Banta

A
Natural History
of
Australia

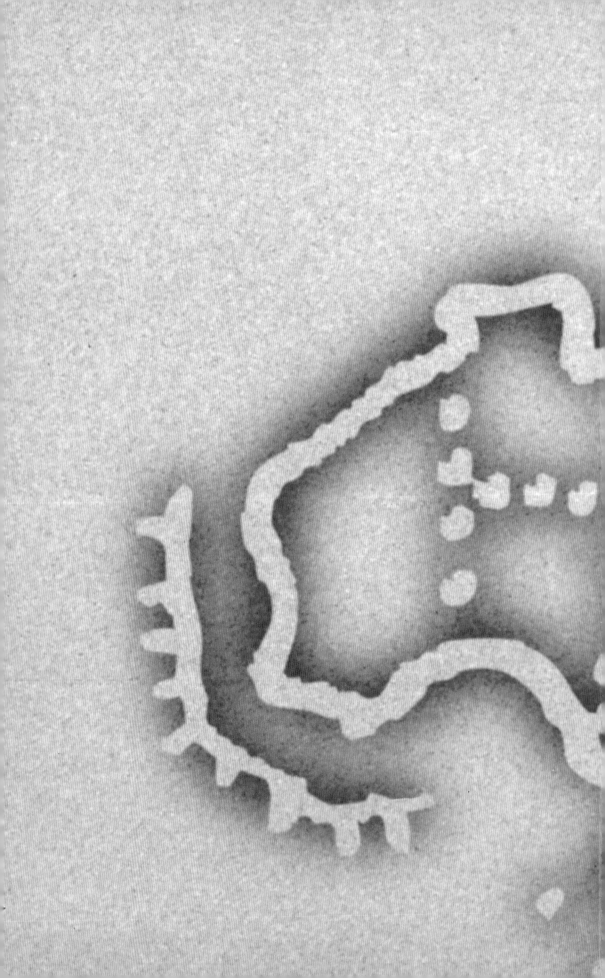

A Natural History *of* Australia

Tim M. Berra, PhD

Professor Emeritus of Zoology

The Ohio State University

With photographs by the author

Published by
Academic Press
525 B Street, Suite 1900
San Diego, CA 92101-4495.

Library of Congress Cataloging-in-Publication Data:
Berra, Tim M., 1943– .
A Natural History of Australia / by Tim M. Berra.
p.   cm.
Includes bibliographical references and index.
ISBN 0-12-093155-9 (alk. paper)
1. Natural history — Australia. I. Title.
QH197.B46   1998   97–42820
508.94 — dc21   CIP.

        1. Natural history - Australia. I. Title.

Available in Australia and New Zealand through
University of New South Wales Press
University of New South Wales
Sydney 2052 Australia

Designer: Mango Design Group
Printer: South China Printing, Hong Kong

Photographs in this book are by Tim M. Berra unless otherwise credited.

By the same author

*Evolution and the Myth
of Creationism*
(Stanford University Press)

*An Atlas of Distribution
of the Freshwater Fish Families
of the World*
(University of Nebraska Press)

*William Beebe: An Annotated
Bibliography*
(Anchor Books)

Contents

Preface

This book is the result of my lifelong love affair with Australia and her biota. The passion was first made possible by a Fulbright Postdoctoral Fellowship in 1969, and the relationship was rekindled by a Fulbright Senior Research Fellowship in 1979. I owe much of my career and the happiness I have gotten from it to the opportunities given me by the Fulbright Program and the foresight of the man after whom it was named, Senator J. William Fulbright.

The two Fulbright Fellowships enabled me to resolve the Murray cod-trout cod taxonomic problem while at the Australian National University in Canberra and to work out the life history of the Australian grayling at Monash University in Melbourne. Subsequently, I spent three months in 1986 and nine months in 1988–89 on sabbatical from The Ohio State University as a Research Associate at the Western Australian Museum in Perth where I unraveled the life history of the bizarre salamanderfish and became involved with the third known specimen of megamouth shark. Research visits to Macquarie University in Sydney in 1994 and 1995 helped explain the world-wide southern hemisphere distribution of the small fish, *Galaxias maculatus*. During all of these visits and their attendant field trips, conference travel, and sightseeing, I've covered over 160 000 km (100 000 miles) throughout Australia. I've driven across Australia east to west and south to north, all the while taking photographs. I hope you enjoy the selection presented here.

The purpose of the book is to explain Australia and the beautifully strange Australian biota to visitors. The book is also intended for the general natural history reader—not least for Australians themselves, who might enjoy seeing how their country has captivated a visiting Yank.

I have included topics that I think an educated reader would want to know about and would expect to find in a book with this title. I have included subjects that are necessary for understanding Australian natural history. One must grasp plate tectonics and continental drift to understand the development of the flora and fauna in isolation from other land masses. This book is also, in part, a reflection and collection of my personal experiences in Australia. I have chosen topics of which I have some personal knowledge, whether it be a formal study or a chance encounter. I have tried to convey something of the Australian character and the land that shaped it. I hope the reader of this book will feel well prepared for a visit to this beautiful and friendly island continent.

Tim M. Berra

Acknowledgments

I am deeply appreciative of the Fulbright Fellowship program of the United States and Australian governments. Without the experience gained while on two Fulbright Fellowships I could never have written this book. I am also grateful for the flexibility and support of my institution, The Ohio State University, in allowing me to dash off to Australia when I was able to secure the funding. My friends, the former Australian Consul-General Terence B. McCarthy and his wife Margaret, made many helpful suggestions on the political and historical material in the manuscript. Dr Gerry Allen of the Western Australian Museum read the entire manuscript and made suggestions for improvements in practically every chapter. Dr Jim Davidson of CSIRO corrected several errors in an early draft of the manuscript, and Dr Weldon Patton of Ohio Wesleyan University was especially helpful with the Great Barrier Reef chapter. Drs Peter and Jean Chesson, formerly of The Ohio State University and Battelle Memorial Institute respectively, commented on various parts of the manuscript. Dr Phillip Habgood of the University of Sydney and Dr Ian Crawford of the Western Australian Museum advised me on the Aboriginal chapter. Dr Mike Tyler, of the University of Adelaide, commented on the frogs section.

I am grateful too to the illustrators whose work embellishes this book: Sue Robinson, Lisa McElravy, Tim Dittmer, Dave Dennis, Lynn Swank, and Connie Fornash. Heidi Shoup verified distribution maps, and made corrections where necessary. Dr Leighton Llewellyn and Dr Gillian Courtice facilitated the photography of several amphibians, reptiles, and birds. I thank the staff and inhabitants of Sydney Aquarium, Oceanworld (Sydney), Taronga Zoo (Sydney), Columbus Zoo (Ohio), and Fort Wayne Children's Zoo (Indiana) for their accessibility and for allowing photographic close-ups of some animals. Marna Utz typed 12 revisions of the manuscript, and compiled the index of scientific names. My wife, Rita, has been my traveling companion and helpmate on most of my trips to Australia.

I appreciate all of this help, which has greatly improved the content of the book. Any errors that remain are, of course, my responsibility.

Jim Jim Falls may be
the most spectacular
point in Kakadu National Park.
The falls are 215 m high, with
a sheer drop of 152 m over
the escarpment.

Jean-Paul Ferrero, AUSCAPE International

Great South Land

●●●●●●●●●●●●●●
EXPLORERS AND SETTLERS

Australia is the only continent occupied by one nation. It is, of course, an island, and is the size of the continental United States. Australia's first settlers were a dark-skinned race of hunters and food gatherers who arrived via raft or canoe from southeast Asia probably 50 000 to 60 000 years ago. At the time of European settlement the Aboriginal population numbered between 300 000 and 1 million people, whose complex and distinctive culture was shaped by a severe environment. They had no agriculture and used only Stone-age tools. (A further discussion on the Aboriginal people can be found in Chapter 3.)

The Greek mathematician Ptolemy produced a map of the world in the 2nd century AD which showed a large unknown land he called *Terra Incognita*. He reasoned that this landmass must be present to balance the northern continents. This argument would not be accepted today. Nevertheless, 1500 years later, European explorers proved that such a continent did exist. In 1606 Dutch explorers, seeking mercantile expansion into Asia, sailed through the strait between Australia and New Guinea—now named Torres Strait after their captain. In the same year the large gulf on the northern coast, the Gulf of Carpentaria, was explored by Willem Janszoon in a Dutch ship, *Duyfken* which made the first known European landing on Australian soil at Cape York. Dirck Hartog explored Shark Bay in Western Australia in 1616, and in 1623 the west coast of Cape York was charted by Jan Carstensz, another Dutchman. In 1642 Abel Tasman visited the small island off the southeast coast of Australia, which he named Van Diemen's Land; it became Tasmania. The Dutch explorers named the newly discovered continent New Holland. In 1688 William Dampier, a buccaneer, became the first Englishman to visit New Holland. He returned in 1698 for further exploration, but his report was not enthusiastic, as he explored the barren northern coast.

It was not until 1770 that Captain James Cook sighted the east coast of the continent. When Cook completed his astronomical observations in Tahiti he sailed the *Endeavour* south to New Zealand and then headed due west. On 20 April 1770 he sighted the southeast Australian coast. He charted the coastline and nine days later landed at Botany Bay, south of the present city of Sydney. Joseph Banks, Cook's botanist, named the place for the variety of botanic specimens he found.

Cook continued his survey 3000 km (1863 miles) north until the *Endeavour* ran aground on the Great Barrier Reef. Repairs took two months; he then sailed north again, passing through Torres Strait and landing on an island off Cape York. There he raised the Union Jack and took possession of the eastern part of the continent for England.

Initially Cook's discoveries had little impact. However, after the American colonies won their War of Independence, an alternative penal colony was needed, for England could no longer transport its criminals to America. In May 1787 Captain Arthur Phillip, with a fleet of 11 ships in his command, sailed from England. There were 1030 people aboard, including 736 convicts. The fleet arrived at Botany Bay on 18 January 1788. Finding it unsuitable for a settlement, eight days later it moved a few miles north to a magnificent harbor, Port Jackson. This small settlement became Sydney, named for Lord Sydney the Home Secretary, and Phillip became the first Governor of the Colony of New South Wales. Phillip took possession of the eastern part of the continent, including Tasmania, on 26 January 1788.

Today, 26 January is commemorated as Australia Day, a national holiday. The 1988 bicentennial celebration marked the 200th anniversary of European settlement of Australia.

Initially, the colonists were supplied with food from overseas, but gradually the land around the settlement was cultivated and better soil was found west of Sydney town, near today's Parramatta. Soon exploration to the west was begun.

In 1802–1803 Captain Matthew Flinders circumnavigated the island-continent for the first time. This closed the era of European discovery which had lasted almost 200 years. Some hydrographic charts made by Flinders, from soundings taken by lowering a weighted line into the water, are still in use today. These old maps are being replaced with new ones made with aircraft-mounted laser technology, a method far beyond Flinders' wildest dreams.

In 1813 a passage was found through the Blue Mountains west of Sydney. This opened the hinterland for further exploration. In 1827 Captain Charles Sturt traced the course of the Lachlan and other inland rivers, and reached the Darling River in 1829. While exploring the Murrumbidgee River he discovered that it flowed into the Murray River. Major Thomas Mitchell became Surveyor-General in 1828 and led many expeditions in South Australia and Victoria. Edward Eyre walked across the entire southern coast—including the Nullarbor Plain—in 1841 and reached the west coast. Ludwig

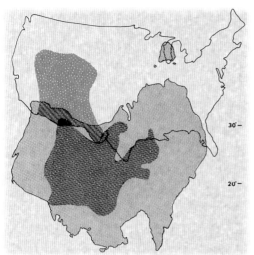

Figure 1.1
Equal area projection of the United States and Australia with west coasts and latitudes aligned. The most arid regions of the two countries are shown by hatching. (Modified from Schall and Pianka, 1978.)

Figure 1.2
Australia and the contiguous United States are nearly identical in area.

Leichhardt walked from Darling Downs in Queensland to Darwin. He and his party vanished in the 'Never-Never Land' (the desert country) in 1848 while attempting to cross the interior from east to west. The Burke and Wills Expedition ended in death for both men in 1860 as they attempted to cross Australia from south to north.

Settlements were established at Hobart in 1803, Brisbane in 1824, and Perth (Swan River) in 1829. Melbourne was founded on Port Phillip Bay in 1835 and Adelaide in 1836.

Captain John MacArthur began experimenting with Spanish merino sheep from South Africa in 1796. By skillful breeding he and others developed a sheep of superior wool-growing quality suited to Australia's dry climate. The wool industry flourished and exporting began.

The population grew rapidly, prompting early governors to grant land to anyone willing to employ and care for convicts. This helped open up new areas for settlement. Immigration and reproduction pushed the population from 34 000 in 1820 to 405 000 in 1850. The increasing number of free settlers forced an end to transportation of convicts in 1840, except to Tasmania and later Western Australia. Tasmania continued to accept convicts until 1853, and Western Australia, which previously had not accepted convicts, received quotas between 1850 and 1868 in order to overcome a labor shortage.

The Australian population soared to 1 140 000 in 1860 and to 1 640 000 in 1870. The reason for this was the goldrush. Gold was discovered in 1851 at Bathurst, New South Wales, and at Ballarat and Bendigo, Victoria. By 1881 the population had reached 2.25 million. In 1892–93 extremely rich goldfields were discovered in Western Australia. Today, Kalgoorlie is still Australia's largest goldfield.

New South Wales achieved self-government via legislation of the British Parliament in 1855, and the other colonies, except Western Australia, became self-governing by 1859. Western Australia achieved self-government in 1890. Federation of the independent colonies followed on 1 January 1901—the birthday of the Commonwealth of Australia.

By 1921 Australia's population had reached 5.41 million, and at the end of World War II in 1945 there were 7.4 million Australians. Today, Australia's population is 18 348 700 (1996 est., ABS).

• • • • • • • • • •

GEOGRAPHY

Australia is distinguished by being the largest island in the world and the smallest continent. Its area is 7 682 300 km² (2.9 million miles²), which is only 2 percent smaller than the continental United States minus Alaska (Figures 1.1 and 1.2). Yet Australia's population is only approximately 18 million compared to America's 266.5 million (1997 census projection). Australia is the sixth largest country in area after Russia, Canada, China, the United States, and Brazil. The coast of Australia is washed by the Indian Ocean on the west and the southern Pacific Ocean on the east. The shoreline extends 37 000 km (23 000 miles).

About 40 percent of Australia is north of the Tropic of Capricorn. It lies 10 347 km (6467 miles) from the west coast of the United States, 1920 km from New Zealand, and 640 km from the closest Asian neighbor, Timor. It is also one of the oldest and flattest of the continents. Australia's average elevation is about 274 m (900 ft), compared to about 700 m as a world average. Rocks from the Murchison River area of Western Australia date to 4.2 billion years ago, and rocks from the northwest corner of Western Australia contain the oldest remains of life, fossil bacteria-like organisms 3.5 billion years old. Australia is tectonically rather quiet, without much volcanic or severe earthquake activity.

Landforms

There are three major structural divisions of the Australian continent. These are the stable Western Shield, the gently warped Central Basin, and the Eastern Uplands (Figure 1.3).

The Western Shield is one vast ancient crustal block about 300 m above sea level. It emerges from Western Australia's coastal plain and covers about half of the continent. A famous outcrop which interrupts the monotony of the plateau is Ayers Rock (now known by its Aboriginal name, Uluru) in the central desert. It is a huge monolith, commonly regarded as the largest in the world. The Western Shield makes up most of the 2.5 million km² (1 million miles²) of Australia that is covered by windblown sand. Much of it, including the Nullarbor Plain, is riverless.

The Central Basin is a great lowland belt extending from the Gulf of Carpentaria in the north to the coast of Victoria in the south. It averages 150 m above sea level. Much of the drainage is internal, to Lake Eyre in South Australia, a 1.4 million km² (540 000 miles²) tectonic depression that dates back 200 million years and was a much larger lake in the Tertiary and Pleistocene. Usually Lake Eyre is a dry saltpan, but it fills with water after heavy rains, most recently in 1988. Its basin at the lowest point is 15 m below sea level. Cretaceous fossils indicate that the Central Basin was covered by the sea relatively recently in geological history. Gibber-strewn plains cover 250 000 km² (100 000 miles²) around the Simpson Desert.

The Eastern Uplands (or Highlands), which average 910 m in elevation, extend along the eastern rim of the continent from Cape York to Tasmania. The Highlands were differentially uplifted in the Tertiary and are more commonly referred to as the Great Dividing Range. This range hugs the edge of the continent 50–400 km from the eastern coast. Rivers to the east drain into the Pacific Ocean, whereas the poorly watered western slopes drain internally or into Australia's largest river system, the Murray-Darling. The southeastern area of the Great Dividing Range is known as the Australian Alps and includes Australia's highest peak, Mount Kosciuszko, 2228 m (7308 ft). The region contains the only areas of Australia to be glaciated during the Pleistocene.

Figure 1.3
Landforms of Australia.
Note: 300 m = 984 ft.

Rainfall and Runoff

Australia is the driest continent, excluding uninhabited Antarctica. Mean annual rainfall is 465 mm (18 in), but it varies from less than 150 mm (6 in) in the central desert to more than 3.6 m (142 in) in the rainforests of the tropics and Tasmania. On average, about 12 percent of the total rainfall becomes surface runoff. The rest is lost in evaporation. As a comparison, North America's annual runoff is about 16 times that of Australia's and equals about 52 percent of the total rainfall.

Spectacular sunsets occur throughout Australia. This scene was taken near Dubbo in New South Wales.

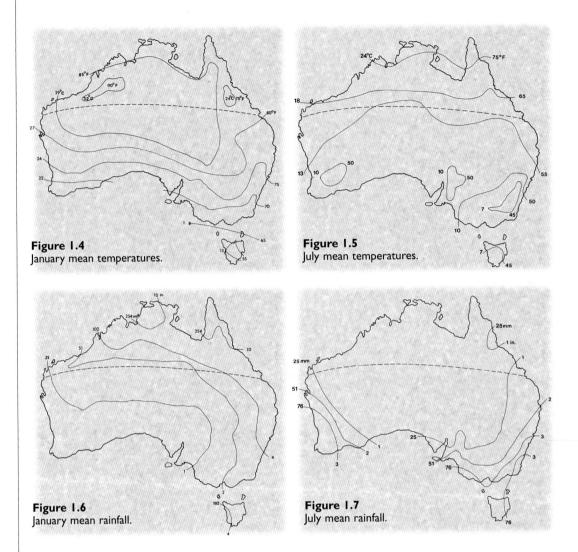

Figure 1.4
January mean temperatures.

Figure 1.5
July mean temperatures.

Figure 1.6
January mean rainfall.

Figure 1.7
July mean rainfall.

The Murray-Darling river system, 2575 km long, drains a catchment of 1.1 million km² (410 000 miles²) in southern Queensland, New South Wales, and Victoria. The Murray-Darling runoff is 22 700 million cubic meters, which is small by world standards. For example, the Mississippi River system runoff is 25 times, and the Amazon River runoff 250 times that of the Murray-Darling.

Climate

The climate of Australia is varied, due to the large size of the continent. The continent lies between latitudes 10°S and 44°S, and, as an island, is protected from extremes of climate by the moderating effects of the surrounding oceans. The absence of extensive high mountain ranges also has an ameliorating effect. The seasons are, of course, reversed from those in the northern hemisphere. Spring lasts from September to November, summer from December to February, autumn from March to May, and winter from June to August.

In the north, 50 percent of Queensland, 40 percent of Western Australia, and 80 percent of the Northern Territory are above the

Tropic of Capricorn, where a generally tropical (hot and humid) climate prevails with temperatures ranging from 27°C to 33°C (80°–91°F). The wet or monsoon season runs from November to April and the dry season extends from April to October. As one travels inland, the climate becomes drier and hotter. Winter temperatures in the arid interior average 23°C (73°F) and summer temperatures reach 43°C (118°F) or higher. Rain rarely falls in these regions. The remainder (about 60 percent) of the continent, including southern Queensland, Western Australia, New South Wales, Victoria, Tasmania, and South Australia lies in the temperate zone and has either a subtropical, a Mediterranean, or a temperate climate. (Figures 1.4–1.7).

The high elevations in the southeastern mainland and Tasmania have the coldest winter weather, with freezing temperatures and regular snowfall.

Much of Australia's weather is characterized by low rainfall and clear skies. Periodic droughts are a regular feature of most areas, with droughts often lasting several years. The latest five-year drought, from 1991 to 1995, was estimated to have cost the Australian economy some $A5 billion.

• • • • • • • • • •
STATES AND TERRITORIES

Politically, Australia is divided into six states and two territories: New South Wales, the Australian Capital Territory (including the Jervis Bay area), Victoria, Tasmania, South Australia, Western Australia, the Northern Territory, and Queensland (Figure 1.8). There are also seven external territories.

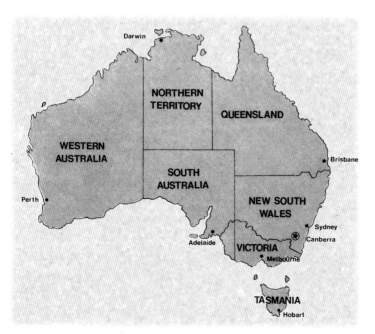

Figure 1.8
States, territories, and capital cities of Australia.

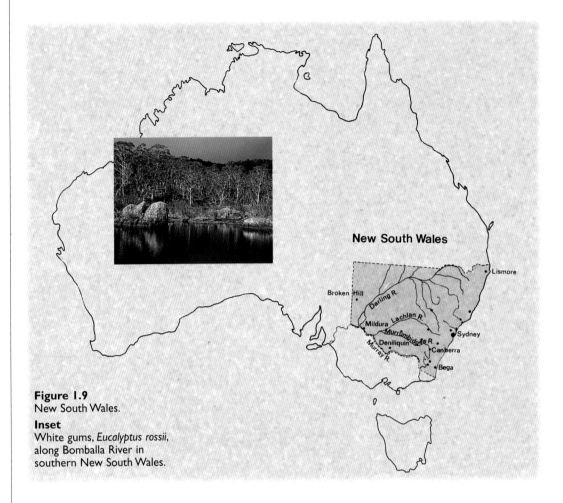

Figure 1.9
New South Wales.

Inset
White gums, *Eucalyptus rossii*,
along Bomballa River in
southern New South Wales.

New South Wales

Imagining the continent of Australia as the face of a clock, we begin at four o'clock with New South Wales and proceed clockwise. The state occupies most of southeastern Australia (see Figure 1.9). It is the most populous (6 307 900) and heavily industrialized state, with a highly urbanized population. It is home to about 35 percent of all Australians and to about 27 percent of all Aboriginals. Sydney is its capital and is the largest city in Australia, with 3 821 400 people (1996 est., ABS). It is one of the world's great seaports.

Natural Features

New South Wales, covers 801 428 km² (309 351 miles²) or 10.4 percent of Australia. Its natural features can be divided into four main zones, extending from north to south:

- Seaboard and coastal lowlands. The coastline is 1460 km (907 miles) long, with few inlets. The major inlets are Port Jackson (Sydney) and Port Hunter (Newcastle). Many fine surfing beaches are strung along this coast.

Sydney Opera House

One of the most architecturally striking and recognizable buildings in the world is the Sydney Opera House. This beautiful building, with its billowing sail-like roofs, commands a majestic site at Bennelong Point on Sydney Harbour within view of the equally famous and striking Sydney Harbour Bridge. The Opera House, a multipurpose performing arts complex, was designed by the Danish architect Jørn Utzon who was selected in an international competition in 1957. At the time, engineers were unsure whether the design was structurally possible. Even Utzon had no firm idea on how to build the roof shells he had sketched so that they would not collapse. Nevertheless, construction began in March 1959. Delays and cost overruns plagued the project and contributed to the toppling of the New South Wales government. Utzon quit the project in controversial circumstances in 1966 and left Australia.

The Sydney Opera House guards the entrance to a bustling ferry dock at Circular Quay and to the city center.

The Opera House was expected to cost about $A7 000 000; however, when the massive complex was completed 14 years later in September 1973 the New South Wales government had spent $A102 million. This huge sum was raised by the introduction of a state lottery—which is still going strong. Queen Elizabeth II officially dedicated the complex on 20 October 1973.

The Opera House is truly among the wonders of the modern world, along with such structures as the Suez Canal, the Eiffel Tower, the Golden Gate Bridge, and the Empire State Building. The entire complex covers about 1.8 ha (4.4 ac) on a site that is among the most spectacular pieces of real estate in the world. The building is about 183 m long and is 118 m wide at its widest point. The highest roof shell towers 67 m above the water. The roofs are constructed from 2194 concrete sections that weigh up to 15.3 tonnes (15 tons) each, held together by 350 km of tensioned cable. The roofs are covered with more than 1 056 000 individual Swedish-made cream and white tiles. The effect of these white 'sails' against the electric blue sky above Sydney Harbour is a sight to behold. About 2000 panes of glass in 700 different sizes, covering 6223 sq m, fill in the mouths of the shell roofs. The glass was made in France and one layer is tinted topaz.

Inside, finished in beautiful native woods, there is a concert hall that seats 2690 with near perfect acoustics, an opera theatre (1547 seats), a drama theatre (544 seats), and a cinema (419 seats). There are many other facilities, such as a recording hall, restaurants, bars, library, and offices.

The Sydney Opera House ranks as one of the world's busiest performing arts centers. More than 53 000 performances were staged in its first 20 years. There are as many as 1.5 million attendances at events each year.

- The Tablelands are an almost unbroken series of plateaus formed by the Great Dividing Range. They form the main watershed which gives rise to both coastal and inland rivers. The northerly New England range averages 750 m (2460 ft) and reaches up to 1200 m. The southern tableland includes Mount Kosciuszko.
- The Western Slopes are a fertile and undulating region. The rich plains along the rivers have regular and adequate rainfall, leading to extensive cultivation.
- The Western Plains cover about two-thirds of the state. Rainfall is sparse and much of the land is used for grazing.

There are two categories of rivers in New South Wales: the short, fast-flowing streams along the coast, which drain only about one-sixth of the state but carry two-thirds of its water; and the sluggish, meandering inland rivers of the Murray-Darling system (Figure 7.1, page 126) which lose much water via evaporation and seepage. Principal coastal rivers include the Hawkesbury (472 km), Hunter (462 km), MacLeay (402 km), Clarence (394 km), Shoalhaven (332 km), Snowy (270 km within New South Wales), and Richmond (262 km). Major inland rivers are the Darling (2617 km within New South Wales), Murray (1936 km within New South Wales), Murrumbidgee (1759 km), Lachlan (1484 km), Macquarie Bogan (950 km), Namoi (847 km), Gwydir (668 km), and Castlereagh (549 km).

Climate

New South Wales is entirely within the temperate zone. The highest temperatures are recorded in the northwest. Bourke, along the Darling River, has recorded a shade temperature of more than 50°C (124°F). The Snowy Mountains in the southern tablelands are the coldest part of the state and have winter frost and snow. Sydney has a beautiful climate which tends to be warm and humid toward the end of

Sydney Harbour Bridge at night.

summer in February and into March. The mean summer temperature of Sydney is 21.4°C (70.5°F) and the winter average is 12.6°C (54.7°F).

Rainfall varies widely over the state, gradually increasing from an annual average of about 200 mm (7.8 in) in the far northwest to about 1800 mm (71 in) on parts of the north coast.

Cities and Towns

The principal cities and their populations (1996 ests, ABS) are Sydney (3 821 400), Newcastle (471 000), and Wollongong (255 700). The populations of the following smaller towns are (1996 ests, ABS): Wagga Wagga (57 800), Coffs Harbour (57 460), Maitland (52 700), Lismore (45 900), Albury (49 210), Tamworth (36 900), Dubbo (36 900), Orange (35 260), Broken Hill (23 720), Armidale (23 450), Goulburn (22 220).

Industry

The state's primary production includes cattle and dairy products, wool, wheat, alfalfa, oats, rice, corn, fruit and vegetables, and fishing, oyster farming, and forestry products. The coastal area supports mixed farming, especially dairying. In the subtropical north, sugar and bananas are grown. The tablelands are a grazing area for sheep and cattle, while the western slopes and the eastern part of the western plains are wool and wheat producing areas. The Riverina district is a large area irrigated by the Murrumbidgee River which yields orchard crops, grapes, rice, and sheep. The arid western plain is suitable only for wool production.

High quality black coal is mined in the central coast, Blue Mountains, and Hunter River Valley. Lead and zinc are mined at Broken Hill.

Machinery, electrical goods, clothing, chemicals and processed foods are major secondary industries in New South Wales.

The Australian Capital Territory

The Australian Capital Territory (ACT) is analogous to the District of Columbia in the United States. It is the site of the federal capital and is not under any state's control. The ACT was carved out of New South Wales in 1911 to provide a location for the capital (Figure 1.10). It has an area of 2357 km^2 (910 miles^2).

The capital city is Canberra (accent on the first syllable) which is an Aboriginal word for 'meeting place'. An international competition was launched in 1912 for the design of the city, which was won by an American architect from Chicago, Walter Burley Griffin, an associate of the renowned Frank Lloyd Wright. He laid out an impressive city plan around a large artificial lake which was eventually created by the damming of the Molonglo River. In 1927 the Parliament and some government departments were transferred from the temporary capital in Melbourne. Canberra is a beautiful and thoughtfully planned city with numerous greenbelts and more than 11 million trees. It boasts broad boulevards with numerous, confusing

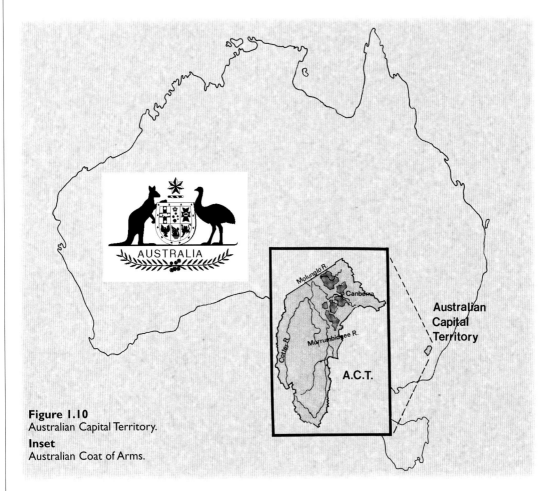

Figure 1.10
Australian Capital Territory.
Inset
Australian Coat of Arms.

Fog shrouds a *Eucalyptus* forest near the Mitchell River in Victoria.

roundabouts. The city nestles about 575 m (1886 ft) above sea level, between Lake Burley Griffin, with its 35 km (22 mile) shoreline, and the surrounding mountains. Canberra is Australia's largest inland city with a population of 344 900 (1996 est., ABS). (This figure includes

A partially cleared valley in the Great Dividing Range at 755 m (2477 ft) elevation south of Omeo in Victoria.

the population of Queanbeyan, a nearby New South Wales town.) It is home to the Commonwealth Parliament, government buildings, and the diplomatic missions of about 58 countries.

About 60 percent of Canberra's wage earners work for the government. Others are employed by many important establishments such as the Australian National University, the Australian Academy of Science, the headquarters of the Commonwealth Scientific and Industrial Research Organization (CSIRO), the National Library, the National Gallery, and the High Court.

When the national capital was created, it was decided that Canberra should have access to the sea. In 1915 the southern headland of Jervis Bay was ceded by New South Wales to the federal government, and is now administered by the ACT. The Jervis Bay territory is an area of 73.6 km^2 (28.4 miles^2), about 150 km northeast of Canberra with about 66 percent of its area being a public park. It is the site of the Royal Australian Navy College.

Victoria

Victoria occupies the southeastern corner of the continent and is the second smallest state, after Tasmania, but it has a larger population than any other Australian state except New South Wales. Its area is 227 600 km^2 (87 884 miles^2) which is only 3 percent of mainland Australia (Figure 1.11). It's about the same size as Utah or the United Kingdom. Victoria's population is 4 552 000 (1996 ests, ABS), most of which lives in the capital, Melbourne (3 248 800), the second largest city in Australia. Victoria is the most densely settled state, with more than 15 persons per square kilometer.

The great majority of Victoria's population, like that of Australia in general, are of Anglo-Saxon descent. However, more than 1 million of Melbourne's citizens were born outside Australia or are the children of migrants, mostly from Europe. Italians are the largest

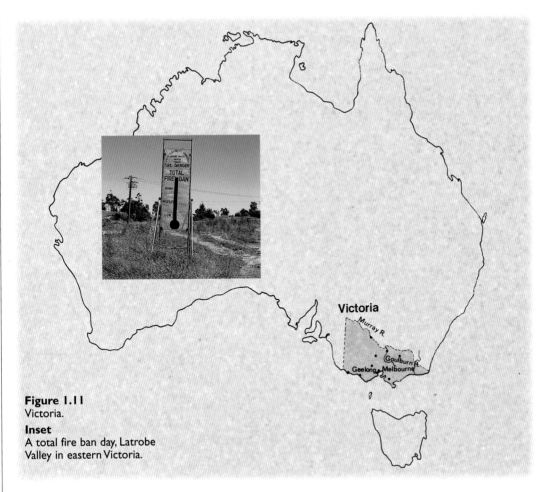

Figure 1.11
Victoria.

Inset
A total fire ban day, Latrobe
Valley in eastern Victoria.

single ethnic community with over 300 000 people. Among the
hundred or more nationalities represented, there are also sizeable
Greek and Yugoslav communities and a Chinese population, some of
whose ancestors came during the goldrush of the 1850s.

Natural Features

Over one-quarter of Victoria is forested and the remainder of the
terrain is open rolling country. Mount Bogong, 1986 m (6514 ft) and
Mount Feathertop, 1922 m (6304 ft) are the highest peaks. The main
coastal rivers are Glenelg (457 km), Yarra (241 km), Hopkins
(209 km), Latrobe (233 km), Tambo (145 km), Mitchell (129 km), and
Snowy (129 km, in Victoria). The principal inland river is the Murray
(Figure 7.1, page 126), stretching 1930 km (1198 miles) from its source
in the Alps to the South Australian border. It forms Victoria's border
with New South Wales. Other major inland rivers are tributaries of
the Murray, such as the Goulburn (566 km), Loddon (338 km), Mitta
Mitta (269 km), Campaspe (249 km), and Ovens (212 km). Artificial
lakes include Lake Eildon (3 390 000 ML) on the Goulburn River
and Lake Hume (3 059 000 ML) on the Murray. Victoria has an 1800
km (1118 mile) coastline on Bass Strait which separates the mainland
from Tasmania. The most important harbor is Port Phillip Bay, on
which Melbourne and Geelong are situated.

Climate

The climate is temperate with hot summers and cool to mild winters. Snow can cover the Victorian part of the Australian Alps from June to September. The Great Dividing Range and country south of it have moderate to cool temperatures and good rainfall throughout the year, while the plains to the north and west are drier and warmer. January and February are hottest. Average maximum temperatures are under 20°C (68°F) in the high land, 24°C (75°F) along the coast and over 32°C (89°F) in the Mallee (scrub bush) in the northwest. Winter temperatures drop below 10°C (50°F) in the ranges. In summer temperatures can rise to above 40°C (104°F).

Cities and Towns

Melbourne, the capital, is located on the Yarra River. It is a major financial and communications center. The city is well known for the Melbourne Cricket Ground which can hold more than 100 000 spectators, and the Flemington Racecourse which is the site of the famous Melbourne Cup horse race held on the first Tuesday of November.

Geelong is the second largest city after Melbourne with 154 000 people (1996 ests, ABS). It is a port and an industrial center. Ballarat (77 600), which was a major goldfield in the 1850s, today is a manufacturing and woolgrowing district. Bendigo (75 000) was also a goldrush town, but today its main activities are agriculture and commerce.

Industry

Primary production in the state includes wool, lambs, cattle, dairy products, wheat and grain, and forestry products. Large oil and natural gas fields off the eastern coast in Bass Strait are in production or under development. These fields provide 90 percent of Australia's domestic oil production and 60 percent of its crude oil needs. The Latrobe Valley, about 150 km east of Melbourne, is a major coalmining area which supplies electricity for all of Victoria. Victoria is also the major center of Australia's motor vehicle and aircraft industry.

Tasmania

Tasmania, a triangular-shaped island south of Victoria, is Australia's smallest state. Its area, including the satellite islands of King, Flinders, and Cape Barren is 68 331 km² (26 376 miles²) (Figure 1.12). Tasmania is separated from the mainland by Bass Strait, which on average is 240 km (150 miles) wide. Originally called Van Diemen's Land, Tasmania was the second colony to be founded. Its population is 473 400 (1996 est., ABS), of whom about 40 percent (195 300) live in the capital, Hobart.

Figure 1.12
Tasmania.

Inset
Much of Tasmania is still forested.

Natural Features

Tasmania is Australia's most mountainous state, but its peaks rarely exceed 1500 m (4920 ft). The Central Plateau slopes southeastward from 1065 m (3493 ft) in the north to 610 m (2000 ft) in the south. Most of the mountainous area on the western edge is sparsely populated. The midlands are a rich agricultural area. Lying between 40° and 44°S, Tasmania has no point that is more than 115 km from the sea.

A farm property carved out of *Eucalyptus* forest along the Esk River in northeastern Tasmania.

Climate

Tasmania has a temperate, maritime climate. Its rainfall is the highest of
any Australian state, with annual averages from 3600 mm (142 in) in
the west to 500 mm (20 in) in the east. The higher elevation means
that winter minimums can fall to 0°C (32°F), while summer
maximums in the lowland area exceed 23°C (74°F). Snow is common
in late winter and early spring, but there is no permanent snowline.

Mist rises from the North Esk
River near Launceston,
Tasmania.

Cities and Towns

Hobart, the capital, spreads along both sides of the Derwent River
and has Mount Wellington (1269 m) for a backdrop. It is Tasmania's
main port and a home to several industries. The first Las Vegas-style
gambling casino in Australia opened in Hobart in 1973. Hobart is also
well known because of the Sydney–Hobart Yacht Race, a popular
annual event which starts in Sydney on Boxing Day (26 December)
each year. The other major city in Tasmania is Launceston in the
north, with a population of 98 000 (1996 est., ABS).

Industry

Agriculture is important to Tasmania's economy, involving one-third
of the state's total area. Meat, dairy, and orchard products form
significant industries. Tasmania is known as the Apple Isle because of
its production of that fruit. Approximately 41 percent of the state is
forested. Timber production is an important factor in the island's
economy, as is mining. Tasmania produces two-thirds of Australia's

The Nullarbor is a vast, dry, virtually treeless plain that covers about 200 000 km² (80 000 miles²) of the south of the continent. The Eyre Highway near the coast is the only road across the Nullarbor.

tin and 10 percent of its copper, zinc, silver, and gold. The countryside around Queenstown on the west coast became denuded of vegetation by early logging to supply fuel for the copper smelters and by fumes from the smelters. Local tourism operators hope that the area will not become revegetated—its strange landscape attracts many visitors.

Because of the abundant rainfall and high country, many hydro-electric power plants have been built, at some considerable expense to the wild rivers and their inhabitants.

South Australia

South Australia (Figure 1.13) is the driest state of the world's driest continent. It is about one and a half times the size of Texas, with 984 000 km² (380 000 miles²) and 1 480 200 inhabitants (1996 est., ABS). Eighty percent of South Australia gets less than 250 mm (10 in) of rain per year—the definition of a desert.

Natural Features

There are two major gulfs along the southern coast: Spencer and St Vincent's. The three large, low-lying areas in the center are Lakes

Limestone cliffs along the Eyre Highway at the Great Australian Bight, South Australia. This area is rich in Miocene marine fossils, and occasional rainfall has carved several large caves.

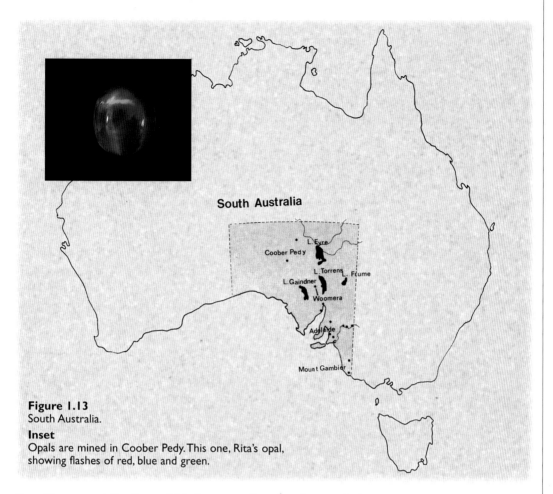

Figure 1.13
South Australia.

Inset
Opals are mined in Coober Pedy. This one, Rita's opal, showing flashes of red, blue and green.

Eyre, Torrens, and Gardiner, which form vast glistening deserts of salt as they dry out. Occasionally they fill with rare rainfall to produce spectacular aggregations of plant and animal life.

The Murray River is a major geographic feature. Its last 700 km (435 miles) flow through South Australia, and its water is extensively used for irrigation. The highest point in South Australia is Mount

Road signs warning of camels, wombats, and kangaroos near Ceduna in South Australia. This is the last 'major' town and the last sighting of trees before crossing the Nullarbor from east to west.

Piles of excavated material dot the desert around Coober Pedy, South Australia, as residents mine for opals formed on the bed of an ancient, dried-up sea.

A kitchen in an underground home in Coober Pedy.

Lofty, 728 m (2384 ft). Most of the state is desert, but the southern part is well-watered and supports agriculture and industry. Kangaroo Island lies 113 km southwest of Adelaide and retains much of its rugged character and wildlife.

The Nullarbor Plain is a vast, treeless, arid, almost uninhabited plateau, across which a section of the Trans-Australian Railway Line runs, linking Sydney with Perth. North of the Nullarbor lies the Great Victorian Desert, an extensive area of sand.

Cities and Towns

Adelaide is the state's capital city with a population of 1 086 500 (1996 est., ABS). It is situated along the Torrens River on the east side of St Vincent's Gulf. Adelaide receives about the highest rainfall in South Australia, 533 mm (21 in) per year, yet it boasts 245 rainless days annually and seven hours of sunshine a day. Its daily maximum temperature averages 22.3°C (72°F). There are no slums, few traffic jams, and more restaurants per inhabitant than any other capital city in Australia. It is a delightful city, well planned around parklands. Adelaide hosts a three-week Festival of Arts every two years which attracts high quality performances and lends a carnival atmosphere to the city.

Industry

Because most of South Australia is a desert, only about 15 percent of the land is devoted to cropping or permanent pasture, and primary industry is the state's biggest export earner.

The Pinnacles Desert, 259km north of Perth, Western Australia. The Pinnacles, formed from limestone, are up to 4 m high.

An hour's drive to the north of Adelaide is the Barossa Valley, a major home of Australia's wine industry. Wineries in the Barossa Valley produce about 60 percent of Australia's annual 414 million liters (108 million gallons) of excellent wine.

Roxby Downs, in the desert about 500 km north of Adelaide, could be classified as the world's single richest new discovery of minerals in 50 years. It probably contains the biggest copper deposit in existence as well as the largest uranium deposits in the world. Motor vehicles and domestic appliances are made in Adelaide.

Opal, a gemstone form of silica similar to quartz but with 6–10 percent water within its mineral structure, is mined at Coober Pedy, 966 km north of Adelaide, where nearly half the 2500 inhabitants live underground to escape the heat that can reach 50°C (122°F) in summer. About 75 percent of the world's supply of this brilliantly colored gemstone comes from Australia. The background color of opals can be white, black, or blue, while internal refraction of light on impurities results in dazzling flashes of red, green, and yellow. The name Coober Pedy is derived from the local Aboriginal words 'Kupa Piti'—meaning 'white man's hole'. Title to about 10 percent of South Australia's area (formerly the Northwest Reserves) is held by Aboriginals.

Western Australia

Western Australia covers about one-third of Australia and is the largest state with 2 525 500 km² (974 843 miles²). It is bounded by the Great Australian Bight on the Southern Ocean, the Indian Ocean on the west, and the Timor Sea on the northern tip. It has a long coastline of 12 500 km (7800 miles) (Figure 1.14). The Great Plateau covers more than 90 percent of the state and averages about 600 m (2000 ft) above sea level. Elevations twice that are found in the northwest, with Mount Meharry reaching 1244 m (4081 ft). The state's population is approximately 1 771 000 (1996 est., ABS), which includes about 47 000 Aboriginals and about 350 000 new Australians, two-thirds of whom are of European origin.

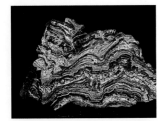

Top: Cross-section of a stromatolite in the Western Australian Museum, showing the layers of sediment formed by cyanobacteria. Such stromatolites are some of the earliest life forms on earth. Below: Stromatolites at low tide at Hamelin Pool, Shark Bay, Western Australia. These stony cushions lie under water at high tide.

Nature's Window in Kalbarri National Park, Western Australia.

Natural Features

Lying opposite the mouth of the Swan River is Rottnest Island, about 19 km off the coast. It became isolated from the mainland only about 6000 years ago. This low-lying island is a public reserve and wildlife sanctuary, and home to the small kangaroo known as the quokka. Rottnest, so named because the Dutch explorer William deVlaming (1696) mistook quokkas for rats, is a popular holiday resort and weekend retreat for Perth residents. No cars are allowed on the island. The coral, *Pocillipora damicornis,* forms small reefs in shallow water close to the southern shore of Rottnest Island.

Major rivers of the state (Figure 7.1, page 126) are the Gascoyne (838 km), Murchison (708 km), Ashburton (644 km), Fitzroy (520 km), Drysdale (443 km), and the Swan with its tributary the Avon (386 km). There are vast expanses of salt lakes near Kalgoorlie (Lake LeFroy) and near Port Hedland (Lake McLeod).

The quartzites of Mount Narryer in the Murchison region of Western Australia contain some of the earliest minerals. Zircon crystals were part of the Earth's first crust and were derived from sediments 4.1–4.2 billion years ago (BYA). The rocks in which these crystals became incorporated date back to more than 3.6 BYA. These are some of the oldest rocks on Earth.

Fossil remains of the oldest lifeforms on earth are recorded from the hot Pilbara region west of Marble Bar (known ironically as the North Pole). These are stromatolites, radiometrically dated at about 3.5 billion years old, and contain four different types of filamentous microfossils. The rock formation is known as the Warrawoona Group. A stromatolite is a stony cushion or dome of rock composed of layers of sediment trapped, along with precipitated limestone, by thin layers of primitive, filamentous, single-celled microorganisms called cyanobacteria. Incredibly, living stromatolites are still being formed today and can be found in a few places on earth where

Ken Griffiths

Monkey Mia

A much better known place in Shark Bay is Monkey Mia, 26 km from the most westerly town in Australia, Denham, which is 911 km north of Perth. Denham has a resident population of about 450 people, but as many as 160 000 tourists visit the area annually. This number is expected to increase as the roads improve. The region is desolate and difficult to reach, but it is becoming increasingly popular as word spreads about the wild bottlenose dolphins,

Wild dolphins come into shallow water at Monkey Mia, Shark Bay, near Denham in Western Australia, and allow human contact.

Tursiops truncatus, that allow visitors to swim with them and even touch them. About three decades ago local fishermen interacted with and probably fed a pod of local wild dolphins, the descendants of which continue to frequent the shallow waters at Monkey Mia and seem to solicit interaction with wading humans.

When I visited Monkey Mia early one morning in February 1986 there were no dolphins to be seen and only two people, who were working on their boat. This was before road improvements made tourist access easier. I walked into the waist-deep water, and within a minute two dolphins appeared and swam around me and between my legs. It was a thrilling experience. They allowed me to gently scratch their belly and sides. Some dolphins don't like to be touched on their heads, fins, or tail, and they especially dislike people reaching over their heads. They usually give a warning by jerking their heads if someone is doing something to upset them. Dolphins are generally quite tolerant, but if the warning is not heeded a dolphin may bite the offender.

I enjoyed the dolphins' company for about an hour and did not feed them. Eventually other people arrived, and children were still playing with the dolphins when I left. I didn't get the impression that the dolphins were waiting for food. They seemed to enjoy the human stimulation—but that is just my subjective, anthropomorphic impression. However, since the surrounding water is teeming with fish, it might well be argued that the dolphins come into Monkey Mia for human contact at least as much as for food. Occasionally a dolphin will present a human with a still living, freshly caught fish. There may be more than 300 dolphins in the herd offshore. Sometimes as many as six females and their calves will come inshore at one time. The dolphins remain in the general area throughout the year, but May through October is the best time for humans to visit.

Survival of first year dolphins is twice as high if the mother has not fed at the beach. It has been argued that young dolphins nursing from handfed females are malnourished and vulnerable to disease, polluted waters, and shark attack. The mother may be spending too much time with the tourists waiting for a handout, and not enough time catching fish and attending to the young. Biologists from the Western Australian Department of Conservation and Land Management are studying this problem and may eventually prohibit hand feeding. Rangers at Monkey Mia keep an eye on the visitors to limit feeding and to prevent any harm to the dolphins or humans.

The social structure of the Monkey Mia dolphins has been the subject of study by scientists from the University of Michigan since the 1980s. Each dolphin can be individually identified by its body scars and marks on fins. For a popular account of the dolphin research at Monkey Mia see Smuts et al. (1991) and for technical details see Smolker et al. (1993) and the references cited therein.

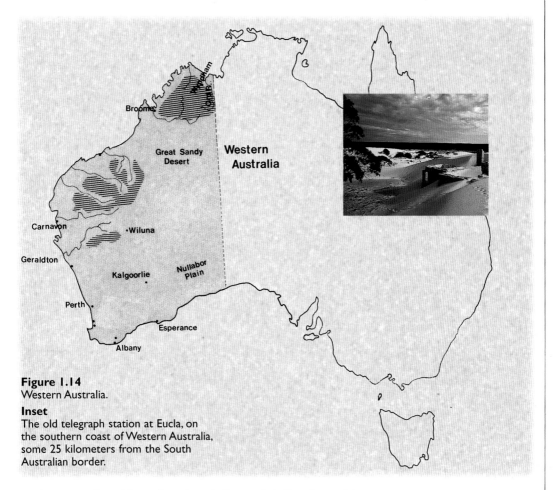

Figure 1.14
Western Australia.

Inset
The old telegraph station at Eucla, on
the southern coast of Western Australia,
some 25 kilometers from the South
Australian border.

conditions are favorable, such as the salt flats and shallow bays of the
Persian Gulf, the Bahamas, and the west coast of Mexico, and at
Hamelin Pool in Shark Bay, Western Australia. In Shark Bay, a sandbar
has blocked one small arm of the bay, and thus impedes water
movement. This restricts tidal action and allows evaporation to
produce hypersaline waters (about twice the salinity of normal sea
water), which few organisms can tolerate. This prevents grazing
organisms such as molluscs from destroying the surviving
stromatolites. Freshwater runoff into Hamelin Pool is negligible, and
Hamelin Pool has now been declared a marine nature reserve.

In the absence of strong influences of winds and currents,
stromatolites tend to orient toward the sun. In the southern
hemisphere this heliotropism results in a northward inclination.
These living fossils grow very slowly, about 0.4 mm per year. Some
individual stromatolites in Hamelin Pool may be 1000 years old,
according to radiocarbon dating. As they photosynthesize and grow,
stromatolites leave a daily record of sediment and calcium carbonate
accumulation. By sectioning a stromatolite, scientists can calculate
the number of days in a year at the time of formation. This also gives
the length of a day from the past. For example, stromatolites from
850 million years ago reveal a year of about 435 days. A day would
then be about 20 hours long. A recent study utilizing fractal

Wave Rock near Hyden in
Western Australia.

geometry by Grotzinger and Rothman (1996) suggests that some
stromatolites may form by abiotic processes.

Western Australia is becoming famous for another marine
visitor—the whale shark, *Rhincodon typus*. This is the largest fish in
the world. It may reach a length of 18 m or so (60 ft), although
12.2 m (40 ft) is the largest specimen recorded. It is a harmless
plankton feeder. These huge fish congregate by the hundreds in the
months of March and April, following the full moon, at Ningaloo
Reef off Exmouth. The aggregation of whale sharks probably occurs
in response to the enriched food resource that follows mass coral
spawning. Divers have come from all over the world to photograph
and swim with these gentle giants, but government regulations
forbid riding on or touching the whale sharks.

Two of the many interesting and unusual landforms in Western
Australia are Pinnacle Desert and Wave Rock. The Pinnacles are part
of Nambung National Park 17 km south of the coastal fishing village
of Cervantes, 259 km north of Perth. The stark, lunar-like landscape
of shifting yellow sand is punctuated by thousands of limestone
pillars that range in size from a few centimeters to 4 m high and 2 m
in diameter at their base. The pinnacles were formed by the leaching
effect of acidic rainwater on soft limestone, which created solution
pipes eventually exploited by plant roots. As the vegetation died, and
windblown sand eroded the debris between the more resistant
materials, the pinnacles were left behind. These surviving columns

The Murchison River Gorge in
coastal central Western
Australia is 80 km (50 miles)
long and is surrounded by
Kalbarri National Park. The
Murchinson River floods once
or twice a year and at other
times is reduced to a string of
billabongs. This view is from
Hawk's Head Lookout.

were protected from erosion by a cap of cement-like calcrete formed on top of the softer limestone.

Wave Rock is the 2.7 billion year old face of a 65 ha (161 ac) granite formation known as Hyden Rock, near the town of Hyden, 340 km north of Perth. Wind and water acting over this immense period of time have undercut the rock's base and produced a curved, smooth, breaking-wave shape. Various minerals have been deposited on the wave face by rainwater cascading over the edge of the rock, resulting in beautiful vertical banding of red, yellow, and black. The wave itself is 14 x 40 m (46 x 131 ft). A low concrete dam wall has been built over the top of Hyden Rock to enable it to serve as a water catchment area for the town.

East Alligator River in Kakadu National Park, Northern Territory, is home to 62 species of fishes, as well as crocodiles and introduced water buffalo.

Climate

This huge state has three major climatic zones which give rise to three biotic divisions: the Northern or Kimberley Division with a tropical biota adapted to reliable monsoonal summer rainfall and dry winters; a temperate biota of the Southwest Division adapted to reliable winter rainfall and dry, hot summers typical of a Mediterranean climate; and an extensive arid zone separating the Kimberley and Southwestern Divisions. Most of the inland and northern areas are desert (Great Sandy Desert, Gibson Desert, Great Victoria Desert, and Nullarbor Plain). There is a tropical coastal belt across the far north and northwest, where the highest mountains are found.

Cities and Towns

Perth is a beautiful capital city situated on the Swan River. It is the fourth largest city in Australia, with a total population of 1 282 800, including Fremantle, a quaint port town, 18 km downstream from Perth (1996 est., ABS). The nearest major city is Adelaide, 2400 km away. Perth has a Mediterranean-type climate of wet winters and long, hot summers. It is the sunniest and windiest of the Australian capitals. It receives an average of eight hours of sunshine per day and 875 mm (34 in) of rain per year. Average maximum temperature is 23°C (73.4°F) and the minimum average is 13°C (55.4°F). King's Park, a 400 ha (988 ac) native bushland, botanical garden and public park, commands a spectacular view of the Perth skyline and Swan

River. It's a beautiful place for bushwalking, birding, cycling, and picnicking within a few minutes of the city center.

Industry

Western Australia is mineral rich. The Kalgoorlie goldrush of the 1880s peaked in 1903 and then declined, but gold remains a major resource. The state now produces 90 percent of Australia's iron output. Other important mining interests are bauxite, uranium, coal, mineral sands, nickel, and copper. Gas fields are also being developed. Other primary industries include cattle, sheep, and grain, especially wheat.

The Northern Territory

The Northern Territory covers one-sixth of the Australian continent (Figure 1.15). Its area of 1 346 000 km² (520 000 miles²) is equal to the area of Italy, Spain, and France combined, and the distance between the northern and southern extremities of the Territory is almost as great as the distance between New York and Miami. About 80 percent of the Northern Territory is in the tropics).

The total population of the Territory is 171 100, of which 46 000 (27 percent) are of Aboriginal descent (1996 est., ABS). This figure represents about 15 percent of Australia's Aboriginal population.

There are three main highways in the Territory. One links Darwin with Alice Springs and goes on south to Adelaide. Another runs from Katherine to the Ord region of Western Australia, and the third connects Tennant Creek with Mount Isa in Queensland. A new standard-gauge railway joins Alice Springs to Adelaide.

The Northern Territory became self-governing in 1978. It has one member in the Australian House of Representatives and two senators in the Upper House. The federal government retains control over uranium mining and Aboriginal land rights, but federal legislation in 1977 enabled the Aboriginal people to regain ownership over their traditional lands, and Aboriginal reserves now total about 33 percent of the Northern Territory, with a further 13 percent under claim.

Natural Features

The coastline is flat, with swamps, mudflats, and mangroves. In the central region the east-west ridges of the McDonnell Ranges reach more than 1200 m (4000 ft).

The Todd River at Jessie's Gap near Alice Springs is a river of sand most of the time. Boat races are held in the Todd River by crews running in the dry river bed while holding their beer-can boats.

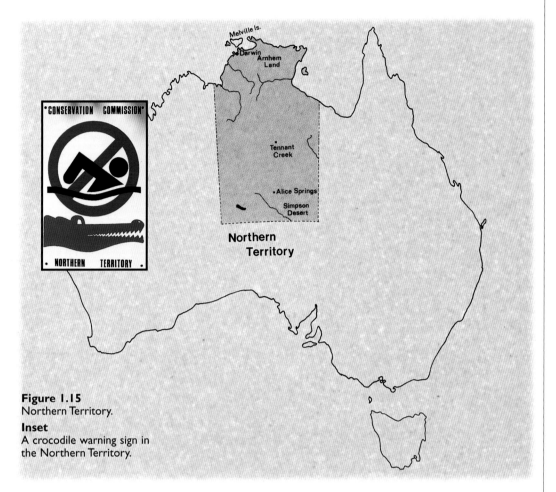

Figure 1.15
Northern Territory.
Inset
A crocodile warning sign in
the Northern Territory.

The southern three-quarters of the Territory is desert or semi-arid
plain, while the northern quarter, called the Top End, contains pockets
of tropical rainforests and savannah. The Arnhem Land Plateau, in the
northeast, rises abruptly from the plain and extends to the Gulf of
Carpentaria. Inland rivers are dry for most of the year and their
irregular flow is lost in the Simpson Desert. Coastal rivers (Figure 7.1,
page 126), listed from west to east, include the Victoria, Daly,
Adelaide, Mary, South Alligator, East Alligator, Roper, and McArthur.

Climate

About two-thirds of the Territory receives less than 500 mm (20 in)
of rainfall per year, with the far north averaging 1575 mm (62 in).
Monsoonal influences in the north produce a wet season (November
to April) and a dry season (May to October). Cyclones (hurricanes)
can be very dangerous in the wet season. The temperature in Darwin
is almost constant the year round. The January maximum and
minimum averages are 31.8°C (89.2°F) and 24.8°C (76.6°F) respectively,
and in July the averages are 30.3°C (86.5°F) and 18.9°C (66°F).

Cities and Towns

In 1839 the commander of HMS *Beagle,* J.C. Wickham, discovered and
named Port Darwin after Charles Darwin, the English naturalist who

Uluru

Uluru is the exposed tip of a 5 km deep arkose sandstone mass about 500 million years old. It is regarded as the world's largest monolith, being 348 m (1141 ft) high, 2.6 km (1.6 miles) long, and 9.4 km (5.6 miles) around its base. It is located some 450 km southwest of Alice Springs. The terracotta color of the Rock is due to oxidized iron in its sandstone, and it changes hue daily with the position of the sun and moisture content of the rock. At sunset the Rock can be brilliant red. Its surface is spectacularly weathered, with erosion holes formed by the heavy flow of periodic desert rainfall over many millions of years. Climbing the Rock along a 1600 m (1 mile) trail is a thrilling adventure and no mean feat (or feet).

Uluru is a sacred site to the local Aboriginal people, and should be treated with respect. Uluru National Park, in which the Rock is situated, is owned by the Aboriginal community and leased to the Australian National Parks and Wildlife Service.

The 36 rounded domes of conglomerate granite of Kata Tjuta (highest point 546 m [1791 ft] above the surrounding plain) are visible from Ayers Rock, 36 km away. Kata Tjuta is also a sacred site.

Top: Erosion from rainfall acting over many millions of years has produced interesting weathering patterns in the sandstone of Uluru.
Below: A trail has been marked out leading to the top of Uluru. It is a taxing climb that takes about an hour.
Inset: Uluru (Ayers Rock) at sunrise.

Kakadu National Park

Kakadu National Park ranges between 150 and 270 km (by road) to the east of Darwin. This huge park, 20 000 km² (7700 miles²) in the Alligator Rivers region, has a wealth of wildlife that inhabits four major landscape features: tidal flats, floodplains, lowlands, and plateau. Sandstone escarpments with spectacular waterfalls mark the southern and eastern edge of the Arnhem Land plateau. Massive unspoiled waterways, floodplains, and billabongs swell with runoff during the wet season and teem with fishes and crocodiles. Tidal communities, sedgelands, swamp forests, grasslands, eucalyptus woodlands, rainforests, and heathlands are all found within this huge heritage area. Such habitat diversity houses over 1500 species of plants, 50 mammal species, 275 bird species, 25 frog species, and at least 60 different species of fishes.

A treasure of Aboriginal rock paintings can be seen on the rugged cliff faces of Nourlangie Rock. Aboriginals have inhabited this area for at least 25 000 years. Some of the scenes in the 1986 movie, *Crocodile Dundee,* were shot within the park.

Below left: A *Pandanus* sp. with aerial roots in Kakadu National Park.
Below: One of several rocky outcrops in Kakadu National Park. These serve as Aboriginal art galleries, decorated with rock paintings that describe beings of the Dreamtime.

sailed aboard the *Beagle* from 1831 to 1836 on a voyage around the world that eventually led to the development of the theory of evolution.

Darwin, the capital of the Territory, was the only Australian city bombed by the Japanese in World War II. This first occurred on 19 February 1942. Darwin was destroyed, 243 people died, and over 300 were wounded. Sixty more air raids followed. Darwin was also devastated by cyclones in 1878, 1897, and 1937, and by Cyclone Tracy on Christmas morning in 1974, which demolished the city and killed 65 people. The swirl around Tracy's eye exceeded 259 km/h (155 miles/h). The population of Darwin dropped from 46 656 to fewer than 12 000 in the aftermath, but after reconstruction in 1978 it increased and is now 80 900 (1996 est., ABS). As a result of the reconstruction, Darwin is a modern tropical city.

Alice Springs, the Territory's second largest city with a population of 25 700 (1996 est., ABS), has a much wider temperature range. In January the average maximum and minimum temperatures are 36.6°C (98°F) and 22.2°C (72°F), but in July the figures are 19.3°C (66.7°F) and 4.5°C (40.1°F). The Alice, as it is called by the locals, lies along the Todd River, which is dry most of the time. It is a truly remote location, 1500 km (900 miles) south of Darwin and 1650 km north of Adelaide. The town was named for a nearby spring, where a telegraph station was erected in 1888. Alice Springs is a major tourist base for visitors to Uluru (Ayers Rock) and nearby Kata Tjuta (known also as the Olgas).

Other population centers are (1996 ests, ABS): Nhulunbuy (3915), Katherine (8809), Tennant Creek (3103), and Groote Eylandt (2563).

Industry

Beef cattle production, tourism, and mining of uranium, manganese, copper, gold, bismuth and bauxite are the major industries.

Queensland

Queensland, also known as the Sunshine State, covers 22.4 percent of Australia, or 1 727 200 km^2 (666 699 miles^2), 54 percent of which is in the tropics. In area, it is the second largest Australian state (Figure 1.16). Queensland's population totals 3 373 200, and is growing about 1 percent faster than the national average. Approximately one-fourth of Australia's Aboriginal population lives in Queensland.

Natural Features

The Great Barrier Reef, which is more than 2000 km (1250 miles) long and from 16 to 160 km off the east coast, is within the Queensland boundary. (For more information on the Barrier Reef, see Chapter 5.)

The Great Dividing Range extends along the entire N-S axis of the east coast. It runs from Cape York Peninsula through New South Wales to Victoria. Queensland's highest point is in the north— Mount Bartle Frere, 1611 m (5284 ft). Between the coastal ranges and the Great Dividing Range is a patchwork of rolling agricultural land.

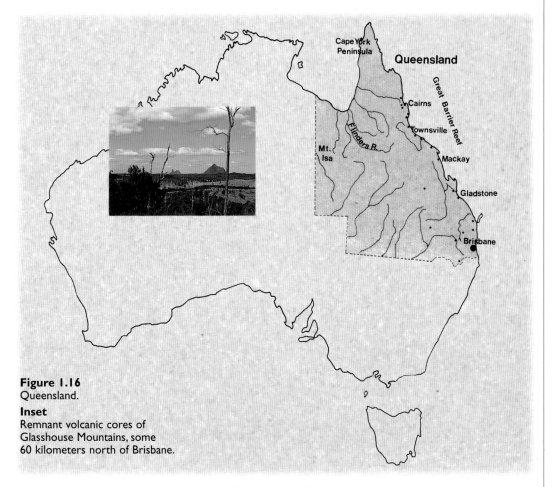

Figure 1.16
Queensland.

Inset
Remnant volcanic cores of
Glasshouse Mountains, some
60 kilometers north of Brisbane.

Some areas such as Darling Downs have rich volcanic soils as a result
of past lava flows. West of the mountains, plains stretch in all
directions. Under these plains lies the Great Artesian Basin, the
world's largest underground store of water. The basin underlies about
20 percent of the entire continent. This natural reservoir has been
tapped by over 18 000 bores to support cattle and sheep pastures.
It is recharged by rainwater from the eastern highlands which
gradually moves westward, as ground water, through aquifers.

On the north coast and along the southern river valleys there are
dense tropical forests, and beautiful sandy beaches line the coast.
Much of the west of Queensland is part of the arid Outback;
however, the southwest is known as the Channel Country because of
the maze of thousands of waterways which, when filled, become a
shimmery web of streams covering thousands of hectares.

Queensland has four principal drainage systems: the Pacific; the
Gulf of Carpentaria; internal Lake Eyre; and the Darling River. The
major rivers east of the Great Divide (Figure 7.1, page 126), which
empty into the Pacific Ocean, are the Barron, the Burdekin, and the
Fitzroy (with its tributaries, Burnett, Mary, and Brisbane). The streams
west of the Divide are generally only recognizable after wet weather.
The major ones that drain into the Gulf of Carpentaria include the
Gilbert, Norman, and Leichhardt rivers. The Diamantina, Thomson,

A view from Heron Island, Barrier Reef.

and Barcoo Rivers flow toward Lake Eyre in South Australia. The Darling system includes the Condamine and MacIntyre rivers.

Climate

Queensland has hot summers and warm winters, and the interior has the greatest variability in temperature. The heat is most intense in the north. Nighttime frosts sometimes occur in the southeast highlands and downs in winter, but the days are normally sunny, dry and about 20°C (68°F).

The heaviest rainfall is in the northeast coastal area and can average 4550 mm (179 in) annually. The rainfall diminishes as one goes west of the Great Divide. Birdsville, in the southwest corner near South Australia and the Northern Territory, receives only about 150 mm (6 in) annually. Queensland's rainy season is December to March, during which cyclones may move from the Coral Sea along the coast, causing substantial damage.

Cities and Towns

Brisbane is the capital of Queensland, and is the third largest city after Sydney and Melbourne. It has 1 525 500 inhabitants (1996 est., ABS) —roughly 45 percent of the population of Queensland (3 373 200). Brisbane is situated on the banks of the meandering Brisbane River, 483 km south of the Tropic of Capricorn. It receives 7.5 hours of sunshine a day, and its summer temperatures are 19–30°C (67–86°F) and winter temperatures are 6–18°C (43–65°F). Average rainfall is 1150 mm (45 in), half of which falls between December and March.

Other population centers include (1996 ests, ABS): the Gold Coast (339 600), Townsville (126 700), Cairns (104 300), Toowoomba (91 900), Ipswich (119 843), Rockhampton (68 300), and Mackay (61 600).

Industry

The pastoral industry (cattle and sheep) is very important in Queensland's economy, as is the cultivation of grain and dairying. Sugar cane is Queensland's major crop, especially on the northern coastal plains. The coastal areas also supply Australia's tropical fruits such as pineapples, bananas, mangoes, and papaws.

Queensland, like Western Australia, has been a leader in the mining boom since the mid-1960s. Coal is mined and silver, lead, copper, and zinc from the Mount Isa area are extremely valuable. Uranium and iron are also important.

Tourism, especially in the beach areas and the Great Barrier Reef, contributes greatly to the economy.

External Territories

Australia has seven external territories: Norfolk Island; the Cocos (Keeling) Islands; Christmas Island; the Coral Sea islands in the Pacific Ocean; the Ashmore and Cartier Islands Territory in the Indian Ocean; the Territory of Heard Island and MacDonald Island in the sub-Antarctic; and the Australian Antarctic Territory.

Norfolk Island lies in the Pacific Ocean 1676 km (1006 miles) east-northeast of Sydney. It was originally one of the harshest convict settlements —established in 1788. The last of the convicts were removed in 1855. In 1856, 194 descendants of the *Bounty* mutineers moved from Pitcairn Island to Norfolk Island at the invitation of Queen Victoria. Today, the permanent population of the island is about 1400, half of which is related to the original mutineers. In 1979 Norfolk Island became self-governing, with a Legislative Assembly and Executive Council.

The Cocos (Keeling) Islands are in the Indian Ocean, 2768 km (1660 miles) northwest of Perth. There are 27 islands composing two atolls with a total land area of 14 km² (5.4 miles²). There are about 400 permanent residents who are descendants of Malays brought to the islands as indentured laborers for the wealthy Clunies-Ross family. The Australian government bought the Clunies-Ross interests in the islands in 1978, and the Cocos Malays became Australian citizens. In 1984 the inhabitants voted for integration of the islands with Australia. There is a local government council and an Australian Government Administrator.

Christmas Island, 2623 km (1573 miles) northwest of Perth, is about 135 km² (52 miles²) in area. It was once the site of intense phosphate mining, an environmentally destructive activity which ceased in 1984. Currently there are about 1300 inhabitants of Chinese and Malay descent. The Australian government is attempting to find alternative industries for the island.

Other Australian external territories are uninhabited.

Spectacular views are common along the Great Ocean Road on Victoria's southwest coast.

The
Gondwana
Legacy

•••••••••••••
CONTINENTAL DRIFT

During the Permian Period, which began about 286 million years ago (MYA), a supercontinent known as Pangaea was forming by the coalescing of the early continents of Gondwana, Laurasia, and Angaraland (Figure 2.1). At the end of the Permian, roughly 248 MYA, half the families of shallow-water marine invertebrates disappeared from the fossil record in the short span of a few million years. Trilobites completely died out, and many other groups were decimated. This Permian mass extinction is probably related to the formation of Pangaea by the early continents. Such a union of separate continents into one large landmass would have eliminated much of the shallow-water habitat. A geological time scale is provided in Table 2.1 to show the relationships of the various periods.

The surface of the Earth today is the result of the division of Pangaea and the movement of the continents into their present positions over a period of time. This, of course, has implications for the biota carried on these continental blocks. Knowledge of how this continental drifting came about is dependent on an understanding of the 'anatomy and physiology' of the Earth, an understanding that has developed in the last three decades. The following discussion is somewhat technical, and if you do not wish to read all of it, you may prefer to study the illustrations only and then skip to the Wallace's Line section on page 46.

Anatomy of the Earth

The Earth's outer shell, or lithosphere, contains a crust about 30 km (19 miles) thick (Figure 2.2). Crustal thickness under the oceans may be as thin as 8 km and under mountain ranges as thick as 60 km. The density of the crust relative to other components of the lithosphere is 2.2–3.0. The next layer is the mantle, which is plastic rock about 2900 km thick, with a relative density of 3.5–5.7. The outer *core* of the Earth, about 2000 km thick, is made up of iron and nickel, probably in a liquid state with a relative density of 9.4–14.2. The solid center or inner core of the earth is about 1360 km thick, and has a relative density of 17.0.

The patches of thick, lightweight crust are the continents, which are separated from one another by larger areas of thinner, denser crust known as the ocean basins. About 70 percent of the Earth's surface is covered by water, including 10 percent which is submerged continental shelf. The light, thick continents can be considered as 'passengers' riding on the back of the thinner, denser oceanic crust.

Figure 2.1
End of the Permian Period, c. 250 MYA. The stippled areas are continental landmasses exposed above sea level, and the thick line represents submerged edges of the continental masses.

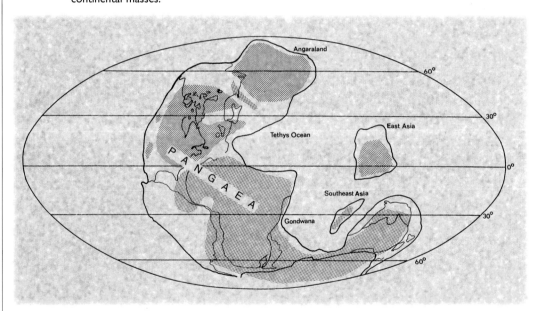

Figure 2.2
Structure of the Earth.

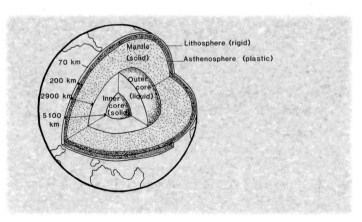

TABLE 2.1   **GEOLOGICAL TIME SCALE**

ERAS, PERIODS AND EPOCHS	SPAN OF TIME COVERED (MYA)*		EMERGING GROUPS OR EVENTS
	From	**To**	
Cenozoic			
Quaternary			
Recent	0.01	today	Retreat of last ice age (epoch covers only last 10 000 years)
Pleistocene	2	0.01	*Homo* (humans)
Tertiary			
Pliocene	5.1	2	*Australopithecus* (ape-man)
Miocene	24.6	5.1	Apes
Oligocene	38	24.6	Monkeys
Eocene	54.9	38	Lemurs, lorises, tarsiers
Paleocene	65	54.9	Origin of primates
Mesozoic			
Cretaceous	144	65	First flowering plants (period closes with of extinction dinosaurs)
Jurassic	213	144	First birds; abundant dinosaurs
Triassic	248	213	First dinosaurs, mammal-like reptiles; origin of mammals
Paleozoic			
Permian	286	248	Radiation of reptiles
Carboniferous	360	286	First reptiles
Devonian	408	360	First amphibians
Silurian	438	408	First jawed fishes
Ordovician	505	438	Spread of jawless fishes
Cambrian	590	505	Trilobites and other hard-bodied invertebrates
Precambrian			
Ediacaran	4600	590	First multicellular fossils of jellyfish, sea pens, wormlike animals, and algae

* Million years ago.

Source: Based on Harland et al. (1982) and Berra (1990).

Plate Tectonics

There are 13 major crustal plates, each of which is independently movable and continuously formed along a mid-ocean ridge. The continuous formation is balanced by continuous destruction of crustal material as it descends into the mantle along a sea floor trench. The destructive process is called subduction. The oceanic crust is like a conveyer belt system arising at mid-oceanic ridges and descending at deep oceanic trenches. This causes the plates and their continental passengers to move. This concept, called the plate

Figure 2.3
Rising convection currents in the Earth's mantle form new oceanic crust at the oceanic ridges. This new, denser material is carried away from the mid-ocean ridges by the convection currents and is returned to the mantle at subduction zones in the oceanic trenches. Volcanic islands and mountain belts are formed over the descending oceanic crust, which eventually becomes new continental crust. The spreading sea floor acts as a tape recorder, recording the magnetic field reversals and thus preserving a record of polar direction and rate of sea floor spreading. (Modified from Tarling, 1980.)

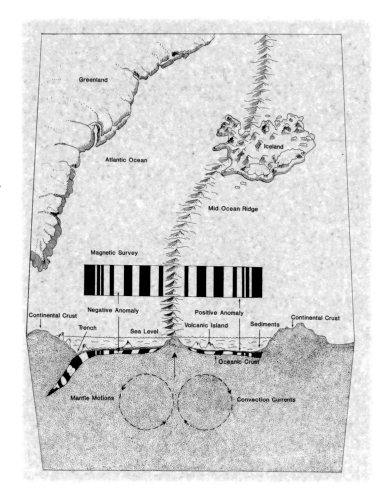

tectonic theory, was not known in 1915 when a young German meteorologist Alfred Wegener suggested that the continents could be fitted together like pieces of a giant jigsaw puzzle. This notion provoked a great deal of controversy at the time. Today, the concept of continental drift and its parent theory, plate tectonics, are firmly established in geophysics.

The Evidence

In the last 30 years a great deal of evidence has been gathered in support of the plate tectonics theory, some of which is outlined below.

Geometric Evidence

If one considers the continental slope at about the 500 m (1640 ft) contour as a boundary, a very good fit can be obtained by rotating and pushing the continents together. For example, the east coast of South America and the west coast of Africa fit rather well together (see Figure 2.14 on page 47).

Geological Evidence

Similarities in the rocks of what are now separate continents can be recognized as part of one landmass. For example, the former unity of the Caledonia-Appalachian Mountains, which are 420 million years old, reflects the union of North America, Greenland, and Europe as Laurasia. Even more striking is the similarity of the Roraima Formation in the Guyana region of South America, with an identical formation in western Africa. These billion-year-old pink rocks contain diamonds in the lower parts of the sediments, and were transported toward what is now the Atlantic Ocean. When a geometric reconstruction is made, both sides, from South America and Africa, match to produce a coherent pattern of sedimentation. Likewise, age discontinuities along the margins of South America and Africa match from side to side, and demonstrate that they were a single continent and that the South Atlantic Ocean began to form about 120 MYA.

Geophysical Evidence

Rock magnetism. As igneous rocks form, their iron ores become magnetized in the direction that parallels the Earth's magnetic field at that time. Likewise, as sediments are deposited, the magnetized grains line up under the influence of the prevailing geomagnetic field. Thus a record of the Earth's magnetic field at that time is preserved (Figure 2.3). This 'fossilized magnetism' shows that the north and south magnetic poles have alternated with each other (polarity reversal) and that the positions of the poles have slowly changed (polar wandering). When the magnetic pole was far from an area, the sediments and organisms indicate a tropical climate; and when the magnetic pole was near an area, there is evidence of glacial conditions.

Magnetic polar wandering reflects the motion of the continents relative to the Earth's rotational pole. Information so generated can be used to determine the ancient latitudes of an area. These data produce a continental pattern the same as that indicated by the geometric, geological, and paleoclimatic evidence. The idea that the Earth's magnetic field reverses its polarity provides a major proof for continental drift by explaining the process of sea floor spreading.

Sea Floor Spreading and Subduction

New ocean rocks form at mid-ocean ridges such as the Mid-Atlantic Ridge, the East Pacific Rise, the Indian Ocean Ridge and the Austral-Antarctic Ridge. Underwater volcanic eruptions release their lava at these ridges. As the lava cools, it becomes magnetized and flows to each side of the ridge. These magnetic ribbons in the ocean floor contain records of the time when the Earth's magnetic field was as it is today and when the field was reversed (Figure 2.3). These records are mirror images on both sides of a ridge. The age of the oceanic igneous rock increases from the site of the ridge. In other words, new oceanic material is being added at the site of the ridge and pushed laterally by convection currents. The oldest rocks are found at the greatest distance from the ridge. By moving the continents with

Figure 2.4
A *Glossopteris* leaf from the late Permian Period of Gondwana, about 248 MYA.

Figure 2.5
Lystrosaurus, a lower Triassic, terrestrial, egg-laying reptile of Antarctica, from about 240 MYA. Such a creature could not exist in the present-day climate of Antarctica—a point which supports the idea of continental drift and polar wandering.

their magnetic anomaly patterns until a particular pair of patterns from opposite sides of a ridge overlap each other, one can determine the position of the continents and oceans when the rocks of the anomaly were being formed at the oceanic ridge.

Since the Earth is not expanding, the increases in the area of the ocean floor must be offset by a loss of oceanic rock. This happens in the deep sea trenches where descending mantle rocks subduct (carry down) the dense oceanic crustal rocks into the mantle. (Figure 2.3.)

Oceanic crust is dense and the continental crust is much lighter—too light, in fact, to be carried into the mantle. Therefore, when the continental rocks of two plates collide at a subduction zone, mountain belts are formed, such as the Himalayas. As oceanic rock descends, volcanic island areas and mountain belts may rise over the descending oceanic crust, out of which new continental crust is formed.

The propulsion force that moves the 13 or so major continental plates is the heat of radioactive decay that generates convection currents in the mantle. Thus plate tectonics and sea floor spreading make it possible to accept Wegener's idea of continental drift.

Maps of the Past

The continents have drifted throughout the entire 4.5 billion year history of the Earth, but the best evidence available for reconstruction is of the continents' former position during the last 300 million years. During the Permian Period, ending about 248 MYA, the continents were joined as a large horseshoe-shaped supercontinent called Pangaea (Figure 2.1, page 38). The Tethys Sea occupied the space between the arms of the horseshoe. The continents of the southern portion, Gondwana (South America, Africa, India, Australia, and Antarctica) were linked in a continuous distribution, unbroken by epicontinental seas or high mountain ranges. Thus plants and animals had a broad area for dispersal, and climate, as influenced by latitude, was the major factor determining distribution of the plants and animals. Much of Gondwana lay in polar or high latitudes with attendant glaciation.

Cold floral and faunal groups appeared in Australia in the mid-Carboniferous and Permian periods. These were dominated by a remarkable seed plant, *Glossopteris* (Figure 2.4), which achieved wide distribution in Gondwana—indicating that the southern continents were joined. Its seeds probably could not have tolerated transport over wide bodies of water. The *Glossopteris*-bearing sandstones gave way to Triassic sandstones and conglomerates with fossil reptiles such as *Lystrosaurus* (Figure 2.5) in South Africa, India, China, and Antarctica. This adds supporting evidence to the concept of Gondwana as a giant southern continent.

The northern continent (Laurasia) was in tropical latitudes. Much of Europe was covered by epicontinental seas, while the equatorial and tropical regions developed a coal-forming flora. Angaraland,

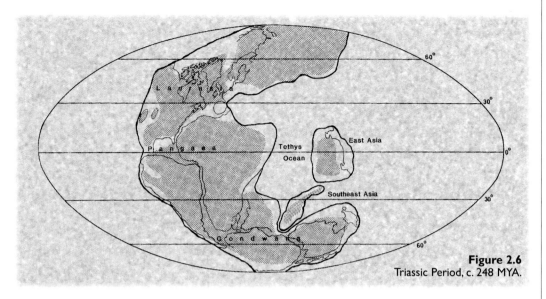

Figure 2.6
Triassic Period, c. 248 MYA.

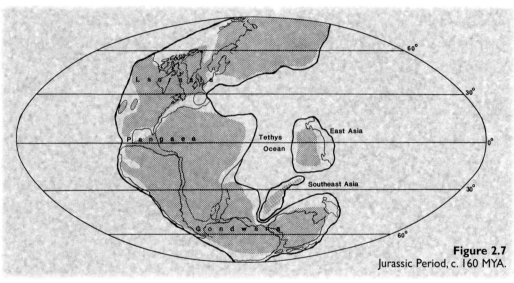

Figure 2.7
Jurassic Period, c. 160 MYA.

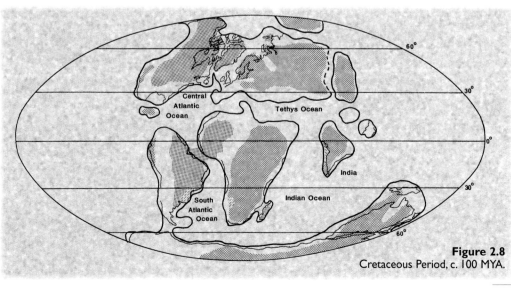

Figure 2.8
Cretaceous Period, c. 100 MYA.

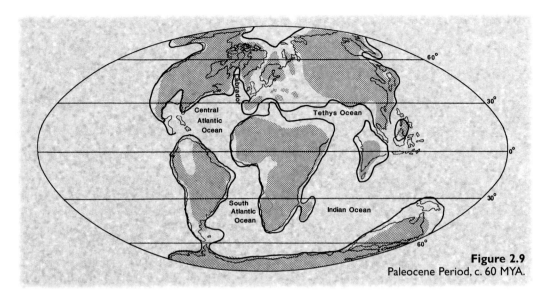

Figure 2.9
Paleocene Period, c. 60 MYA.

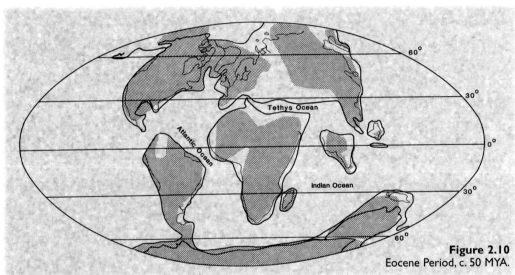

Figure 2.10
Eocene Period, c. 50 MYA.

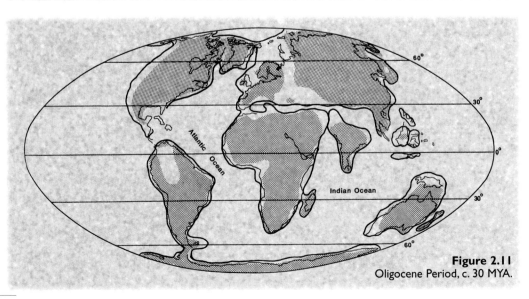

Figure 2.11
Oligocene Period, c. 30 MYA.

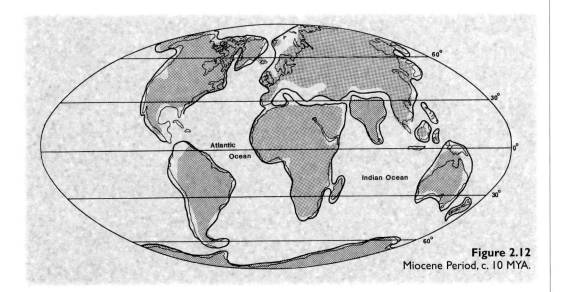

Figure 2.12
Miocene Period, c. 10 MYA.

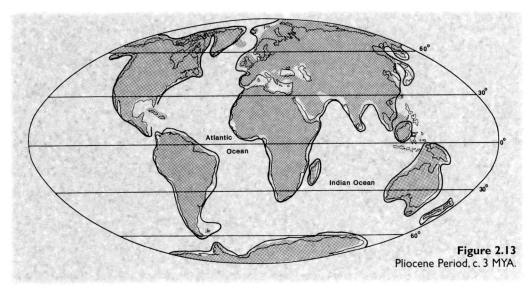

Figure 2.13
Pliocene Period, c. 3 MYA.

now Siberia, was disjunct from Laurasia and was temperate. The relationship of East Asia is not yet known.

Changes in sea level caused short-term effects such as flooding or increased aridity during the early Triassic Period, 248 MYA. During this period, Angaraland became connected to Laurasia as the sea level receded (Figure 2.6). Pangaea began to break up during the late Triassic/early Jurassic, around 213 MYA, with the first split occurring in the central Atlantic between North America and Africa. Gondwana began to split in the Jurassic, about 160 MYA (Figure 2.7). India separated from Africa, and lines of weakness appeared in Australia, Antarctica, Arabia, and southern Africa. This marked the beginning of the Indian Ocean.

During the Cretaceous Period, around 100 MYA (Figure 2.8), India drifted northward at a rapid rate (20 cm/yr; 200 km/million yrs), which isolated the Indian biota. The landmass passed from the high southern latitudes into temperate and then tropical conditions, finally joining with the main Asian landmass. The resulting collision produced the Himalayas. The rest of Gondwana remained linked via Antarctica until about 50 MYA.

Although Antarctica was increasingly in southern latitudes, its climate was still not a major obstacle to life. South America and Africa remained linked at Brazil but the South Atlantic was forming off Argentina. About 80 MYA New Zealand separated from Australia.

During the Paleocene Period, 60 MYA (Figure 2.9), Africa was separated from South America and Australia, which were still linked by Antarctica. The Central Atlantic Ocean was opening up but North America, Greenland, and Europe were united. Australia separated from Antarctica about 53 MYA, in the Eocene Period, and Europe drifted away from Greenland (Figure 2.10). Finally, South America broke from Antarctica around 30–35 MYA, in the Oligocene Period (Figure 2.11). Thus the Gondwanan continents began their independent existence.

Australia's northward movement is about 7 cm (2.75 in) a year. At that rate Hobart will reach the latitude of Sydney, and Cape York will be on the Equator, about 20 million years from now.

During the Miocene Period, around 10 MYA (Figure 2.12), Africa and India were joined to Europe and Asia, while South America remained an island. During the Pliocene (3 MYA) (Figure 2.13) the continents achieved their present positions and extensive glaciation began at both poles. At about that time, South America became joined to North America via the Isthmus of Panama.

· · · · · · · · · ·
WALLACE'S LINE

Alfred Russel Wallace (1823–1913) was a British naturalist and explorer of the Amazon (between 1848 and 1852) and of the Malay Archipelago (between 1854 and 1862). He arrived at essentially the same explanation of evolution—natural selection—as did Charles Darwin. Papers on the subject both by Darwin and Wallace were read at a meeting of the Linnean Society on 1 July 1858. With the publication of his monumental *On the Origin of Species* in 1859, Darwin secured his place in history as the principal originator of the cornerstone of modern biology, but Wallace is remembered more for his contribution to zoogeography than for his work on evolutionary theory.

It was in a letter to Henry Bates (naturalist-explorer of South America) in 1858 that Wallace first indicated that the world's landmasses could be divided into zoological regions. On the basis of mammal, bird, and insect distributions, Wallace recognized six zoogeographical realms, which he named as follows: Nearctic (North America except tropical Mexico), Neotropical (South and Central America with tropical Mexico), Palearctic (nontropical Eurasia and the northern tip of Africa), Ethiopian (Africa and South Arabia), Oriental (tropical Asia and nearby islands), and Australian (Australia, New Guinea and New Zealand) (Figure 2.14).

From 1860 onward, Wallace published a series of articles developing this topic, which culminated in his two volume *Geographical Distribution of Animals* (1876). Wallace drew a line between Bali and Lombok and between Borneo and Celebes (now known as Sulawesi) which passed to the east of the Philippines (Figure 2.15). The area to the west of this line was the Oriental faunal zone, and the area to the east was the Australian zone.

Thomas Henry Huxley named the boundary Wallace's Line. Huxley also modified it on the basis of bird studies, placing the Philippines east of the line, within the Australian realm. Wallace, however, ignored this change in later papers. Wallace's Line represents a major faunal break separating the rich Oriental continental fauna from the depauperate Australian island fauna.

Figure 2.14
Zoogeographical realms and major biomes of the world.

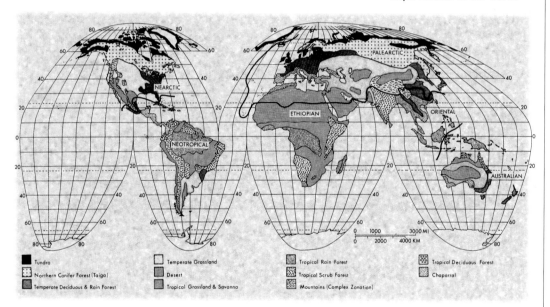

Freshwater fish distribution most closely follows Wallace's Line. Of the 23 families of primary division freshwater fishes (fishes with little or no salt tolerance, whose distribution has not depended upon dispersal through the sea) occurring in Borneo, only the Osteoglossidae has managed to cross into the Australian realm unaided by humans. Three genera *(Anabas, Ophicephalus, Clarias)* have probably been transported across the Makasar Strait to Celebes by humans. A few salt-tolerant species of *Oryzias* (Adrianichthyidae) have reached Celebes, Lombok, and Timor, and the endemic secondary division subfamily Adrianichthyinae occurs only on Celebes. The dominant freshwater fish family of the world is the Cyprinidae; however, only two cyprinid genera, *Puntius* and *Rasbora,* occur on both Bali and Lombok. *Rasbora* reaches Sumbawa. These two cyprinid

Figure 2.15
Lines of zoogeographical
importance in Wallacea.

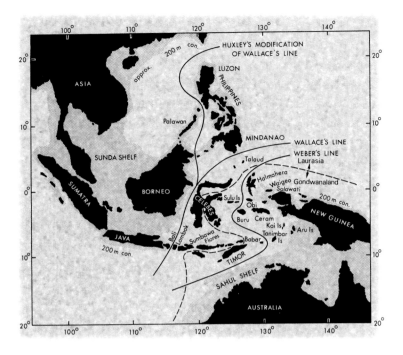

genera and a few others also occur in the Philippines, along with an endemic silurid and clariid catfish. With these few exceptions, the Asian and Australian freshwater fish faunas do not mix but, rather, end abruptly, unlike the faunas of North and South America which mingle in a Central American transitional zone.

Unlike the freshwater fishes, to which salt water is a serious barrier, some terrestrial vertebrates have managed to cross Wallace's Line (Table 2.2). Similar calculations show that about 68 percent of the Passerine bird fauna of Celebes comes from the west, as does 80 percent of the butterfly, amphibian, and reptilian faunas.

A second zoogeographical line, Weber's Line, represents the line of faunal balance that separated the islands which have a majority of Oriental groups from those having an Australian majority (Figure 2.15). Weber's Line is closer to Australia and reflects the fact that the Oriental fauna has spread more to the east than the Australian fauna has to the west. Which line, if any, one chooses to defend really depends on the taxonomic group in question and how many exceptions one is willing to tolerate. Furthermore, these lines are less relevant to plant geographers, whose subject is much more directly influenced by climate. These zoogeographical boundaries are discussed in detail by Mayr (1944) Darlington (1957), and Whitmore (1981).

Wallacea, the island area between Wallace's and Weber's lines, is one of the most geologically complex areas of the world. Geophysicists suggest that the first collision between Australia-New Guinea and the Asian islands occurred about 15 MYA (mid-Miocene), near Celebes. By the late Miocene or early Pliocene a land connection between Australia and Celebes was exposed. This could have provided a corridor for plant and animal migration.

The Makasar Strait between Borneo and Celebes may have been bridged in the mid-Miocene and again in the late Pliocene.

Recent evidence suggests that Celebes is of composite origin, from both Laurasia and Gondwana. Its biota, although predominantly Asian, is highly endemic but has elements of both zones. Modern consideration of Wallace's Line should circumscribe Celebes, which can be considered as a separate province. Whitmore (1981) provides further information on the relationship between Wallace's Line and plate tectonics.

TABLE 2.2 **PERCENTAGE OF SPECIES OF ASIAN ORIGIN IN THE LESSER SUNDA ISLANDS**

	Bali	Lombok	Sumbawa	Flores
Amphibians	92	78	75	70
Reptiles	94	85	87	78
Birds	87	73	68	63

Source: The Earl of Cranbrook (1989).

Aboriginal rock art in the
Australian Capital Territory,
depicting a kangaroo and a turtle.

The **3**
First
Australians

THE ABORIGINAL PEOPLE:
WHERE DID THEY COME FROM AND WHEN?

The original human inhabitants of Australia were named Aborigines by the European explorers. This name means 'from the beginning'. The prevailing anthropological opinion today is that Australian Aboriginals had a single racial origin, and arrived in northwest Australia about 50 000–60 000 years ago from Sundaland in southeast Asia. This would have been during the late Pleistocene—the Wurm or Wisconsin Ice Age. Sea levels were much lower, and the land areas were about 30 percent greater than today. The exposed Sunda Shelf united Asia and Sumatra, Java and Borneo (Figure 2.15, page 48). The gaps between the various islands of Wallacea would have been perhaps no more than 50 km (31 miles)—less than one-third of what they are today. The early Aboriginal ancestors might have been able to see each landfall as they set out on their rafts or outrigger canoes. Nevertheless, the ancestors of the Aboriginals were the world's first open-ocean sailors. Some reached the Andaman Islands, the Philippines, Timor, Sulawesi (Celebes), and New Guinea. The most likely entry point was from Timor into the northwest shelf of Australia. About 53 000 years ago, this would have been a 90 km sea voyage.

Once established in New Guinea, these Stone Age people had access to Australia and Tasmania, as all were connected above sea level by the Sahul Shelf at that time. Although each other's closest relatives, the considerable morphological differences between Australian Aborigines and New Guineans suggest that various localities of the Sahul Shelf were populated by different groups of people who subsequently adapted to their local environment. These landing sites are now 100 m (328 ft) below sea level, which makes it very difficult to recover traces of these earliest coastal settlers.

Stone tools and artifacts from Upper Swan Bridge Camp, near Perth, have been radiocarbon dated to between 37 100 and 39 500 years ago. Early stone tools found at Lake Mungo in southwestern New South Wales have been dated at 32 000 years old, proving that humans had reached the interior of Australia by that date. Human female bones from a cremation burial at Lake Mungo have been radiocarbon dated at 26 000 years. This is the earliest evidence of cremation practices anywhere in the world. Other sites show positive evidence of human presence about 20 000 years ago in the Northern Territory and the Nullarbor Plain. It is accepted that Aboriginal people have been established in the western half of Australia for at least 40 000 years and in the eastern half for the last 32 000 years. More discoveries may push these dates back further. A small skull fragment from Lake Eyre has been tentatively dated at 60 000 years ago by fluorine analysis of hemoglobin from the bone. The significance of this recent discovery and the validity of its date are still being assessed. Recent studies by Rhy Jones, archaeologist at the Australian National University in Canberra, of rock shelters in northern Australia reveal a date of 60 000 years as determined by thermoluminescence and optical dating. This is all the more amazing when one considers that the Neanderthals did not vanish from the European scene until about 35 000 years ago. The discovery close to the Hopkins River estuary near Warrnambool, Victoria, of seemingly cooked (blackened) shellfish remains may push the date of human occupation back to between 60 000–85 000 years ago if confirmed.

The ancient culture and the technological age. An Aboriginal man in traditional garb pauses while playing the didjeridoo for tourists in Sydney in order to take a call on his cellular phone.

An even earlier series of dates puts human habitation of Australia at 116 000 and 176 000 years ago. (See Wilford (1966) and Dayton and Woodford (1996) for this story, and Fullagar et al. (1996) for the scientific reference.) This is before most paleoanthropologists think that modern *Homo sapiens* left the African continent. Anthropologist Richard Fullagar, from the Australian Museum, and geologists Lesley Head and David Price of the University of Wollongong applied the thermoluminescence technique, which measures electrons trapped in quartz crystals, to the sediments containing red ochre and stone artifacts from a rocky habitation site in the Northern Territory just across the border from the town of Kununurra, Western Australia. If these 100 000+ dates hold up under the intense scrutiny that is sure to follow—and that is a very big if—paleoanthropologists will have to reassess the later phases of human evolution. Was it archic *Homo sapiens* that first arrived in Australia, probably from Indonesia? Such a boat journey would have involved a high degree of social organization and sophistication. Or did anatomically modern *Homo sapiens* leave Africa earlier than the 100 000 or so years ago currently accepted. In addition to pushing back the arrival of humans in Australia, Fullagar and colleagues report rock art at a date of 75 000 years. Thousands of 30 mm (1.2 in) circular depressions (cupules) about 15 mm deep are reported to have been carved into sandstone walls at a locality called Jinmium. This is twice as old as European cave paintings and 15 000 years older than previously reported Australian rock art. As exciting as these discoveries are, for the moment and until these new, very ancient dates are confirmed with another dating technique, it is prudent to accept an estimate of 50 000–60 000 years ago as the arrival time of the ancestral Aboriginals in greater Australia (Australia plus New Guinea).

Aboriginal rock art of a kangaroo in Kakadu National Park, Northern Territory, showing the anatomically correct position of the testes anterior to the penis.

Paleoanthropology

The cradle of the human family is the African continent. Darwin correctly predicted this in *The Descent of Man,* published in 1871. There was virtually no hominid fossil record known at that time. Darwin's reasoning was biogeographical and elegantly simple. He wrote:

> In each great region of the world the living mammals are closely related to the extinct species of the same region. It is therefore probable that Africa was formerly inhabited by extinct apes closely allied to the gorilla and chimpanzee: and as these two species are now man's nearest allies, it is somewhat more probable that our early progenitors lived on the African continent than elsewhere.

The earliest known member of our family (Hominidae) is *Ardipithecus ramidus* from Ethiopia, the fossils of which date back to 4.4 million years ago (MYA). This ape-like creature may be a side branch or an ancestor to the australopithecines. In fact, *Ardipithecus* was originally assigned to the genus *Australopithecus,* but later moved to its own genus as more fossils became available. The genus *Australopithecus* dates back to about 4.2 MYA and is only found in Africa. This lineage is most likely derived from the as yet undiscovered common ancestor of apes and humans. So far, there are six known species, all of which are small-brained, upright walkers. The recently described *A. anamensis* (4.2-3.9 MYA) is a possible ancestor of *A. afarensis,* or 'Lucy', whose species lived from about 3.8 to 2.8 MYA. Other australopithecines include *A. africanus* (2.8–1.9 MYA), who may or may not be in the direct line of human evolution, and the more robust australopithecines such as *A. aethiopicus* (2.5 MYA), *A. robustus* (2.0–1.5 MYA), and *A. boisei* (2.0–1.2 MYA).

A. afarensis (or *A. africanus* according to some paleoanthropologists) gave rise to the first known species of our genus, *Homo habilis,* meaning 'handy man'. This species was the first to make and use stone tools. *H. habilis* lived in Africa from 2.1 to 1.5 MYA and gave rise to *H. erectus* (1.8–0.25 MYA), which evolved in Africa and rapidly spread to Asia. *H. erectus* was the first hominid to leave the African continent. Specimens of *H. erectus* that date back to 1.8–1.7 MYA have been found in Java. *H. erectus* descendants eventually became recognized as *H. sapiens,* the earliest form of which has many *H. erectus* features and is referred to as archaic *H. sapiens.* These fossils date to roughly 500 000 years ago.

Around 150 000 years ago another form appeared—the Neanderthals. They had heavy brow ridges, elongate skulls and very weak chins. Neanderthal people inhabited Europe and Western Asia, and are considered by some anthropologists to be a subspecies of *H. sapiens* and by others to be a distinct species, *H. neanderthalensis.* Neanderthals disappear from the fossil record at about 35 000 years ago.

A bark painting of Wandijina or Windjedda, a mouthless spirit, from Western Australia.

Somewhere around 100 000 years ago, anatomically modern
H. sapiens appeared. According to the Out of Africa Hypothesis, these
species first evolved from *H. erectus* in Africa and spread to the rest of
the world. However, the Regional Continuity Hypothesis states that
H. erectus in each region evolved into *H. sapiens* of that region. Some
prominent Australian anthropologists support the latter view due to
the high incidence of *H. erectus* traits in Aboriginal fossil. (See the
April 1992 issue of the *Scientific American* for more details of these
two ideas.)

A study by Swisher et al. (1996) redated some *H. erectus* fossils
from Solo, Java and concluded that this species did not become
extinct about 250 000 years ago, but was still extant as recently as
27 000–53 000 years ago. If their redating holds up and if the
specimens are actually *H. erectus* rather than *H. sapiens*, then three
species of humans were contemporaneous in various parts of the
world—*H. erectus*, *H. neanderthalensis*, and *H. sapiens*. This might pose a
problem for the Regional Continuity model because the *H. erectus*
people from Solo may actually have been younger than their
presumed descendants, the first Australians. However, Regional
Continuity proponents point to a host of morphological characters
that are similar in Solo skulls and the earliest humans from Australia.
Time and more fossils will tell. This is why it is so important to
continue paleoanthropological studies in Australia.

Physical Anthropology

Skulls from Kow Swamp in Victoria show robust archaic features,
suggesting that some *Homo erectus* genes may have persisted in
Australia as recently as 10 000 years ago. Modern Aboriginals have
more *H. erectus* features than any other modern race. This suggests
regional genetic continuity of *H. erectus* through specimens from Solo,
Java, to Australian Aboriginals. This may be a reflection of an early
breeding line going back to *H. erectus* populations of Java during the
Pleistocene. The Lake Nitchie skeleton, dated at about 6820 years
ago, has the highest number of archaic traits of any known Aboriginal
skull. On the other hand, Lake Mungo specimens (c. 30 000 years
ago), as well as fossils from Keilor (c. 15 000 years ago) and Green
Gully (8000 years ago), are very gracile (lightly built) in contrast
with the Kow Swamp specimens (c. 13 000 years ago). However,
individual features of fossil skulls found in Australia fall within the
range of variation of modern Aboriginals. It remains to be explained
why modern-looking gracile specimens from Lake Mungo predate
the archaic-looking robust specimens from Kow Swamp. Some
authorities consider that the robust and gracile remains reflect a dual
origin for Aborigines, while other experts suggest that the robust
specimens were males and the more gracile forms were females.

The oldest example of a *H. sapiens* skull in southeast Asia is from
Niah Cave in Borneo. Its surrounding charcoal is dated at 40 000

years old. It can be speculated that the Aboriginals were the earliest generalized *H. sapiens* to arrive at their ultimate geographical destination. This could account for the slightly higher occurrence of *H. erectus* traits in Aboriginals compared to New Guineans, Orientals, and Europeans. Older references consider Aboriginals as a separate racial group along with four other major racial categories of humans—Caucasoid, Khoisanoid (San Bushmen), Mongoloid, and Negroid. However, it should be kept in mind that humans are a highly variable species, and that the vast majority of human genetic variation occurs within any given race, not between races. A large enough random sample of people from any race will contain about 85 percent of all the genetic variation of our species. The remaining 15 percent of the genetic variation has a geographical pattern, with about 9 percent of the variation being due to differences among linguistic and cultural groups *within* any given race. The remaining 6 percent of human genetic variation reflects inherited differences between races, with many of these differences being primarily adaptations to the local environment.

The ability to learn from Aboriginal material may soon be sharply curtailed or eliminated altogether as a result of some Aboriginal groups insisting that university and museum holdings of skeletal material be reburied or cremated. In some cases, this has already happened. Laws are being contemplated that may bring an end to physical anthropology in Australia. There are few Aboriginals trained in physical anthropology to explain the value of research to contemporary Aboriginals. A similar phenomenon is sweeping North America, where federal laws that put native rights and religion ahead of science are affecting anthropology and archaeology, and causing invaluable material to be lost forever.

Full-blooded Aboriginals are characterized by a dark skin, broad nose, prominent brow ridges and curly to wavy hair. The male height ranges from 145–190 cm (4.8–6.2 ft). The range of skin color is from dark bronze to chocolate, but not as dark as some African Negroes. Hair color is black or brown, but blonde hair is common until puberty among central Australian Aboriginals. The hair is never woolly. The body frame is slim with long legs and arms, which is an efficient adaptation for radiating body heat to the environment. The linear body gives a graceful appearance to walking and running movements. Aboriginals from cooler southern areas tend to be stockier, this being an adaptation for conserving body heat. Brow ridges range from moderate to highly prominent above dark brown, deeply-set eyes. Full-blooded Aboriginals have the largest jaws and teeth of any living population of humans. The noses of full-blooded Aboriginals are short and broad with a high bridge. Most Aboriginals belong to blood groups O and A, but there are a few individuals with B and AB in Arnham Land and Queensland. The Rh negative factor is absent among full-blooded Aboriginals. Blood group N is the world's highest at 70 percent.

Aboriginal school children in Alice Springs, Northern Territory.

Aboriginal women and child in Alice Springs, Northern Territory.

The skulls of male and female Australian Aboriginals tend to be thicker than the cranial vaults of other races. This thickness is not correlated with size. Peter Brown and Graham Knuckey of the University of New England suggested that the ritual of settling disputes by blows to the head with digging sticks exerted selective pressure for increased skull thickness. Of 409 Aboriginal skeletons examined from South Australia, 57.4 percent had depressed fractures. Also common were healed fractures of the distal forearm, used to parry blows. Skeletons more than 11 000 years old show similar fractures indicating that this method of conflict resolution is ancient. It is still practised today and continues to be a social problem. The effects are even more detrimental today, as more European genes have developed thinner vaults in the Aboriginal population.

Past and Present Hypotheses of Origin

Until recently, anthropologists favored a *trihybrid hypothesis* of Aboriginal origin, developed in the 1930s by Joseph Birdsell, an American anthropologist. According to this hypothesis there were three separate and successive waves of colonization by three distinct racial types. The first wave was by the oceanic Negritos, called Barrineans. They colonized Australia and eventually dispersed to Tasmania before it became isolated from the mainland about 12 000 years ago due to rising post-glacial sea levels. Full-blooded Tasmanian Aboriginals are now extinct, having succumbed to European disease and persecution. The last full-blooded Tasmanian died in 1905, but mixed-blooded descendants still exist. The second wave was that of the Murrayians, who were dispersed by European settlement but some survive today. The Murrayians were replaced in northern coastal areas by the third wave, that of the Carpentarians, who were darker and had less body hair. The older anthropological literature developed complicated, elegant explanations of these three successive waves, but in recent years the *homogeneous hypothesis* has come to prevail. Genetic (blood groups, fingerprints, serum proteins and enzymes) and linguistic evidence seem to support the concept that Aboriginals are of homogeneous origin and distinct from other ethnic groups. The fact that the very early remains are similar to modern Aboriginals and the suggestion that various physical differences evolved as adaptations to varying environments all seem to disprove the trihybrid hypothesis, which is rejected today by most anthropologists.

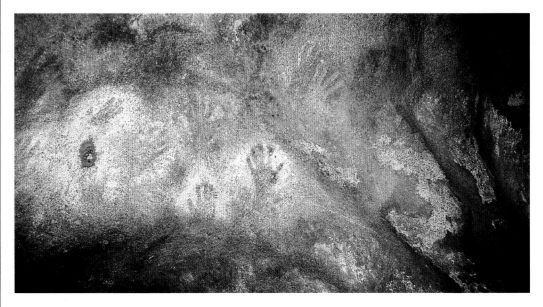

Aboriginal hand prints created by spitting pigment around a hand placed on a rock. This example is from Wave Rock, Western Australia.

How Many Aboriginals Were There, and Who is an Aboriginal Today?

It is thought that at the time of European colonization, 200 years ago, there were at least 300 000 and possibly 1 000 000 Aboriginals throughout the continent. They were organized into about 500 tribes speaking about 260 distinct languages. A tentative population estimate of 750 000 is reasonable and was probably the carrying capacity that could be supported by their hunting-gathering culture, which did not include agriculture or animal husbandry, although protoagriculture in the form of yam planting existed. Small populations amid an abundance of game and a diverse larder of plant foods did not want, in spite of the lack of agriculture. The population most probably remained near the 750 000 figure for thousands of years until the arrival of the Europeans.

By the 1920s the Aboriginal population had been reduced to approximately 60 000. Most fell victim to diseases brought by Europeans (such as measles, whooping cough, influenza, tuberculosis, syphilis, and gonorrhea) to which the Aboriginals had no immunity. Others succumbed to alcohol or were killed by the new settlers. The 1986 census indicated that there were then some 227 645 Aboriginals and Torres Strait Islanders, of whom perhaps 40 000 were full-blooded, although census figures did not differentiate between full-blooded and mixed descent. The total 1986 number is an increase of 42 percent over the 1981 census figure and reflects an increased desire among Aboriginals to identify themselves as such. The most recent survey (30 June 1994) estimated that there were 303 300 Aboriginal and Torres Strait Islanders. Aboriginals form about 1.7 percent of the Australian population.

An Aboriginal is defined as 'a person of Aboriginal descent who identifies as an Aboriginal and is accepted as such by the community in which he or she lives'. The 1994 estimates for the Aboriginal and Torres Strait Islander population of each state are:

An Aboriginal man in Alice Springs, Northern Territory.

New South Wales (including Australian Capital Territory)	80 500
Queensland	79 800
Western Australia	47 300
Northern Territory	46 000
Victoria	19 200
South Australia	18 400
Tasmania	10 100

Aboriginal and Torres Strait Islander people make up about 27 percent of the total population in the Northern Territory, but less than 3 percent in all other states. About 28 percent of Aboriginals were living in capital cities (1994 data), and 40 percent were under 15 years old (1991 data).

TORRES STRAIT ISLANDERS

Australia's other indigenous people are the Torres Strait Islanders, a Melanesian minority of some 21 000 whose culture is basically Papuan. The Torres Strait Islands are scattered across the shallow sea separating Papua New Guinea from the northeastern tip of Australia. In fact, the international border between Australia and Papua New Guinea passes through the islands which are an Australian possession administered by the Queensland government.

More than half of the Torres Strait Islanders live on the Australian mainland along the coast of north Queensland, especially in Townsville and Cairns. The rest live on 15 islands between the tip of north Queensland and the southern coast of Papua New Guinea, in communities ranging from 30 to 400 people. More than 2000 live on Thursday Island, the administrative and business center, and nearby islands.

Torres Strait Islanders in the west and central island groups reflect Melanesian origins, while those in the east show some Polynesian heritage. The area is also an amalgam of Aboriginal and Papuan stock. There is even evidence of European, Malay, Chinese and Japanese ancestry. The western islanders, who inhabit islands located only a few kilometers off the Papuan coast, speak a distinct

Aboriginal language, while the eastern islanders speak a Papuan language. The physical characteristics of Torres Strait Islanders are Papuan (Melanesian). Interestingly, the hair type of Aboriginals from northern Cape York is also Melanesian, suggesting that cultural and trading links across Torres Strait have undoubtedly led to some gene flow.

Unlike Australian Aboriginals, agriculture was practiced extensively by Papuans and the eastern Torres Strait Islanders, perhaps because the eastern islands, which are of relatively recent volcanic origin, have rich soils and less marine resources than the older, more elevated western islands. Likewise, the bow and arrow were used in Papua and the Torres Strait Islands, but not by Australian Aboriginals who used spears and spear throwers. Cape York Aboriginals' main items of trade were spears, which were eagerly sought by islanders who used them for harpooning dugong.

Today, Torres Strait Islanders are Australian citizens, with the attendant rights and benefits; they have been eligible to vote in federal and state elections since the 1960s. Fishing and pearling are their traditional sources of income, which remains the mainstay of Torres Strait Islands economy today.

• • • • • • • • • •

ABORIGINAL CULTURE

For thousands of years Aboriginal people used simple implements made of stone, bone, wood, and shells. As few Aboriginals today live a fully traditional lifestyle, the use of such implements is now comparatively rare. Most tools served more than one function. The world's oldest known ground-edge hatchets date to about 22 000 years and come from sites near Oenpelli in Arnhem Land, in the Northern Territory. Aboriginals were using similar stone hatchets only 200 years ago. Digging sticks, wooden spears, and scraping stones were also used in very early times. The spearthrower, or woomera, was a later development. This clever device was a length of thin wood with a wooden tooth lashed to one end. The spear shaft was laid along the woomera with its end against the tooth, and the woomera acted as a lever to impart a great deal of thrust to the spear. Bows and arrows were never utilized by Aboriginals.

Spearthrower, or woomera, and a coolamon used for carrying food and water.

Boomerangs have been in use for at least 10 000 years. There are many different designs for different purposes. Those that do not return are more properly called killing sticks. These were hunting and fighting weapons which could kill kangaroos, birds, and reptiles or cause serious injuries in warfare when thrown or used as a club. The spinning action caused the tips to travel at high speed, imparting a more severe blow than a non-revolving object such as a rock. A boomerang that ricocheted off the ground into the body of an opponent could cause a serious injury.

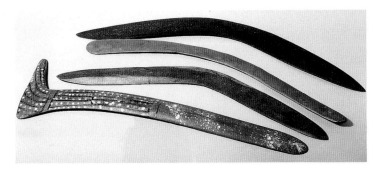

Left: Boomerangs. The top three are returning boomerangs, and the bottom one is a non-returning killing stick used as a club. The boomerang second from the bottom shows the two colors of mulga wood, *Acacia aneura*. Below: A didjeridoo carved from a hollow tree limb.

The returning boomerang is thin and crescent-shaped. It was used for amusement and for bird hunting. With a crude net stretched across a flyway, a hunter would throw the boomerang over a flight of oncoming ducks, parrots, or pigeons, and at the same time make a call of a hawk. The shadow of the boomerang would cause the birds to dive and some would be trapped in the net, while the boomerang would return to the thrower. Boomerangs were also thrown *at* flying birds; if the target was not knocked out of the sky, the boomerang would return to the hunter to be thrown again. Boomerangs were used, too, for skinning kangaroos, digging cooking pits, and making rhythmic clapping sounds for dancers. The aerodynamics of boomerangs are discussed by Hess (1968) and Thomas (1983).

Another implement used by the Aboriginals is the bullroarer. This thin wooden slat was whirled around by a cord. The roaring sound it made represented the Rainbow Serpent and warned off women and uninitiated males.

The didjeridoo is a well-known Aboriginal wind instrument consisting of a length of hollow wood, usually ranging in size from 0.5 to 2 m (1.6–6.6 ft). It is often ornately painted. The player blows into one end and vibrates his lips to produce a sophisticated musical tone. An Aboriginal gathering for singing and dancing or for religious ceremony is called a corroboree.

Aboriginal people's lifestyle stood in relative harmony with their environment, even though they modified their surroundings, increasing the extent of grasslands by the use of fire. The people saw themselves as part of the natural world, attributing human characteristics to natural phenomena, such as rocks, plants, and animals. Their complex systems of beliefs were expressed in songs,

Aboriginal rock art at Burrunguy (Nourlangie Rock) in Kakadu National Park, Northern Territory. This composition of people and mythical beings at the Anbangbang shelter was painted in 1964 by artist Najombolmi ('Barramundi Charlie'). Depicted in the painting are the evil spirit Namandi (largest figure at top); the Lightning Man Namargon, who is encircled by a bolt of lightning; and his wife Barging, whose left foot is almost touching a barramundi, *Lates calcarifer*. A group of people (Binin) are painted on the bottom. Water flowing down the face of the rock and rubbing by water buffaloes have obliterated some of the painting.

dances, and oral literature passed from generation to generation. Their complex mythology was centered on the so-called Dreamtime, or 'golden age of the past', in which spirit ancestors travelled through the land and gave it physical form. The term 'Dreamtime' is an inadequate translation of the Aboriginal concept, which does not refer to ideas obtained as a dream but rather is a continuing philosophy of life—one to which many Aboriginal people still adhere. The activities of the Great Rainbow Snake or the Fertility Mother still survive in stories, paintings, and ceremonies. The ancestral creative beings of the Dreamtime imparted the kinship system, marriage rules, and social norms of Aboriginal culture. Sex and age were the major determinants of status, and kinship rules governed social interaction.

There was little recognition of the physiology of procreation in Aboriginal culture. Intercourse and semen were less relevant to conception than the spirit-child who entered the mother and grew within her stomach. Circumcision was widely practised as a ritual initiation to manhood. Some Western Desert people also practiced subincision, in which the underside of the penis was slit open from the urethral opening to the base. This was done as a badge of full manhood and was believed to make men resemble Dreamtime beings. It also emulated emus, who have a grooved penis. This practice resulted in 'splashy' urination, but did not seem to reduce virility.

A man could have more than one wife. His first wife was usually much older and a widow. A second wife might be a very young girl betrothed to him at his circumcision. These arrangements helped to cement relationships between families and groups. When a visitor came to a camp, a man might 'lend' his wife in order to show hospitality, and she was expected to comply.

The family was a self-sufficient economic unit and labor was rigidly divided by sex. The women gathered grain, roots, berries, grubs, and other small insects, prepared the meals, and carried the eating utensils. The men hunted game and located water. Much of the people's life was spent moving from place to place seeking food and water. Fire was known for thousands of years and was made by friction and percussion. Food was cooked in an earth oven.

Several groups of families made up a tribe, which occupied a given area and claimed hunting rights in that area. The tribe had a common language and ranged in size from 100 to 1500 people. There were no chiefs or rulers, only elders who enforced the social and religious rules.

There were three kinds of visual art and craft:

- Painting was done on tree bark strips, sand, rocks, and caves. The now famous 'x-ray' paintings show the internal organs and bones of animals. Other styles include the application of hundreds of dots of color. Earth pigments such as red and yellow ochers, charcoal, and white clay were used, with feathers, human hair, and chewed twigs serving as brushes.
- Carving and engraving. Rocks and wood were carved and engraved, and human skin was incised. Spears, woomeras, boomerangs, didjeridoos, and wooden bowls were decorated by carving and painting.
- Weaving. Fiber baskets, mats, and bags were woven from grasses, and elaborate string headdresses were made for ceremonies. Tree bark was used for making shelters, and was folded and tied for carrying anything from babies to water.

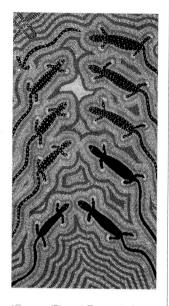

'Goanna/Pirenti Dreaming', an acrylic dot painting on canvas by Janet Forrester of the Luritja/Arrernte language group. Moving from the bottom to the top, the painting tells the following story: Goanna and Pirenti met and started talking. Pirenti wanted to paint Goanna, and he put dots on Goanna. Pirenti wanted to be painted the same and Goanna painted him. They went to the waterhole. Goanna got wet, and the paint smudged. Pirenti chased him, but Goanna went into a hole and got away. (Goanna can be seen disappearing into the hole at the top left, where only his hind legs and tail are visible.) [From the private collection of the author.]

ABORIGINAL RIGHTS

In the 1960s, legislation was purged of discriminatory provisions, and today Aboriginals and Torres Strait Islanders have the benefits of other Australian citizens, such as the right to vote and access to social security benefits. A national referendum held in 1967 gave the Australian government the power to pass special laws for Aboriginals. Government policy is based on the right of Aboriginals to retain their traditional lifestyle and racial identity, or to adopt a partially or wholly European lifestyle. The Australian government provides education, housing, and health care for Aboriginals. Since the passage of the *Aboriginal Land Rights Act,* 428 127 km² (165 257 miles²) of the Northern Territory (33 percent of its total area) has been granted to Aboriginals and 12 percent more is under claim. In South Australia Aboriginals have freehold title to over 183 146 km²

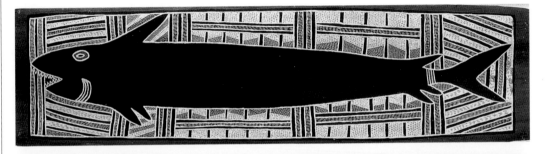

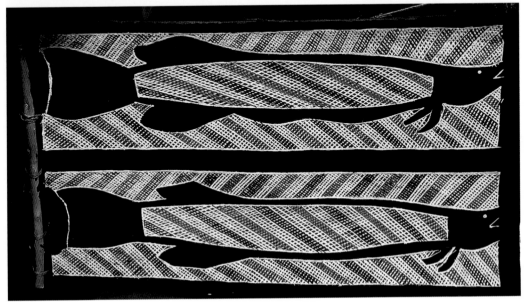

(70 600 miles²), which is about 19 percent of the total area. Much of the Aboriginal land in Western Australia is rich in minerals and may some day result in increased wealth for the indigenous people, whose reserves and leases currently occupy about 227 788 km² (87 975 miles²) or 9 percent of the total land area of Western Australia. About 12 percent of the total area of Australia was Aboriginal freehold, leasehold, or reserve land as of November 1987, much of it in isolated, semi-arid areas.

In 1992 the High Court ruled that the concept of *terra nullius* (in effect, unowned land), under which the European occupation of Australia had proceeded, was not valid—and that a form of 'native title' existed prior to European settlement. This native title could still exist on land where the government had not issued title. *The Native Title Act*, passed by the Commonwealth Parliament in December 1993, established a system of courts and tribunals to deal with native title matters. Aborigines and Torres Strait Islanders must prove continuing association with the land claimed to gain native title. The Act also set up a Land Acquisition Fund designed to help Aboriginals to buy land to which they proved native title.

An appreciation of Aboriginal culture is slowly developing among non-Aboriginal Australians, although a great deal of prejudice is still evident. Aboriginals are worse off than any other group of Australian

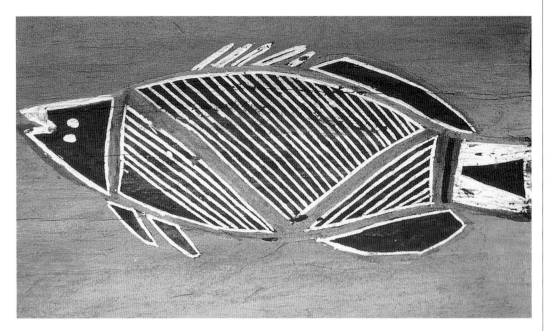

citizens. Alcoholism remains a devastating problem for them. Their life expectancy is 20 years less than that of non-Aboriginal Australians, infant mortality is three times higher, unemployment is six times the national average, and average income levels are half those of other Australians. However, since the 1960s increased government spending on health, housing, education, and other services has improved prospects for a better life and has enabled Aboriginals to hold more responsible positions, especially in official agencies.

Above and opposite: Aboriginal bark paintings of a shark, bonytongue, *Scleropages leichardti,* and barramundi, *Lates calcarifer.*

Sturt's desert pea,
Clianthus formosus.
White or pink flowered
forms also occur occasionally.

A Floral Mosaic: The Plants

4

••••••••••••••
EVOLUTION OF THE PLANTS

The present Australian flora is a mosaic of the world flora that has developed under conditions of isolation and varying degrees of aridity. It shows many Gondwanan affinities to Africa, South America, and New Zealand. It is closely related to formerly geographically connected land areas where climates are similar. For example, the southern beech *(Nothofagus)* forests, are similar in Australia, New Zealand, and South America. Tropical rainforests of Queensland and New Guinea were once connected, and the alpine floras of Tasmania, New Zealand, and South America reflect past connections. It is the drier parts of Australia that have the characteristic Australian elements of *Eucalyptus,* acacias, casuarinas, spinifex, *Banksia,* and so on. These archetypical Australians are derived from ancestral Gondwanan elements. There are about 15 000 species of Australian flowering plants, of which approximately 80 percent are endemic. From Gondwanan beginnings the Australian flora evolved in isolation for about 30 million years without significant foreign input. In the last few million years invaders from Asia have migrated to Australia, especially in the north.

Development of the Flora

In *The Flowering of Gondwana,* Mary White (1990) provided us with insights into the historical development of the Australian flora. Around 400 MYA, at the end of the Silurian Period and beginning of the Devonian Period, small, primitive vascular plants colonized the land. The lycopods (club mosses) are examples which still exist today. Giant club mosses, such as *Lepidodendron,* dominated the landscape in tree-like form into the Carboniferous Period.

During the Devonian Period, secondary growth evolved which allowed plants to increase in diameter, resulting in thicker trunks that could support greater heights. A vascular system capable of transporting nutrients and water to all parts of the increasingly larger plants developed. The retention of female gametes accompanied the increase in size, as plants now grew at a greater height from the ground where wind dispersal of spores was facilitated. This eventually led to seed formation. Giant horsetails, similar to the modern genus *Equisetum,* were typical of that period. In the late Devonian and early Carboniferous times, fern-type foliage appeared in the fossil record. The giant club moss flora died out as Australia drifted closer to the South Pole by the mid-Carboniferous, around 325 MYA. Primitive seed-ferns with early leaf structure and venation—part of the *Racopteris* flora—dominated the late Carboniferous Period.

The climate became more moderate during the Permian Period, giving rise to the *Glossopteris* flora, a group of gymnosperms that flourished about 286–248 MYA. Also during this time ginkos originated, tree-ferns were abundant, and coal deposits were accumulating as the altered remains of cold, swamp-plant communities. Fossilized tongue-shaped *Glossopteris* leaves (Figure 2.4, page 42) are widespread over all of the southern continents, reflecting the landmasses' connection during the Permian—as Gondwana. *Glossopteris* vanished from the fossil record in the Triassic Period, 248–213 MYA. During this time of increasing drought, many plants evolved adaptations which enabled them to survive in hot, dry environments. The mid- to late-Triassic flora was characterized by forked-frond seed-ferns called *Dicroidium,* which were associated with tree-ferns, ginkos, cycads, conifers, and horsetails.

All of these groups prospered into the hot and wet Jurassic Period, 213–144 MYA. This was the last geological period to be dominated by ancient plant groups. A more modern flora began in the Cretaceous Period, 144–65 MYA, which was a time of great change. Gondwana was breaking up, sea levels were changing, the climate fluctuated, and the flowering plants (angiosperms) were evolving. These most primitive flowering plants were able to spread to all of the southern continents before Gondwana separated into continental fragments. As the various continents drifted apart, their floras evolved in isolation.

Sea levels receded during the late Cretaceous Period, revealing denuded areas of landscape ready for colonization by flowering plants. Angiosperm pollen, or something very close to it, first appeared in the fossil record of the early Cretaceous (about 140 MYA) in West Gondwana (western South America and eastern Africa) and spread north and south. Australia received its inoculation with angiosperms from West Gondwana across Africa, India, and Antarctica in the late Cretaceous. This explains some of the similarity of the floras of these areas.

The *Nothofagus* (southern beech of the family Fragaceae) flora arose after Africa drifted northeast and became isolated. *Nothofagus* today is absent from Africa but survives in cold regions of South America, the Falkland Islands, southeastern Australia, New Guinea, New Zealand, and New Caledonia. The southern conifers of the families Araucariaceae and Podocarpaceae as well as the cycads and tree-ferns recall Australia's ancient Gondwanan connections. Holly *(Ilex)*, *Nothofagus*, the Winteraceae, and the primitive Proteaceae shed some of the first angiosperm pollen in the Australian fossil record.

The land was relatively flat, with few barriers to plant migration, and edaphic (soil) conditions allowed a pan-Australian flora to exist during the Eocene and Oligocene epochs. The Miocene marked the end of the period of stability. Earth movements began to fragment the old peneplain. The Great Dividing Range arose, and, concomitantly, marine inundation produced limestone deposits inland from the Great Australian Bight. These factors led to greater habitat diversity and geographic isolation which greatly influenced the subsequent flora.

The arid adaptation shown by much of the Australian sclerophyll (leathery leaf) flora was selected for from at least the mid-Eocene. Rainforests contracted and semi-desert and desert conditions prevailed over much of Australia. *Eucalyptus* pollen made its first appearance in the fossil record of the early Oligocene Epoch, about 34 MYA, and *Acacia* followed at the close of the Oligocene, about 25 MYA.

The Miocene earth movements and marine flooding isolated Australian flora into two disjunct areas, southwestern and southeastern. The Miocene limestone presented an edaphic barrier to the mingling of the two floras, and continued isolation resulted in further endemism in these areas. Today, about 80 percent of the Western Australian flora is endemic.

The transition to open grasslands in eastern Australia began around 5 MYA, as shown by the fossil pollen record. During the last two million years of Earth history, the Quaternary Period, the Australian flora developed under conditions of rapid and extreme climatic change. Glaciation came and went, sea levels rose and fell by as much as 200 m (656 ft), and Tasmania was connected to the mainland for a time. The tectonic collision of the Timor plate with the Australian plate about 4 MYA resulted in the addition of

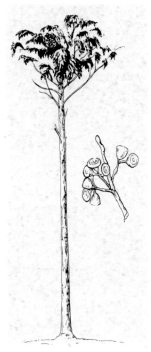

Figure 4.1
Mountain ash, *Eucalyptus regans,* and fruit.

Figure 4.2
Mallee, *Eucalyptus luehmanniana.*

Right: A karri forest, *Eucalyptus diversicolor,* near Pemberton in Western Australia. The fire tower is at the top of a 67 m (220 ft) tall karri.
Far right: An archetypical *Eucalyptus,* from Tasmania.

southeast Asian elements to Australian flora, especially in the north of Australia. It is interesting to note that the drought-adapted, poor-soil flora of Australia made much less of an incursion into Asia than did the Asian flora into Australia. The Australian plants probably could not tolerate the rich volcanic soils of southeast Asia.

Of the 15 000 or so species of Australian flowering plants, the largest groups are the acacias (Mimosaceae), eucalypts (Myrtaceae), grevilleas (Proteaceae), tea-trees (Myrtaceae), heaths (Ericaceae), grasses (Poaceae), and orchids (Orchidaceae).

· · · · · · · · · ·
PLANT GROUPS OF PARTICULAR INTEREST

Eucalyptus: Gum Trees

When one thinks of the Australian landscape, *Eucalyptus* trees immediately come to mind. They dominate the part of the country where rainfall is sufficient for their growth. There are about 700 species of *Eucalyptus* and all but seven are Australian. Most of these seven occur in New Guinea and one reaches the Philippines. There are no native *Eucalyptus* in New Zealand. It is not known whether this great variety of species is derived from a single protoeucalypt or is of polyphyletic origin within the Myrtaceae.

Eucalyptus are the tallest flowering plants (angiosperms) in the world. (The California redwoods are gymnosperms, not flowering plants, and can exceed 100 m [328 ft]). Some mountain ash, *E. regans* (Figure 4.1), exceed 90 m (295 ft) in Victoria and Tasmania, but their

diameters average only about 2 m (6.6 ft). The karri, *E. diversicolor,* of southwest Western Australia also reaches 90 m but has a circumference of 7.3 m (24 ft) at its base.

Eucalyptus thrive in cooler areas with winter rainfall, and in tropical areas with summer rainfall. They grow in rich or poor soil, in the Australian Alps, or along rivers. They are absent from tropical rainforests and rare in extremely arid environments. *Eucalyptus* growing in fertile, well-watered forests are usually straight and tall and have a limited canopy. Isolated trees on open plains are short with a broader canopy.

Mallee (Figure 4.2), various species of *Eucalyptus* in arid areas, usually have multiple slender trunks and a widespread canopy 3–6 m off the ground. Mallee are protected from drought and fire by underground lignotubers from which new stems grow after the above-ground parts are destroyed. Snow gums, *E. pauciflora,* growing on windswept mountains may have flat tops.

The foliage of juvenile *Eucalyptus* is different from that of the mature trees. Floral arrangements utilizing the round powdery-blue, opposite, stalkless (sessile) juvenile leaves are common both in Australian and in America. The mature leaf is long and shiny green. *Eucalyptus* trees produce the characteristic aromatic essential oils from glands on their leaves. The bluish haze in the Blue Mountains near Sydney comes from a cocktail of hydrocarbon gases given off by *Eucalyptus* and other plants. One of these gases, isoprene, may protect the photosynthetic process from excessive temperatures—above 37°C (99°F).

The word *Eucalyptus* means 'well-covered' in Greek, and reflects the fact that the flower bud is tightly sealed by an operculum; the shape and size of which form a useful taxonomic character, as do the anthers of the flower. A woody capsule covers the minute seeds to form 'gum nuts' (Figure 4.3). Valves of the dry capsule open to release the seed for dispersal.

Eucalyptus are the source of practically all of Australia's hardwood. Eucalypts such as ironbark, *E. sideroxylong,* are immensely strong; others are very soft and used for paper pulping. River red gum, *E. camaldulensis* (Figure 4.4), is one of the most beautiful trees in the world and is found growing along streams and rivers in all states; it is used for fence posts, railroad ties (sleepers), and other structures where durability in damp ground is important. Jarrah, *E. marginata,* of southwest Western Australia, has a beautiful red wood not unlike mahogany, which is used for furniture making.

Eucalyptus flowers have conspicuous stamens (brush blossoms) and are pollinated by insects, birds, bats, and some marsupials (Figure 4.5). They yield more honey than any other plant. Beekeepers follow the blossoming of *Eucalyptus* from district to district. Each species produces a distinctive color and taste, with yellow box, *E. melliodora,* being preferred. *E. citriodora,* the lemon-scented eucalyptus, is cultivated for an oil that forms the trade product, Citronella, used as

Figure 4.3
Gum nuts of lemon-scented gum, *Eucalyptus maculata.*

Figure 4.4
River red gum, *Eucalyptus camaldulensis.*

Figure 4.5
Blossom of scarlet-flowered gum, *Eucalyptus ficifolia.*

Flower of a eucalypt.

Figure 4.6
Flower of narrow-leaf bottlebrush, *Callistemon linearis.*

Figure 4.7
Red-leaved wattle, *Acacia rubida.*

Figure 4.8
Juniper wattle, *Acacia ulicifolia.*

Crimson bottlebrush,
Callistemon citrinus, in Victoria.

a mosquito repellent and in the perfume industry. *Eucalyptus* oil is distilled from leaves of several species and used in pharmaceutical preparations such as expectorants and inhalants, and in germicidal products.

Eucalyptus species are of great value in preventing soil erosion and are one of the most aesthetically pleasing objects in the natural world. Consequently, they have been introduced into over 70 countries around the world as ornamentals or for reforestation and erosion control. The Tasmanian blue gum, *E. globolus,* has been widely planted in California. In fact, the presence of the appropriate species of *Eucalyptus* in southern California enables the San Diego, San Francisco, and Los Angeles zoos to permanently exhibit koalas. Koalas feed exclusively on the leaves of only a few species of the 700 eucalypts. Some species such as sugar gum, *E. cladocalyx,* produce a toxic glycoside that releases prussic acid after digestion. This can result in livestock poisoning.

The 700 species of *Eucalyptus* represent but one genus of about 95 in the family Myrtaceae. There are also about 3000 species of myrtles, of which Australia has almost 1200 in 69 of the 95 genera. Other Australian myrtles include the tea-trees *(Leptospermum),* the bottlebrushes *(Callistemon)* (Figure 4.6), and the lillypillies *(Eugenia).*

Acacia: Wattles

Acacia, or wattles as Australians call them, are leguminous plants of the family Mimosaceae. Acacias are smaller and much more drought-tolerant than most *Eucalyptus* are. They do not have the economic status of eucalypts, but their ecological importance and the great beauty of their blossom have earned their presence on Australia's coat of arms. There are about 835 species of Australian acacias and more remain to be described. There are two types, based on foliage. One has pinnate (feathery) leaves (Figure 4.7) and the larger group has a spine-like structure formed from a flattened leaf stalk (Figure 4.8) that takes the place of a normal leaf.

Acacias range in size from low ground mats to forest trees. Silver wattles, *A. dealbata,* may reach 30 m (98 ft) in southeastern wet forests. Mulga, *A. aneura* (Figure 4.9), is very drought-resistant and is widespread throughout arid central Australia. It was used by Aboriginals for boomerangs and spear shafts. Mulga has a dark heartwood and a contrasting pale softwood, from which beautiful carvings can be made. Mulga also forms subsistence fodder for inland stock during severe droughts.

Some acacias are used for timber. Blackwood, *A. melanoxylon,* is ranked with walnut as a prime wood or veneer for furniture making, and golden wattle, *A. pycnantha* (Figure 4.10), is considered one of the world's best barks for the tanning process. Many wattles are cultivated for their profusion of golden blossoms.

Bushfires may assist wattle germination by breaking down the hard coat of the seed pod. The ant-*Acacia* association so characteristic of Africa and Central America is lacking in Australia.

Figure 4.9
Mulga, *Acacia aneura.*

Figure 4.10
Golden wattle, *Acacia pycnantha.*

A wattle, *Acacia* sp., flowering in July in Victoria.

Banksia and the Proteaceae

The genus *Banksia* was named in honor of Sir Joseph Banks, who collected the first specimens in 1770 at Botany Bay while serving as the naturalist aboard Captain Cook's ship HMS *Endeavour.* Banksias are woody evergreen plants in the primitive family Proteaceae. This family has low vagility (dispersal ability) and spread via the continuous landmass of Gondwana. This southern hemisphere family has representatives in South America, Africa, southeast Asia, and Australia. It is a much less important element in the New Zealand flora. In Australia most of the genera are sclerophyllous. About 65 percent of the species are Australian and about 25 percent are South African.

Banksia come in many shapes and sizes, from prostrate shrubs to trees 30 m (98 ft) tall. The tree forms usually have a single, irregular trunk, and the shrub forms may have multiple stems at ground level. About half of the species are fire-tolerant, with a thick corky bark,

Figure 4.11
Flower of *Banksia integrifolia*.

Banksia prionotes from
Western Australia.

and sprout new growth after a fire. The fire-sensitive species are killed by fire, but they regenerate from seeds, the release of which is facilitated by the heat of the fire. However, two severe fires in one year may endanger the survival of the fire-sensitive species. *Banksia* leaves are usually tough and prominently toothed. The flowers are produced in dense spikes on a central, woody, cone-like axis which may contain hundreds to thousands of individual flowers (Figure 4.11). *Banksia* is pollinated by birds, small marsupials, and insects.

There are about 92 species of *Banksia*, all of which are Australian. Only one species, *B. dentata*, occurs naturally outside Australia, in Papua New Guinea, Irian Jaya, and the Aru Islands. The southwest corner of Western Australia has the greatest concentration of Banksias, having 61 species, with southeast and eastern Australia having about 14 species. *Banksia* species also grow in Tasmania.

Banksia are attractive plants and are widely used for landscaping in Australia. Some are grown from nursery stock, but most are grown from wild seeds or wild transplants. The longlasting flower spikes are harvested commercially for cut flowers, with over 1.2 million flowering stems from 29 species being cut each year— and this is a very conservative estimate. The most heavily exploited species include the Western Australian *B. coccinea, B. baxteri*, and *B. hookeriana*. Flowering usually occurs from February to September, with yellows, oranges, and reds predominating.

Banksias are also used in arts and crafts. The large dried fruiting 'cones', especially of *B. grandis* from Western Australia, are cut transversely into coasters or shaped into candleholders, lampstands, and other ornaments.

The Proteaceae family also includes the only Australian native plants to gain acceptance as a food crop. These are the macadamia nuts *(Macadamia integrifolia* and *M. tetraphyla)* from southeastern Queensland. Macadamia nuts were first grown commercially in

Hawaii and later in California. Since demand exceeds supply, these high quality nuts are still very expensive, but extensive orchards have recently been established in Australia and should reach full production soon.

Another important Australian genus of Proteaceae is *Grevillea,* or spider plant.

The Proteacea is thought to be among the earliest flowering plant families to have evolved. Recently, nuts from an undescribed living species from Mount Bartle Frere in northeastern Queensland were examined and found to be virtually identical to a fossilized nut from 60-million-year-old sediments located near Melbourne. This discovery may eventually shed light on the evolutionary history of the angiosperms.

Figure 4.12
Rough tree-fern, *Cyathea australis.*

Other Archetypical Aussie Plants

The she-oaks, *Casuarina,* are represented by about 45 species of the Casuarinaceae in Australia. These wind-pollinated trees are also found in New Guinea and Indonesia, and one species is common to Madagascar and the Mascarenes. Most, however, are endemic to Western Australia. The small, filamentous, needle-like branches resemble the quills of cassowaries resulting in the name *Casuarina*— which means cassowary. Some she-oaks can reach 30 m (98 ft) along the banks of eastern rivers, whereas others are small shrubs. Some species are useful as timber, with an oak-like grain. Pollen of this family appeared in the fossil record with that of *Banksia.*

Flower of spider plant, *Grevillea* sp.

Tree-ferns of the family Cyatheaceae (*Cyathea, Dicksonia, Culcita*) are common in moist forests (Figure 4.12). There are no true Australian pines (Pinaceae) and only about three dozen native conifers. The most important are the hoop pine, *Araucaria cunninghamii,* the kauris, *Agathis* spp., the black and brown pines, *Podocarpus* spp., and the Huon pine, *Dacrydium franklinii.* The Huon pine reaches 30 m (98 ft) in the wet forests of Tasmania. Its yellow wood resists termites and borers and is in much demand for boat building. The family Araucariaceae occurs in South America, the Philippines, New Caledonia, Fiji, Malaysia, New Guinea, and New Zealand. *Araucaria cunninghamii* (hoop pine) is one of the most important native softwoods and reaches a height of 60 m (196 ft) in the coastal forests of northern New South Wales and Queensland. *Araucaria* may represent a stage of development between wet rainforest and dry scherophyll forest. The trees frequently tower over other species in the rainforests. The familiar Norfolk Island Pine, *A. heterophylla,* belongs to this family and is widely cultivated as an ornamental tree along beachfronts and in gardens.

In December 1994, after a four month investigation, the National Parks and Wildlife Service announced that an archaic but previously unknown species of large tree had been discovered in a remote gorge in the Blue Mountains, 200 km (320 miles) west of Sydney. This

The flower spike of *Xanthorrhoea* is made up of hundreds of individual flowers.

Figure 4.13
Baobab tree, *Adansonia gregorii*.

Nuytsia floribunda, the Christmas bush, so called because it flowers during the holiday season. This plant is a member of the mistletoe family and is a hemiparasite.

Xanthorrhoea flower spike, near Keith in South Australia.

rainforest tree is a relic whose fossil record dates back about 160 million years. It has been dubbed the Wollemi Pine, *Wellemia nobilis*, after the Wollemi National Park where it was found. It is a new genus and species of plant in the Araucariaceae. So far 23 adults and 16 juveniles have been located. The biggest specimens reach 40 m (130 ft) with a 3 m (10 ft) girth. Scientists at the Royal Botanic Gardens in Sydney have succeeded in cloning the Wollemi Pine, and now grow small plants in tissue culture.

The baobab tree *Adansonia gregorii* (Figure 4.13), a member of the Bombacaceae, is found in northern Australia, and a related species occurs in Africa. The baobab may reach 24 m (79 ft) in circumference and has a large gourd-like fruit, the pulp of which is edible and is said to taste like cream-of-tartar.

The Christmas bush, *Nuytsia floribunda,* of Western Australia produces a profusion of beautiful yellow blossoms high in the foliage of host trees. This member of the Loranthaceae, an ancient Gondwanan family, is a hemiparasite. It makes its own food via photosynthesis, but its roots are parasitic on host trees' roots.

Spiny-leaved, tussock-forming, highly xeromorphic (drought adapted) grasses of the genus *Triodia* cover the interior plains in northern arid areas for hundreds of kilometers. These grasses are popularly known as 'spinifex' and consist of about 30 exclusively Australian species. Spinifex can grow in deep sand or rocky ground where there is little or no soil. Some species provide forage for livestock during severe droughts. Spinifex is the Australian answer to the succulent euphorbs of arid Africa and the cacti of desert America. Spinifex is highly inflammable. The resin of *T. pungens,* which ranges from the northwest coast to south central Queensland, was used by Aboriginals for hafting axe and spear heads.

Millions of sheep graze in semi-desert areas on various members of the saltbush family Chenopodiaceae, such as *Bassia paradoxa.* These plants have succulent, reduced leaves and are covered with downy hairs. They are especially salt-tolerant and thrive in the scrub country on the margins of deserts.

Other odd plants are the 20 species of grass-trees, or 'blackboys', *Xanthorrhoea* (Xanthorrhoeaceae or Lilliaceae) (Figure 4.14). This group evolved in Australia after its isolation as an island. In spring, assisted by bushfires, their spear-like flower stalks, covered with thousands of white flowers, shoot to a height of 3m (10 ft). They resemble yuccas, with dense basal tufts of long rigid grass-like leaves. When used as a landscaping plant, their trunks are sometimes charred with a blowtorch to stimulate growth.

Sturt's desert pea, *Clianthus formosus*, is named in honor of Charles Sturt who explored Australia's inland river system in 1828–30. This legume (family Papilionaceae) occurs throughout arid areas from the north coast of Western Australia to the western plains of New South Wales. Its red and black flowers can reach a length of 10 cm (4 in), and the leaves are covered with short hairs.

Western Australia is noted for its wildflowers and is estimated to have over 2000 endemic species—among them the official state flower, the kangaroo paw, *Anigozanthos manglesii* (Figure 4.15). Botanists consider this area to have the purest Australian flora.

Figure 4.14
Flowering spike of grass-tree, *Xanthorrhoea minor,* and isolated flower.

Figure 4.15
Kangaroo paw, *Anigozanthos* sp.

Blackboy, *Xanthorrhoea reflexa,* from Mt Chudalup in Western Australia.

An ornamental planting of multicolored kangaroo paw, *Anigozanthos*.

PLANT COMMUNITIES

Rainfall increases as one moves from the center of Australia outward. There is a sequence running from desert through grassland, shrubland, scrub, heathland, woodland, open forest, and ending with closed forest in the wettest areas. Edaphic properties determine the understory. The dominant genus is *Eucalyptus* and the species change with the microhabitat. Specht (1981) lists the structural forms of vegetation in Australia, as summarized in Table 4.1 (page 80).

The dominant plants of the desert are perennial grasses and succulent herbs. After the rare rainfall, the desert may burst into bloom with annual wildflowers. Moving away from the arid edge of the desert one encounters semi-arid savannah—grassland with few scattered trees. This gives way to grassy shrubland and then to scrub (small multiple stem, *Eucalyptus*); the scrub gives way to heathland. The species composition varies from west to east, with the east far less diversified. As rainfall increases away from the center, sclerophyll forests form what most Australians think of as their characteristic landscape. This is not the most extensive area of vegetation, but it is where most people live. The *Eucalyptus* reach the greatest diversity here and form evergreen forests with sclerophyllous understory shrubs. The sclerophyll forests can be either wet or dry depending on the amount of rainfall. In northern Australia a dense assemblage of stunted trees forms a monsoon forest in areas with a marked dry season. In the highest part of the Great Divide in southeastern Australia, snow may cover the ground for several months each year. The trees at high elevation are stunted, and many areas have alpine meadows of grasses and mosses.

On tidal mudflats and estuaries, mangrove forests may develop in all states except Tasmania. There are 40 species of mangroves in Australia, with the grey mangrove, *Avicennia marina*, being the most common. This species, like most mangroves, has a highly modified root system that forms a tangled mass of arches. New shoots emerge from these roots and form a dense network of joined trees that can reach 14 m (46 ft) tall, although most are half that size. Aerial roots, (pneumatophores), project through the anerobic mud and absorb oxygen. Mangroves are, of course, tolerant to saline conditions. They provide an important nursery area for many species of fishes.

True rainforests or jungles, with crowded foliage, buttressed bases, woody vines, epiphytes, ferns, and orchids, develop in areas of high rainfall. Patches of rainforest occur along the north and east coast of the continent. These pockets are remnants of a formerly more widespread distribution. Australia's northern rainforests contain the largest concentration of living primitive flowering plants on the planet. This represents a relict flora. Queensland has about 55 percent of Australia's rainforests. The coastal area of Queensland between Cooktown and Townsville is the largest continuous area of rainforest in Australia. This wet, tropical region contains about 800 000 ha (2 million acres) of rainforest, with rainfall varying

Flower of kangaroo paw, *Anigozanthos* sp., from Western Australia.

between 1300 and more than 3000 mm (51–118 in). The town of Tully on the coast between Cairns and Townsville receives 4267 mm (167 in) of rain on average annually. In the East Gippsland region of Victoria scattered patches of temperate lowland rainforest are found along gullies, where beautiful tree-ferns such as *Dicksonia antarctica* occur. The western third of Tasmania is covered in temperate rainforest. In fact, about one-third of all the rainforest area of Australia is in Tasmania.

TABLE 4.1 **STRUCTURAL FORM**

LIFE-FORM AND HEIGHT OF TALLEST STRATUM	FOLIAGE COVER OF TALLEST STRATUM	
	100–70%	70–50%
Trees* More than 30 m	Tall closed forest	Tall forest
Trees 10–30 m	Closed forest	Forest
Trees Less than 10 m	Low closed forest	Low forest
Shrubs* More than 2 m	Closed scrub	Scrub
Shrubs 0.25–2 m		
Sclerophyllous	Closed heathland	Heathland
Non-sclerophyllous	—	—
Shrubs Less than 0.25 m		
Sclerophyllous	—	—
Non-sclerophyllous	—	—
Hummock grasses	—	—
Herbaceous layer		
Graminoids	Closed (tussock) grassland	(Tussock) grassland
Sedges	Closed sedgeland	Sedgeland
Herbs	Clossed herbland	Herbland
Ferns	Closed fernland	Fernland

*A tree is defined as a woody plant usually having a single stem; a shrub is a woody plant usually having many stems arising at or near the base.
Source: Modified from Specht (1981).

Rainfall, temperature, and soil fertility interact to determine what kind of rainforest will develop. The distribution of rain throughout the year is important. An area that receives all of its rain in a few months will sprout different plants than an area that receives a constant amount every month. Likewise, an area subject to some cold temperatures in winter will give rise to a flora different from that of a uniformly warm area. Botanists classify rainforests in a variety of ways. Adam (1992) presents an excellent review of Australian rainforests and should be consulted for technical details.

Australia is one of the few technologically developed nations to have extensive tracts of rainforest, and there have been numerous battles over their preservation. As environmental consciousness grew, clashes erupted between loggers and conservationists. This came to a head in August 1979, in the Terania Creek region of northeast New South Wales, as protesters lay in the path of bulldozers, climbed trees marked for cutting, and disrupted logging operations. After a lengthy and expensive inquiry logging was permitted to continue, but the movement to preserve rainforests from exploitation had gained momentum.

AUSTRALIAN VEGETATION

50–30%	30–10%	Less than 10%
Tall open forest	Tall woodland	-
Open forest	Woodland	Open woodland
Low open forest	Low woodland	Low open woodland
Open scrub	Tall shrubland	Tall open shrubland
Open heathland	Shrubland	Open shrubland
Open shrubland	Low shrubland	Low open shrubland
—	Dwarf open heathland	Dwarf open heathland
(fell-field)	(fell-field)	
—	Dwarf shrubland	Dwarf open shrubland
—	Hummock grassland	Open hummock grassland
(Tussock) grassland	Open (tussock) grassland	Very open (tussock) grassland
Sedgeland	Open sedgeland	Very open sedgeland
Herbland	Open herbland	Very open herbland
Fernland	—	—

The Commonwealth government nominated various rainforest areas for listing on the World Heritage List, a convention adopted by the General Conference of UNESCO in 1972. This provoked confrontation over states' rights, but later, in 1986, the New South Wales Labor government assumed a leadership role and proposed that the Commonwealth government include much of the state's rainforests on the Heritage List. The state of Queensland continued to oppose the nomination of the Greater Daintree region of undisturbed tropical rainforest, but when a Queensland Labor government was elected in 1989 it agreed to cooperate with the Commonwealth government in the management of the area.

The great turning point came in the early 1980s. The Tasmanian government opposed listing of southwestern areas of Tasmania because it planned to build a hydro-electric dam on the Gordon River, below the junction with the Franklin River. This would have flooded an extensive area of rainforest. The Commonwealth Labor government under Bob Hawke prevented the construction of the dam, an action that was later challenged by Tasmania in Australia's High Court. The court, however, upheld the Commonwealth's right to override the state.

· · · · · · · · · ·

IMPACTS ON THE FLORA

Bushfires

Bushfires are an important ecological factor in the Australian environment. The volatile oils in *Eucalyptus* leaves make them highly inflammable and generate intense heat. Fire can be stimulatory rather than inhibitory to some species, depending on its intensity, frequency, duration, season of occurrence, and other factors. Thick bark protects the cambium of many Australian species, especially *Eucalyptus*. Some species depend on fire-resistant seed, or on underground lignotubers that contain buds from which recovery may be made. All seeds of a given species may not be of the same hardiness. Some may germinate after rain; others require the heat from a fire to induce germination. There is obvious survival advantage in such variation.

Natural bushfires are caused by lightning or spontaneous combustion, but most bushfires are the result of carelessness or acts of vandalism by humans. In an average summer about 400 000 ha (988 000 acres) of forests and grassland are consumed by bushfires. Summer bushfires are promoted by a dry spring that desiccates the vegetation, thereby providing fuel. During the hot dry days of summer, local governments frequently proclaim total fire bans.

Tree fern forest along the Omeo Highway in Victoria.

This means that no fire may be started for any reason—for example, no trash burning nor barbecues are permitted. In wetter times, local governments attempt to reduce the build-up of fuel by controlled burning of small areas of bush.

Furious bushfires fanned by hot northerly winds in drought-desiccated Victoria and South Australia in 1982 killed 72 people, 340 000 sheep, and 18 000 cattle, and caused over A$400 million in property damage. More than 6000 people were made homeless and 2000 homes were reduced to ashes.

Introduced Pine

Many of Australia's native grasses have been replaced with cultivated species from other countries. In addition, grazing by cattle and sheep has reduced much of the native vegetation. Monterey pine, *Pinus radiata,* from California, has been extensively planted in Australia, and a good deal of the native *Eucalyptus* forests have been logged to make room for the introduced pine, resulting in loss of wildlife habitat. Currently about 500 000 ha (1.2 million acres) are in pine plantations. This figure is expected to double by the year 2000. The pine is used in the softwood and pulp industries. Since only about 5 percent of Australia is forested (as compared to 32 percent of continental United States), Australia can ill afford to replace its native forests with pine plantations which are an almost sterile habitat for native wildlife.

At least one species of native bird seems to have benefited from the pine plantations, however. The white-tailed black cockatoo, *Calyptorhynchus latirustris,* has become a regular visitor to the major plantations in southwest Western Australia, where the birds strip and eat the cones from the trees. The pine cones are similar to the fruits of native species that form part of the regular diet of the cockatoos. Foresters consider the bird a nuisance.

Prickly Pear and Cactoblastis

The prickly pear cactus, *Opuntia stricta,* is native to the coastal areas of the Gulf of Mexico. It was brought to Australia by the early European settlers as a source of food for cochineal insects (Hemiptera) which produce carminic acid, from which a brilliant red dye, carmine, was made. This dye was used for soldiers' coats. In 1840–50 property owners planted prickly pear cuttings to form hedges around their pastures, and by 1884, prickly pear had grown to plague proportions. The plants were cut down and thrown out, which only encouraged the discarded cactus pads to take root and spread. By 1900, 4 million ha (9.9 million acres) were made useless by the dense growth of the pest, and by 1920, 23 million ha were infested. The cactus was expanding its range at the rate of 1 million ha per year. At least 26 million ha were affected by 1926. Most of the infestation was in

Queensland, ranging from about Mackay down to Newcastle in New South Wales, a distance of 1450 km (900 miles).

The area involved was too large (and the cost prohibitive) to attempt manual or chemical control. As a result, the Commonwealth Prickly Pear Board was established in 1920 to find a biological control for this problem, which had an enormous financial impact on the local graziers. *Cactoblastis catorum,* a moth from Argentina and Brazil, was chosen. The moth's caterpillar burrows into the flat pads of prickly pears and literally eats them from the inside out. At the end of 1927, after much testing, 9 million eggs of *Cactoblastis* were released at several sites. Within six months the moths had dispersed over a large area and destroyed hundreds of hectares of cactus. More moths were released from 1928 to 1930, when they became generally established.

As the dense growth of *Opuntia* collapsed and rotted, the food source of the *Cactoblastis* moth decreased. This caused a decline in the moth population, which allowed a regrowth of the cactus between 1931 and 1933. (These oscillations are common as predator and prey, or host and parasite, strike a balance.) The secondary growth of *Opuntia* was under control by the end of 1934 and has been completely controlled by *Cactoblastis* since then.

Agricultural Problems

Adding to the environmental problems created by bushfires and introduced species is agriculture, a mainstay of the Australian economy. In the past 50 years as much land has been cleared in Australia as in the preceding 150 years. Much of this land was once virgin bush. Its loss in favor of farming has reduced wildlife habitat and has adversely affected biodiversity.

In the last decade, an average of 500 000 ha (1.2 million acres) has been cleared each year. In addition, there is widespread disturbance of the natural vegetation that remains. The hooves of livestock compact the soil and trample the native plants which evolved in association with the padded, large-surface-area feet of macropods. Australia's soils are thin, only about 10 cm (4 in) deep on average, with little organic matter. Removal of natural vegetation exposes soils to the erosional effects of rain and wind.

Crop irrigation, especially in Australia's richest agricultural region, the Murray-Darling Basin, is resulting in increasing soil salinity. Only about 25 percent of the water that flows into the basin actually reaches the sea. Yet, the demand for water is increasing.

The country's main research agency, the Commonwealth Scientific and Industrial Research Organization (CSIRO), and state governments are aware of the problems and various studies are under way, including studies on the many regions that are relatively undisturbed but the future of which is not certain. One possible alternative could be the farming of native animals such as kangaroos and emus, which might be ecologically less damaging. Much of Australian agriculture is very high tech, efficient, and economically successful. Perhaps it is now time to recognize that this driest of continents should not be asked to support more inhabitants than it already has.

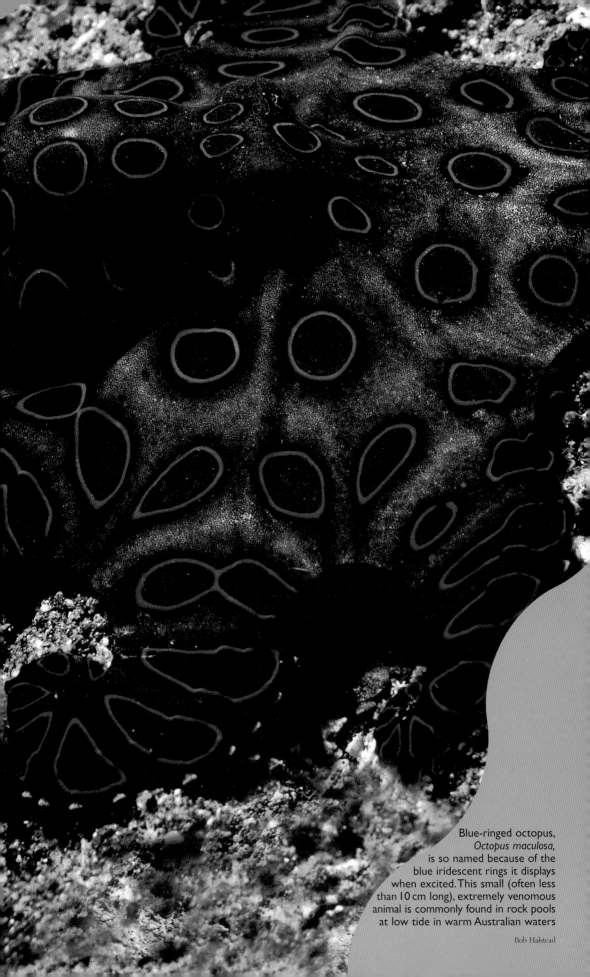

Blue-ringed octopus,
Octopus maculosa,
is so named because of the
blue iridescent rings it displays
when excited. This small (often less
than 10 cm long), extremely venomous
animal is commonly found in rock pools
at low tide in warm Australian waters

Bob Halstead

Dangerous Waters

........·.......
THE GREAT BARRIER REEF

T he Great Barrier Reef is the largest structure built by living organisms — including humans. It comprises a series of reefs extending more than 2000 km (1250 miles) along the northeast coast of Australia, at an offshore distance ranging from 16 km in the north to 160 km in the south, from latitude 9°S to 24°S (Figure 5.1). Thus its length is comparable to the entire west coast of the continental United States, from Washington to southern California. The reef and island area cover some 208 000 km² (80 000 mi²), equalling England, Wales, and Ireland combined. This area is only about one-third of the expanse of the surrounding water within the Great Barrier Reef province. There are some 2500 reefs with areas ranging from a few hundred hectares to more than 52 km², together with many much smaller reefs. The Great Barrier Reef actually consists of many thousands of separate reefs—some mainly dry or barely awash at low tide, some with islands of coral sand (cays), others fringing high islands or mainland coasts. The reef has risen in warm waters on the shallow shelf fringing the Australian continent. The continental shelf has a change of gradient at about 180 m (590 ft).

Each reef is formed from the skeletons of living marine organisms including calcareous remains of the green algae *Halimeda* and coral polyps. So prolific is the calcification of *Halimeda* that 14 cm (5.5 in) of sediment may accumulate in 1000 years. The cement that binds these flakes together is largely manufactured by calcareous red algae called corallines, which form a carpet over the coral banks, especially in surge areas. The joint contribution of calcareous green and red algae accounts for more than half the accretion of carbonates on the Great Barrier Reef.

The first known physical encounter between Europeans and the Great Barrier Reef occurred in 1770, when Captain James Cook ran the 21-meter HMS *Endeavour* aground on it. The work of charting channels and passages through the maze of reefs, begun by Cook, continued during the 19th century when scientists, including the Swiss-American zoologist Alexander Agassiz, made expeditions to this newfound wonder of the world. The contribution of the Great Barrier Reef Expedition of 1928–29 remains unsurpassed in the fields of coral ecology and physiology. A modern laboratory designed to continue reef studies was set up on Heron Island in 1951 and is run by the University of Queensland and the Great Barrier Reef Committee.

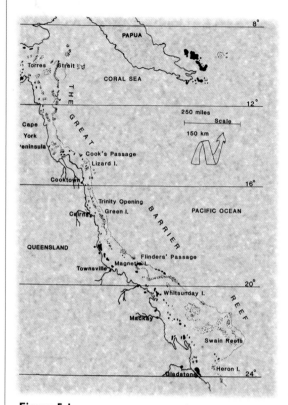

Figure 5.1
The Great Barrier Reef.

White sand of low-lying
Heron Island, as seen from the
Great Barrier Reef at low tide.

Coral Reef Growth

It was no less a biologist than Charles Darwin who, in 1842, proposed the accepted theory of coral growth. Darwin developed his explanation while examining reefs on his voyage around the world in HMS *Beagle* from 1831 to 1836. He suggested that reefs originally became established on the shores of recent volcanic islands (or the continental shelf in the case of the Great Barrier Reef) by colonization of larvae from nearby reefs (Figure 5.2a). As an island gradually subsided under its own weight and the weight of expanding coral growth, the fringing corals along its shores grew upwards. The upward growth more or less matched the rate of subsidence of the island (Figure 5.2b).

Corals can only grow in the surface waters of the sea where light can penetrate because of the presence of photosynthetic algal symbionts. Growth is most rapid on the outer side of the reef where oxygen and nutrients are greatest. As the island submerges, the fringing reef becomes a barrier reef. When the island is completely submerged the reef forms an atoll (Figure 5.2c). As the reef erodes due to wave action, the debris fills in the area enclosed by the atoll and forms a shallow lagoon. A sandy soil is formed from decomposition of upper parts of the reef and from transported organic matter. Palms and other vegetation colonize the coral islands and help hold the soil.

Darwin's theory of coral growth has been confirmed by drilling on atolls which revealed thick layers of reef material over volcanic rock. On Eniwetok Atoll the drill bored through a coral layer over 1000 m (3280 ft) thick before basalt was encountered. Borings have established that the Eniwetok reefs were growing on the continental shelf as early as Miocene times, more than 25 million years ago. An exploratory bore sunk through a coral cay called Wreck Island, in the southern part of the Great Barrier Reef, penetrated 540 m (1770 ft) before encountering underlying volcanic rock. Non-marine sediments underlie the earliest traces of reefs, indicating that the region was then above sea level.

Subsidence of the continental shelf has proceeded with some reversals since the early Miocene Epoch. Platform-like surfaces on the seabed and changes of gradients on the sides of the reef are distinguishable at various depths. These may be related to periods of standstill in the sea level fluctuations associated with the Ice Age glaciers.

Much of the Great Barrier Reef itself is geologically young—2 million years or younger. It was formed in the Pleistocene Epoch during a time of rapidly fluctuating sea levels worldwide. The reef grew during the short periods of high sea level and eroded during low sea level periods. It is a composite of remnant reefs, one on top of another. About 12 000 years ago the margins of the Great Barrier Reef were flooded as northern hemisphere ice sheets melted and raised the sea level. The last 6500 years have seen relative climatic and sea level stability. This has allowed significant horizontal growth and the formation of reef flats. The submarine topography of the reef is complicated by valleys which are products of ancient periods of land erosion that crossed the continental shelf, linking present river mouths with some deeper gaps in the outer reef barriers. The Great Barrier Reef has two major shipping channels: Flinders Passage near Townsville, and Trinity Opening off Cairns.

Figure 5.2
Stages of coral atoll formation.

Reef Conditions, Regions, and Diversity

Great Barrier Reef water shows little seasonal variation. The surface water temperature of this part of the southwest Pacific is high, ranging from 21° to 38°C (70–100°F). The change in temperature with depth is small due to the perpetual stirring action of the southeast trade winds which pound the outer edge of the reef for nine months of the year (February to November) at a steady 30–35 knots for weeks at a time. In the southern hemisphere late-summer, the northwest monsoon season occurs, bringing high rainfall. Cyclonic disturbances are more likely to occur from January to March. Average salinity on the reef is about 3.5 percent, and oxygen content is high at about 90 percent saturation most of the time.

Reef tides have a daily cycle, ranging from 2.4 m (8 ft) to 9 m (30 ft). The draining waters often form foaming white falls at the reef edges. Waters are generally crystal clear, with submarine features clearly visible to depths of 30 m. The further the reef is from the mainland the more luxuriant and prolific are the coral growths. Inshore, the water clarity and coral growth are reduced by the detrimental effects of sediments from river mouths, especially during floods. These conditions cause the abrupt ending of the reef opposite the Gulf of Papua, probably due to the siltladen freshwater discharge of the Fly River in Papua New Guinea.

Fruiting *Pandanus tectorius,* showing prop roots that trap sand and help build islands on the Great Barrier Reef.

There is a gradual sloping of the sea bottom from north to south, which permits a reasonably clear-cut division of the Great Barrier Reef into three regions:

- Northern Region: All reefs to the north of 16°S latitude, characterized by the shallowness of the water covering the shelf; the depth is less than 36 m (120 ft).
- Central Region: From 16°S to 21°S; the depth over the greater part of this region ranges from 36 m to 54 m (120–180 ft).
- Southern Region: From 21°S to 24°S; the deepest part of this reef province reaches 144 m (480 ft).

The Great Barrier Reef has a very high coral diversity—at least 350 species—and is justly famous for its profusion of color and form. These corals show a greater affinity with eastern New Guinea and the western Pacific than they do with Indonesia.

It has become almost a cliché to say that rainforests and coral reefs are the two most productive, species-rich ecosystems on Earth. Phenotypically these two biomes look very different. In rainforests the plants are the most visible structure and the animals are relatively hidden. On a coral reef, the reverse applies. All one sees is the animals, while plants seem to be virtually absent. However, this is not actually the case. Like all ecosystems, except deep-sea thermal vents, coral reefs are supported by photosynthesis.

A portion of the Great Barrier Reef near Heron Island, showing coral which can be seen through the crystal clear water.

Zooxanthellae and Coral Biology

The reef-building (or hermatypic) corals have within their tissues symbiotic algae—microscopic dinoflagellates known as zooxanthellae. The name 'zooxanthellae' reflects the animal nature of corals (*zoo*) and the yellow-brown pigment (*xanthos*). There may be up to three million algae cells per square centimeter of coral surface, and each coral animal cell may have an algal cell within it. The algae provide much of the color seen in corals. One such species is *Symbiodinium microadriaticum*. Because these minute plants require sunlight, reef-building corals are restricted to shallow tropical waters and flourishing growths are not found below about 54 m (180 ft). Maximum reef building takes place between 4.5 m and 27 m.

The symbiotic relationship benefits both corals and algae. The algae gain a secure place to live within the coral tissue, and the zooxanthellae consume the nitrogenous waste and the carbon dioxide of their hosts while releasing oxygen. Some 95–98 percent of all carbon fixed by the algae is delivered to the host and becomes a food source for the coral. An enzyme in the coral promotes the leakage of nutrients from algae to coral. This supplements the coral's ability to deposit its calcium skeleton. Because of this photosynthetic help, corals deposit calcium several times faster during the day than they do during the night. This enables growth to exceed erosion. Thus light is the most important abiotic factor in coral nutrition and structural growth. Hermatypic corals grow only in relatively shallow and clear waters.

Some coral growth can occur down to the 1 percent light level, where corals are often flattened. Presumably, this increases surface area to light exposure. In shallow waters, corals have special compounds in their tissues which act as filters that protect the corals from damaging ultraviolet radiation, and their growth form is usually more convoluted. Some corals have highly plastic growth forms which range from branched to plates or mounds to sheets, depending on physical factors in the environment such as current and sunlight. The genus *Acropora*, staghorn coral, has many different growth forms from table-like to the finely divided branches of its namesake.

Figure 5.3
A tangle of staghorn coral,
Acropora humilis.

Optimal temperatures for coral growth occur between 25°C and 29°C (77–84°F). Nutrients and highly oxygenated waters rich in calcium are, of course, necessary. The temperature in the Caribbean and around the Hawaiian Islands is on average less than that of the tropical Pacific, and this is thought to be the reason why the growth of reef corals is slower in those areas.

A reef is a conglomerate of living and dead plants and animals; and the living reef may be only a thin veneer, perhaps a meter thick, making up less than 10 percent of the total mass. Staghorn corals are known to have the largest number of species, the fastest growth, and the highest rate of oxygen consumption. They are also the most sensitive to temperature and salinity. Staghorns are often the dominant corals on a given area of reef—73 species of *Acropora* have been listed for the Great Barrier Reef region (Figure 5.3). This genus has about 250 species described from the Indo-Pacific as a whole, many of which will, no doubt, prove to be growth forms rather than different species. About 60 genera and 350 species of reef-building corals have been recorded from the Great Barrier Reef waters. Of these, only about 25 play a significant part in reef building. These and others are listed in Table 5.1.

Corals are colonial polyps in the phylum Cnidaria, and are related to sea anemones and jellyfish. They feed by capturing zooplankton with their nematocyst-covered tentacles. The nematocysts are stinging cells that fire a tiny harpoon into the victim. This immobilizes the prey. (The same mechanism is responsible for the stings of jellyfish.) Food is moved to the mouth by the tentacles and digested within the coral polyp. The food captured by any one polyp nourishes the entire colony. This feeding is augmented by nutrients provided by the resident zooxanthellae. Hard corals tend to feed at night.

If you time your visit to the Great Barrier Reef just right, you may see a spectacular orgy of coral spawning. In late spring or early summer, one or two nights after a full moon, many coral species spawn simultaneously. Millions of colorful pink, red, or orange eggs and sperm bundles are released to the sea, usually within a few minutes of one another. Most corals are hermaphroditic, containing both male and female gonads, while others have separate male and

TABLE 5.1 **PRINCIPAL HERMATYPIC CORALS OF THE GREAT BARRIER REEF**

PHYLUM: CNIDARIA; CLASS: ANTHOZOA; SUBCLASS: HEXACORALLIA;

ORDER: SCLERACTINIA

SUBORDER	FAMILY	GENUS	SUBORDER	FAMILY	GENUS
ASTROCOENIIDA	Pocilloporidae	Seriatopora	FAVIIDA	Faviidae	Favia, Favites
		Stylophora			Goniastrea
		Pocillopora			Leptastrea
					Plesiastrea
	Acroporidae	Acropora			Leptoria
		Montipora			Platygyra
		Astreopora			Echinopora
	Agariciidae	Pavona		Oculinidae	Galaxea
		Pachyseris			Acrhelia
		Leptoseris			
				Merulinidae	Merulina
	Siderasireidae	Coscinaraea			Hydnophora
		Psammocora			
				Mussidae	Lobophyllia
					Symphyllia
FUNGIIDA	Poritidae	Goniopora			
		Porites		Pectinidae	Pectinia
		Alveopora			Oxypora
					Mycedium
	Fungiidae	Fungia			
		Podabacia	DENDROPHYLLIIDA	Dendrophyliidae	Turbinaria
		Herpolithia			Tubastrea
		Polyphyllia			
		Sandalolitha			

Source: Bennett (1971) and Veron (1986).

female polyps. Spawning intensifies on the fourth to sixth night littering the surface of the ocean with millions of the brightly colored reproductive bundles. These packets rupture and release the sperm that then fertilizes an egg of its own species. The fertilized egg hatches into a free-swimming larva that drifts in the plankton for a few days, and eventually settles on the bottom to begin the formation of a new coral.

Corals also reproduce asexually by budding off daughter corals which attach themselves near the parent and grow into a new colony that is a clone of the parent. This can take place at any time of the year, and it avoids the problems faced by larvae in the plankton.

Encrusting Algae and Plants

Among the reef organisms, calcareous marine algae (stony seaweeds or nullipores) are as important as the corals. The encrusting red algae *Paragoniolithon* on the windward side and *Porolithon* on sheltered areas form the fortifying purplish-red algal rim that is one of the Great Barrier Reef's most characteristic features. Other species such as the calcareous green algae, *Halimeda,* are commonly present. A thin veneer of algal turf supports a wide variety of herbivorous grazing fishes. Above the surface the plant life of the cays is rather restricted, consisting of only 30–40 species, practically all of them characteristic of the Indo-West Pacific region. About half of the plant species were dispersed to the islands of the reef by sea currents, and about one-quarter were carried by sea birds in their gut or on their feathers.

Erosion

Reef building results from the interaction of the forces of growth and destruction. The role of calcareous algae and corals in growth of the reef have already been discussed (see page 90). Destruction of the coral is caused by physical factors such as storms, wind, and wave action, and biological factors, including the borers such as sponges, polychaetes, and bivalve, as well as urchins and parrotfishes. The sediments produced by erosion form a base for colonization by reef-building organisms, which will ultimately be eroded again.

Crown-of-Thorns Starfish

Acanthaster planci is a large seastar with 7–23 arms and sharp spines that can reach 45 cm (18 in) in length. Adults average about 30 cm (1 ft) in diameter but may exceed 60 cm. After an initial six months of feeding on algae, their diet switches to the tissues of hard corals. In the 1960s there were outbreaks of these animals in the central third of the Great Barrier Reef, off Cairns, as well as outbreaks in the Ryukyus, Okinawa, Micronesia, Tahiti, Fiji, and Samoa. This resulted in the local destruction of about 80 percent of the corals on some reefs in the affected areas. Complete recovery of a damaged reef may take 20–40 years or longer.

The tall beach oak, *Casuarina equisetifolia,* and the low-growing *Messerchmidia argentea* on Heron Island are typical of the tropical vegetation found on the islands of the Great Barrier Reef.

It is not clear whether the proliferation of the crown-of-thorns seastar is a periodic natural phenomenon occurring in response to favorable conditions or a result of perturbations in the environment caused by humans. One hypothesis claims that the increase in *Acanthaster* populations may be due to the removal by shell collectors of the triton, *Charonia tritonis,* a gastropod known to feed on seastars. A more plausible hypothesis states that survival and growth of larval seastars is favored by a combination of low salinities and high temperature. This hypothesis claims that such disturbances are a consequence of cyclones near land areas subject to rapid development that increases runoff. Heavy rains after a drought wash high levels of nitrates and phosphates into coastal waters. This fertilizer enhances larval seastars' survival and growth by enriching their planktonic food organisms. Since a large female seastar can produce 20–50 million eggs, even a fraction of 1 percent increase in survival would result in thousands more offspring per female. These animals are reproductively active for several years. Most die before they reach eight years of age.

A new outbreak of this coral-eating seastar occurred in early 1996 in the northern part of the reef and is expected to last about eight years. The middle reef, 15–20 km (9–12 miles) off shore, is most effected because that is where currents deposit the seastar larvae. No outbreaks have been reported on the inner and outer reefs. These outbreaks occur roughly every 15 years.

**Crown-of-thorns starfish,
Acanthaster placi.**

Neville Co

Other Reef Organisms

Unfortunately there is no list of the fauna of the Great Barrier Reef waters. It is so vast and diverse that a complete study of all groups for all reefs has not been possible. The following are a few of the more interesting animals.

Sponges are among the more conspicuous organisms exposed at low tide, yet the sponge fauna on the Great Barrier Reef is almost unknown. The West Indies has about 300 species, but it is not known how many are in the Great Barrier Reef province. There is no commercial sponge fishery in Australian waters. Hydroids such as stinging coral, *Millepora* (two species), are present and their massive calcareous skeletons act as reef builders. Their nematocysts can produce a severe burning sensation if touched.

The Alcyonaria or Octocorallia are anthozoans called 'soft corals' which may form massive colonies. Sea whips, sea fans, and black corals inhabit the reef base in water too deep and cool for sun-loving corals. Their proteinaceous supporting tissue is called gorgonin. This material from black coral takes a high polish, which is why it is widely sought after for jewellery. Black coral has a bushy or corkscrew growth form in deep water below 20 m (66 ft). Sea fans grow at right angles to the current, enhancing plankton collection by the fan's polyps. Many soft corals possess chemical defenses in the form of terpenes, which are distasteful compounds in their tissue that discourage predators.

All sorts of marine worms, various tube dwellers and other polychaetes are also present in the Reef. Crustaceans and many beautiful Mollusca, especially gastropods, have been collected, perhaps excessively in spots.

Another noteworthy animal is the needle-spined sea urchin, *Diadema setosum* (Figure 5.4). Needle-spined sea urchins would appear to be almost predator-proof; however, some triggerfish flip them over, break the urchin's hard shell with their beaks, and eat the animal's soft insides. The spines of the needle-spined sea urchins

Figure 5.4
Needle-spined sea urchin, *Diadema setosum*.

Above: *Tridacna maxima*, a giant clam with a beautiful blue mantle, in which, like other giant clams, it carries microscopic dinoflagellates known as zooxanthellae.
Left: Razorfish, *Aeoliscus stigatus*, taking refuge among the needle spines of the sea urchin, *Diadema setosum*.

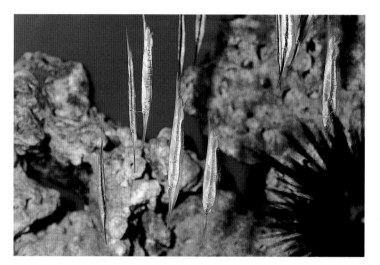

A sign on a beach in the Northern Territory warning swimmers about box jellyfish or sea wasps, *Chironex fleckeri,* the sting of which can cause death within minutes. Increased publicity of the dangers of this creature has, however, drastically reduced the number of fatalities.

offer refuge to the small razorfish, *Aeoliscus stigatus.* This elongated, 120 mm (5 in) fish hides among the spines. Its translucent body, with a black stripe, provides camouflage among the spines of the urchin, where it feeds on small crustaceans. *Aeoliscus stigatus* is related to pipefishes, and an integument of thin plates encases the body.

Giant clams, *Tridacna,* have zooxanthellae in their multi-colored mantles. The largest of the seven species in the giant clam family, *Tridacna gigas,* grows to about 140 cm (56 in), and its shell may weigh 260 kg (572 lb). The deposition of heavy metals in the kidney of giant clams makes an excellent indicator of trace metal pollution of the reef. Asian poachers have nearly driven the giant clam to extinction, and in an attempt to reverse this trend an artificial breeding program has been established off Orpheus Island, Queensland. In 1993 the Australian navy transferred 3000 giant clams, weighing nearly 20 tonnes, from the breeding grounds to a secret location on the Great Barrier Reef. Hopefully, this effort, together with increased anti-poaching patrols, will allow the giant clam to survive well into the future.

• • • • • • • • • •
VENOMOUS MARINE ANIMALS

Sea Wasp

Chironex fleckeri is probably the most venomous animal known. Fortunately it does not usually occur on the Great Barrier Reef, which sees over a million tourists each year. Nor does it occur on the popular beaches of Queensland's Gold Coast or further south in Australia. It does occur in the north—in the Australian tropics—from about October to May. Also known, misleadingly, as the 'box jellyfish', it is actually a cubomedusa of the Cnidarian class Scyphozoa, and has been responsible for at least 65 deaths in the last 100 years.

Its venom can kill a human in less than four minutes, and an adult sea wasp has enough venom to kill 60 people. For comparison, the most dangerous snake in Australia, and perhaps the world, is the taipan. Its venom load can kill 30 humans, usually taking several hours for death to occur. While the snake bite is usually localized in one spot and is not terribly painful, the sting of a sea wasp can spread out over much of the body area as the tentacles contact flesh. The pain is horrible.

The square-shaped body of an adult sea wasp may be as big as a basketball and have 60 dangling tentacles 4.6 m (15 ft) long. The surface of the tentacles is studded with tiny stinging cells called nematocysts which have a mechanical trigger. When the tentacles brush up against a prey, such as shrimp or a human, a minute coiled harpoon is injected along with a tiny quantity of venom. Contact with just 3 m (10 ft) of tentacles can result in death as millions of nematocysts fire into the victim. A smaller amount of venom can produce very painful welts on the body.

Chironex produces two unstable, high molecular weight toxins; both are cardiotoxic and one is also hemolytic. Effects include an initial rise in arterial pressure followed by hypo-hypertensive oscillations and by cardiac irregularities. The victim usually experiences excruciating pain that occurs immediately on contact with the tentacles. The person may become confused, act irrationally and lose consciousness, which may result in drowning. Pain diminishes within 4–12 hours. If death occurs, it usually does so within the first ten minutes, and after one hour survival is likely. First aid includes pressure bandaging and immobilization, removal of the nematocysts with vinegar (not water), and artificial respiration if needed. There is a sea wasp antitoxin prepared from hyper-immunized sheep which restores normal breathing and provides pain relief within minutes. Beaches in northern Queensland, the Northern Territory and Western Australia are posted with warning signs during the summer sea wasp season. Diehard surfers often wear pantyhose with the crotch cut out to accommodate their head, so that their arms are covered by the legs of the pantyhose. Another pair of pantyhose is worn in the usual way. The thickness of the nylon is enough to prevent the nematocyst's harpoon from reaching the skin.

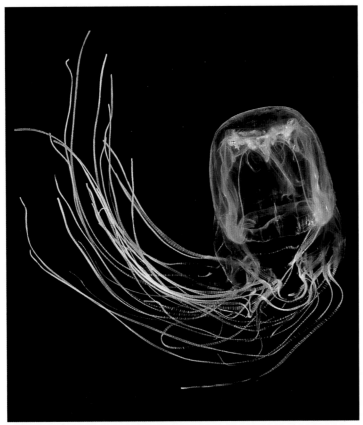

The northern Australian sea wasp, *Chironex fleckeri,* named after Dr Hugo Flecker whose writings are among the first reports of its lethality. It is one of the most venomous animals in the world, and is particularly dangerous because it is very difficult to spot in water.

Sea wasps tend to move inshore during calm weather and converge in the vicinity of stream and river mouths following rains, where adults spawn and die usually during late summer. The planula larvae formed from the fusion of sperm and egg settle to the bottom, develop into polyps and anchor on rocks. The polyps produce more polyps by asexual reproduction. The polyps transform into medusae in the spring and move seaward where they feed on small shrimp and fishes. Amazingly, sea turtles feed on sea wasps, apparently immune to the stings. It seems that the tough lining of the turtle digestive tract prevents damage from the nematocysts.

Blue-ringed Octopus

Octopus maculosa is a small animal, usually less than 10 cm (4 in) in length. It is found in rock pools at low tide on the Great Barrier Reef and elsewhere in warm Australia waters. It displays iridescent blue rings when excited. Handling this animal has resulted in death of the handler within a few minutes. Its saliva contains tetrodotoxin which is neurotoxic and of low molecular weight at 319. This toxin is also found in pufferfish. The bite is painless and may go unnoticed until tissue becomes swollen 15 minutes later. If the patient survives the next hour he or she will notice a local stinging sensation for about six hours, but muscular twitching may last for weeks. The victim may also have difficulty breathing and may be paralyzed for 4–12 hours. First aid is similar to that for snakebite—pressure bandaging and immobilization—with artificial respiration and external heart massage if necessary. No antitoxin is available.

Cone Shells

There are at least eight dangerous species of cone shells (*Conus*), with about 25 percent of the stings being fatal. Cone shells inhabit shallow waters and reefs, may reach 10 cm (4 in) in length and are fish eaters. They deliver their venom by thrusting a minute harpoon through their proboscis. The sting is painful, causing inflammation and swelling. It is followed by numbness, tingling in lips, muscular paralysis, blurred vision, respiratory paralysis, and other neurotoxic damage. Treatment is similar to that for the blue-ringed octopus.

Stonefish and Firefish

Synanceia horrida is a member of the scorpionfish family, Scorpaenidae. It grows to about 30 cm (1 ft); its mouth is vertical and the eyes look upward. It lies quietly in shallow waters, buried in sand, indistinguishable from coral rocks. *S. horrida* has 13 dorsal spines capable of piercing material as thick as that of a tennis shoe (Figure 5.5). Two venom glands discharge along ducts on each spine when pressure is applied. The venom is a protein of molecular weight 150 000. It is a myotoxin that acts on skeletal, smooth and cardiac

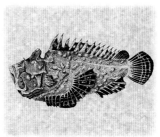

Geographer cone, *Conus geographus,* showing its siphon (projecting from the pointed end) and one of its eyes below it. Reputed to be the most dangerous of cone shells, it is known to have caused at least 12 human fatalities.

muscle, resulting in muscular paralysis, respiratory depression, and cardiac arrest. Immediate pain is experienced, and it increases in intensity over the next 10 minutes to an unbearable level. The affected limbs become swollen, hot, and red. The pain then spreads to lymph nodes. First aid includes immersion of the limb in hot water up to 50°C (122°F) for 30 minutes. If done early, this may produce rapid relief, with the heat denaturing the proteinaceous, high-molecular-weight toxin. There is a stonefish antivenom produced from horses. In Australia, more than 80 cases of stonefish envenomation have been reported, but no deaths have been documented.

Figure 5.5
Stonefish, *Synanceia horrida.*

The firefish, *Pterois volitans,* is another venomous member of the Scorpaenidae. This colorful, feathery-finned beauty is well known to divers and aquarists. When threatened, the fish rotates its body to confront the intruder with its 13 dorsal spines. Puncture wounds from the long dorsal spines are inflicted with a rapid darting motion. The sting soon becomes extremely painful and radiates from the affected part. The area around the sting may become swollen, hot, and cyanotic. In severe cases gangrene may result. Shock and heart

Left: Stonefish, *Synanceia horrida,* is well camouflaged on the sea floor.
Below: The venomous firefish, *Pterois volitans* is deceptively attractive. Even though it is not aggressive, it should be handled with extreme care.

Green moray, *Gymnothorax prasinus,* is distributed from southern Queensland south to Shark Bay in Western Australia. It is not normally a resident of the Great Barrier Reef, but at least 18 other species of *Gymnothorax* are found on the reef.

failure and even death have been known to occur. Immersion of the injured limb in very hot water may relieve the pain, but treatment by a physician may be necessary. This species or close relatives occur throughout the Indo-Pacific. See Halstead (1978) and Williamson et al. (1996) for more information on this beautiful fish and its relatives, and indeed on all poisonous and venomous marine animals.

· · · · · · · · · ·

REEF FISHES

The fish fauna of the Great Barrier Reef is almost infinite in shape, size, and color. The number of tropical marine species recorded from the reef's waters is in the vicinity of 1500, but the surveys are incomplete. Ichthyologists at the Australian Museum have catalogued over 500 species from one single reef, One Tree Island in the Capricorn Group. This gives an idea of the extraordinary diversity of Great Barrier Reef fishes.

The greater number of reef fishes are carnivorous, feeding on smaller fishes and invertebrates and even corals. The large carnivores such as sharks are usually found in the deeper water around the reef edge, but may enter shallow areas as the tide rises. Plankton-feeding species such as sardines may be present in enormous numbers cruising along the shore. Larger pelagic fishes forage in their wake.

Moray Eels

Moray eels (family Muraenidae) are represented by at least 29 colorful species on the Great Barrier Reef, 18 of which are in the largest genus *Gymnothorax.* Their elongated shape is ideally suited for preying on fishes that hide in the nooks and crannies of coral reefs. They lack paired fins and scales. Their danger to humans is overstated in popular mythology, but poking about in a coral crevice is not advisable. The moray may mistake a human hand for an octopus, one of its normal prey items, and inflict a painful bite with its sharp, recurved teeth. Moray eels should never be eaten because of the danger of fatal tropical fish poisoning.

Pinecone Fish

The pinecone or pineapple fish, *Cleidopus gloriamaris* (family Monocentridae), is so named because of the large plate-like scales that encase the body. A prominent spine on each pelvic fin and 5–7 dorsal spines add to the body armor. A light organ on each side of the lower jaw is filled with symbiotic bioluminescent bacteria. During the day this spot appears red, but at night the organ gives off a luminous blue-green glow which may help the fish locate or attract its prey.

Pinecone fish, *Cleidopus gloriamaris*. The red spot on the lower jaw is a site of symbiotic bioluminescent bacteria.

Squirrelfishes

The 25 or so species of squirrelfishes (Holocentridae) found on the Great Barrier Reef are usually red perch-like fishes with very large eyes. They feed at night on crustaceans and hide under ledges during the day.

Pipefishes

Squirrelfish, *Myripristis* sp., is also called soldierfish. Note the large eye of this nocturnal fish.

Pipefishes (family Syngnathidae) have an elongated body encased in a series of bony rings. Pelvic fins are absent. Some species, such as the ringed pipefish, *Doryrhamphus dactyliophorus,* can be very colorful. The ringed pipefish, reaching 18 cm (7 in) in length, has red to black bands around a bright yellow body. It occurs on the Great Barrier Reef and elsewhere in the Indo-Pacific. Female pipefishes and their seahorse relatives deposit their eggs on a special part of the trunk or tail of the male. In some species, this region develops into a pouch. The males incubate the eggs and 'give birth' to baby pipefishes or seahorses.

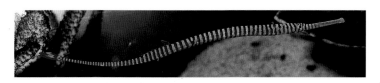

Ringed pipefish, *Doryrhamphus dactyliophorus.*

Weedy seadragon, *Phyllopteryx taeniolatus,* from temperate waters of southern Australia.

The weedy seadragon, *Phyllopteryx taeniolatus,* is a large, colorful pipefish with leaflike flaps that may help camouflage it among kelp and algae beds. This species is a temperate, inshore form found only in southern Australia, from New South Wales to Geraldton, Western Australia. It does not occur on the Great Barrier Reef. It reaches about 45 cm (18 in) and males may carry 100 eggs in a subcaudal brood pouch. Propulsion is provided by the tiny pectoral fins located behind the operculum. Its tail is not prehensile and the animal cannot maintain a grip on vegetation as a seahorse can. It may be found washed up on the beach after a storm.

Groupers

The grouper family (Serranidae) has about 80 species that are commonly found on the Great Barrier Reef. The largest bony fish there is the Queensland grouper, *Epinephelus lanceolatus,* which can reach 270 cm (9 ft) and weigh over 400 kg (880 lb).

Cromileptes altivelis, a secretive serranid that is highly valued as a food fish.

Cardinalfishes

The cardinalfishes (family Apogonidae) are small, 10 cm (4 in) long, brightly colored fishes that represent one of the largest families of coral reef fishes. They resemble squirrelfishes, but possess two short, well-separated dorsal fins, whereas squirrelfishes have a single, long dorsal fin. Like squirrelfishes, cardinalfishes are usually red, have large eyes, and are nocturnal. At least 50 species are recognized from the Great Barrier Reef. Cardinalfishes are unusual among marine families in that the males incubate the fertilized eggs in their mouth for several days. They do not feed during this time. Egg-carrying males can be distinguished by their swollen throats.

Jacks or Trevallies

The jacks or trevallies (family Carangidae) are silvery schooling fishes highly prized as food by anglers and commercial fishermen. They are fast-swimming, voracious predators of small fishes along the outer reef slope. Their speed is reflected by their deeply forked tails, narrow caudal peduncles, and compressed bodies. About 50 species can be seen on the Great Barrier Reef.

Snappers

Snappers (family Lutjanidae) are medium-sized, brightly colored, nocturnal fish predators. Most have enlarged canine teeth on the jaws and terminal mouths. They are among the most important food fishes in warm waters, but some species also carry the tropical fish poisoning disease, ciguatera. This toxin is produced by the dinoflagellate, *Gambierdiscus toxicus.* Snappers can accumulate the toxin by eating herbivorous fishes such as surgeonfishes and parrotfishes. The snappers are not affected by the poison, but human symptoms develop 3–5 hours after eating tainted fish. Abdominal pain, nausea, and vomiting are followed by diarrhea, metallic taste in the mouth, muscular weakness, tingling sensations of mouth, palms and soles, pain in the joints, and reversal of hot and cold sensations. Acute symptoms usually subside in about 8–10 hours. Humans do not acquire an immunity to the poison. Instead, they may become even more sensitive to the toxin. Severe cases can be fatal within ten minutes, but death usually requires several days. Because of this problem it is illegal to sell certain species, such as the two-spot red snapper, *Lutjanus bohar,* in Australia. There is no antidote and treatment is largely symptomatic. Cooking does not denature the toxin. However, many other species are safe to eat and are often sold in markets and restaurants. There is no known way to detect a ciguatoxic fish by its appearance. Larger fishes tend to have more toxin in their flesh than small fishes do. Freshness or staleness of the flesh bears no relationship to toxicity.

Figure 5.6
Long-nosed butterflyfish, *Forcipiger longirostris.*

Butterflyfishes

One of the most beautiful groups of fishes is the butterflyfish family (Chaetodontidae). Many of these disk-shaped fishes have tiny pointed snouts (Figure 5.6) that delicately pluck polyps from the coral skeleton. Others feed on benthic invertebrates, zooplankton and algae. Butterflyfishes occur on isolated patch reefs and defend their territory. At dusk their feeding activity stops, and they take shelter in coral crevices. Yellow is a common color for many butterflyfishes, and their patterns include spots and stripes. Some species become very pale at night, but in the morning they reappear as colorful as ever. They are usually found in shallow water less than 20 m (66 ft) deep.

Below: Forcepsfish, *Forcipiger flavissimus,* a butterflyfish with a delicate snout adapted for probing the interstices of coral reefs.
Bottom: Imperial or emperor angelfish, *Pomacanthus imperator.* Juveniles are colored black with concentric blue and white circles centered posteriorly.

Angelfishes

Equally as beautiful as the butterflyfishes are their close relatives, the angelfishes (family Pomacanthidae). They are also strikingly colored, compressed fishes. Angelfishes have a long spine at the corner of the preopercle bone on the gill cover. This preopercular spine is lacking in the butterflyfish. Angelfishes forage near the bottom in shallow water during daylight hours. The juvenile color pattern is often very different from that of the adult.

Racoon butterflyfish, *Chaetodon lunula,* so named because of the eye mask. It feeds on small benthic invertebrates.

Damselfishes

The damselfishes (family Pomacentridae) are one of the most common coral reef families. Over 100 species of damsels can be seen in the waters of the Great Barrier Reef. They are highly territorial and defend their place on the reef from all others, regardless of size. Their territories often contain the best algae food sources. Anemonefishes are damsels of the genus *Amphiprion.* They enjoy a symbiotic relationship with sea anemones. The colorful orange, black, and white anemonefishes are not harmed by the nematocysts of anemone tentacles. Their mucus coating prevents the stinging cells from firing. Many species of damsels are highly prized by the aquarium trade.

Above: Blue-green chromis, *Chromis viridis (Pomacentridae),* occur on the Great Barrier Reef in large aggregations, but they retreat into isolated coral heads at the first sign of danger. Right: Clown anemonefish, *Amphiprion percula.*

Wrasses

The wrasse family (Labridae) is a large one with over 100 species of colorful fishes on the Great Barrier Reef. Juvenile coloration is often markedly different from the adult's, and sex reversals are common. Many wrasses specialize in the consumption of hardshelled prey such as molluscs, sea urchins, and crabs, which they crush with their powerful pharyngeal teeth. Wrasses are diurnal and disappear into the reef as soon as the evening light wanes, and they are late risers in the morning. Some *Labroides* are cleaner-fishes, specializing in removing crustacean ectoparasites and mucus from larger fishes which encourage this activity by going to 'cleaner stations' on the reef and adopting an 'inviting' posture. This is like going to the dentist — humans too adopt a compliant posture and allow a cleaner to probe our delicate oral tissues with sharp objects.

Some *Labroides* and *Thalassoma* have bizarre sex lives. About ten individuals form a social group, of which one is the male and nine are females. The single male fertilizes the eggs of the females (polygyny). The male is the dominant member of the group and is recognizable by his distinctive color pattern and aggressiveness. If the male dies or is removed experimentally, the most dominant female of the group becomes a male. Her color pattern and behavior become masculine, and the ovaries are replaced with testicular tissue capable of producing sperm. Apparently the social stress caused by the presence of the male inhibits sex reversal by females. When the male is no longer present, the inhibition is removed and a hormonal change occurs in the dominant female, leading to her transformation into a male.

Harlequin tuskfish, *Choerodon fasciatus*, a member of the family Labridae, or wrasses. Its beautiful colors have made it an attractive marine aquarium species.

Figure 5.7
Parrotfish, *Scarus gibbus*, and its beak.

Parrotfishes

Parrotfishes (family Scaridae) are so named because their teeth are fused into plate-like beaks (Figure 5.7), and because of their beautiful color patterns. They look like heavy-bodied wrasses, to which they are related. Parrotfishes feed on algae they scrape from dead or living coral rock. The limestone particles they ingest help to grind the algae. Evidence of coral grazing by parrotfishes is obvious in some places where the massive corals such as *Porites* are covered with marks made by the beak-like teeth scraping algae and coral that is later extruded as a fine sediment. The role of parrotfishes in deposition of sediments over geological time is surely great. Parrotfishes undergo sex and color changes as they mature. The initial phase is usually hermaphroditic, but some are only females. These initial-phase females eventually become terminal-phase males, often exhibiting a vivid green or blue color. Group spawning of initial-phase fishes with many males may take place, or pair-spawning may involve an initial-phase female and a terminal-phase male. Males are territorial and maintain harems. Juveniles are often of very different colors than the adults. At night, parrotfishes secrete a mucus covering around themselves and sleep under ledges or in caves in the reef.

Blennies

Blennies (family Blennidae) are small, colorful, bottom-dwelling, mostly herbivorous fishes. They often live in holes (in the reef or on the bottom), which they back into tail first. Most are under 15 cm

Top: Blue tang or palette surgeonfish, *Paracanthurus hepatus*. A single caudal spine lies in a shallow groove at the leading edge of the yellow V. Above: Moorish idol, *Zanclus canescens*.

Clown triggerfish, *Balistoides conspicillum*. Females guarding a nest may chase intruders including divers.

(6 in). A noteworthy blenny is *Aspidontus taeniatus*. It mimics the color pattern and shape of the cleaner wrasse, *Labroides dimidiatus*, which enables it to nip pieces of fins of unsuspecting fishes that expect to be cleaned. The mimicry extends even to the juvenile color pattern of the model, which differs from that of the adult model. At least 50 species of blennies inhabit the Great Barrier Reef.

Gobies

Gobies (family Gobiidae) belong to the largest family of marine fishes in the world, with about 1900 species, most of which live in the Indo-Pacific region. Gobies are small, elongated, bottom-dwelling fishes with pelvic fins fused into a sucking disk. Most are under 8 cm (3 in) as adults, but some reach twice that size. Some gobies live with shrimp in burrows built by the shrimp. Gobies act as 'watch dogs', while their house mates maintain the burrow. Most gobies are carnivores, feeding on small invertebrates. The careful and patient diver or reef walker can expect to see about 60 species of gobies on the Great Barrier Reef, but many more will probably go unseen because of their small size, secretive nature, and camouflaged coloration.

Surgeonfishes

Surgeonfishes (family Acanthuridae) are so called because they have a sharp spine or spines on the caudal peduncle. They can wound other fishes, or a human hand, with a quick flick of the tail. Many species have bright warning coloration around the peduncle spine, and most have a deep, compressed body and feed on algae during the day. They sleep in coral crevices at night. Up to 35 species can be found on the Great Barrier Reef.

Moorish Idol

The Moorish idol, *Zanclus canescens*, is the only member of its family, Zanclidae. It is common in reef areas, and its elongated dorsal fin filament and striking black, white, and yellow coloration make it a very attractive marine aquarium animal. Unlike its relatives, the surgeonfishes, the caudal peduncle of *Zanclus* is unarmed.

Barracudas

Six species of barracuda (Sphyraenidae) can be found on the Great Barrier Reef, including the great barracuda, *Sphyraena barracuda*. This species may reach 170 cm (5.5 ft) and weigh up to 40 kg (88 lb). Barracuda can be very curious and often come quite close to divers, but an attack on humans is extremely rare and usually only happens in murky waters or after provocation. Large specimens should not be eaten because of the danger of ciguatera poisoning.

Triggerfishes

Triggerfishes (family Balistidae) can lock the stout first dorsal spine in an erect position by bracing the second dorsal spine against it. When frightened, they lock their spines in place while hiding in a crevice in the reef. They cannot be easily removed from their hole while the spine is erect. The swimming motion of triggerfishes is unique. They undulate the soft dorsal and anal fin and gracefully flutter. The caudal fin is used for a quick burst of speed. Triggerfishes lack pelvic fins, have a compressed body, tough skin, and small mouth with chisel-like teeth. They feed on shelled prey with their powerful jaws and sharp teeth. Females, guarding their eggs, may chase and nip divers.

Leatherjackets, (family Monacanthidae), are very similar to triggerfishes, but they can change their color pattern to match their surroundings.

Boxfishes

Boxfishes (family Ostraciidae) are, as the name implies, encased in an inflexible bony container. They swim slowly with a sculling motion of the dorsal and anal fins and an occasional movement of the caudal fin. The cowfish *(Lactoria)* species have a pair of horn-like projections over the eyes. They feed on small, sessile, benthic animals such as sponges, tunicates, and soft corals, as well as algae. The rough skin over their box-like armor can secrete deadly toxin called ostracitoxin. With such a physical and chemical defence, they have few predators.

Longhorn cowfish, *Lactoria cornuta,* measuring up to 46 cm (18 in) in length, is often found in weedy areas near rocks or reefs.

Pufferfishes

Pufferfishes (family Tetraodontidae) can swell their bodies by gulping water into the stomach, making them too large to be swallowed by a predator. They also have a powerful neurotoxic secretion called tetraodontoxin (= tetrodotoxin) in their tissues, especially the liver and gonads. Death can result if this toxin is ingested by humans. In Japan, puffer flesh is a delicacy called *fugu,* which licensed *fugu* chefs prepare in a way that hopefully avoids contamination by poisonous tissues. Nevertheless, several deaths occur each year from eating *fugu.* Captain Cook nearly died of puffer poisoning during his second voyage when he ate a small amount of liver and roe in New Caledonia in 1774.

Porcupinefishes

Related to the puffers are the porcupinefishes (family Diodontidae). When they inflate their bodies, the spines on the skin stand up, making predation even more difficult. Their powerful dental plate can crush hardshelled prey as well as deliver a painful bite to a careless human. As the family name implies, puffers have their beak divided into four 'teeth', while porcupinefishes have two.

Protection of the Reef Fishes

Recreational fishing and tourism is important financially, but commercial fishing is more or less precluded by the irregularities found on the Great Barrier Reef. Nevertheless, coral trout, *Plectropomus leopardus,* a serranid, is overfished in many areas, and tuna are taken in nearby waters, but turtles are protected. The pearl industry is important in the Torres Strait area to the north. Tourism is now well developed at several points along the Queensland coast.

Some years ago, oil companies proposed drilling for oil on the reef, an activity that could spell disaster for the complex reef community. Fortunately, in late 1993, the Queensland government banned oil drilling on the Great Barrier Reef, and today the entire zone between the seaward edge of the Great Barrier Reef and the mainland coast is a closed reserve, protected by law. The reef gives the Australia's east coast natural protection against a seaborne invasion. It can only be hoped that Queensland will continue to protect the reef as well as the reef protects Queensland.

• • • • • • • • • •

SHARK ATTACKS

The world averages about 26 shark attacks per year. The mortality rate for attacks dropped from 46 percent in 1940 to about 16 percent in 1973. This is far below the number of persons who die from bee stings (at least 17 per year in United States). Virtually all the shark

Gray nurse or sand tiger shark, *Carcharias taurus*. Although this ragged-toothed shark is rather sluggish, it can be dangerous and should not be provoked. It can measure up to 2.2 m (7.2 ft) in length.

attacks occurred between latitude 47°S and 46°N, with 54 percent occurring in the southern hemisphere. This pattern follows what might be expected, given the distribution of humans.

Attempts have been made to correlate warm water temperatures (22°C, 70°F) with attacks, but they are most likely an indication of people's preference for swimming in warm water rather than a reflection of shark behavior. Most attacks (62 percent) occurred in water less than 1.5 m (5 ft) deep, but this probably also reflects human recreational patterns. A majority of attacks took place 30 m (100 ft) from shore, but again, this is where most people swim. The data show that one's chances of being attacked increase as one moves further from the shore. Men are about ten times as likely to be attacked as women when present in equal numbers off bathing beaches—males may be more active and swim further out, thus attracting more attention from sharks.

In Australia, popular bathing beaches off Sydney, Newcastle, and Wollongong have been meshed since 1937, and anti-shark measures have protected many Queensland coastal resorts since 1962. In New South Wales, 150 m (490 ft) long 'curtains' made of 60 cm (24 in) gill nets mesh, are set parallel to the beach and touching the bottom some distance offshore. No attempt is made to shield the entire beach, and sharks are caught moving in both directions, that is, to and from the beach. The idea is not to fence off the beaches from sharks, but to reduce the shark population to such low numbers that an attack becomes very improbable.

Meshing has definitely reduced the shark population around the state's beaches. It is most effective against resident species such as *Carcharhinus* spp., as opposed to migratory species like the great white shark, *Carcharodon carcharias*. There have been very few shark attacks at meshed beaches. Whitley (1963) recorded only nine attacks at

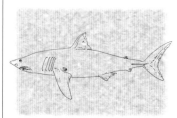

Figure 5.8
Great White Shark,
Carcharodon carcharias.

meshed beaches in Australia, whereas 41 attacks were recorded prior to meshing. There has been only one fatal shark attack in New South Wales waters (1982) since 1967.

Conservationists are now concerned that shark populations have become dangerously depleted. Meshing also threatens harmless species such as dolphins, dugongs, turtles, and rays.

An analysis of sharks caught in the New South Wales beach meshes showed that significantly higher catches of dangerous whaler (*Carcharhinus* spp.), great white (*Carcharodon carcharias*), and tiger (*Galeocerdo cuvieri*) sharks occurred when deeper water was closer (12–18 m) to the beach. The following species and their numbers were recorded from October 1972 to December 1990:

hammerhead sharks (*Sphyrna* spp.)	2033
whaler sharks	831
angel sharks (*Squatina australis*)	753
Port Jackson sharks (*Heterodontus* spp.)	435
white sharks (the largest of which was 5 m)	185
tiger sharks	176
gray nurse sharks (*Carcharias taurus*)	65
seven-gill sharks (*Notorynchus cepedianus*)	50
thresher sharks (*Alopias* spp.)	34
shortfin mako sharks (*Isurus oxyrinchus*)	17
wobbegong sharks (*Orectolobus* spp.)	16

Worldwide, of the 350 or so living species of sharks only 27 have been implicated in attacks on humans. Some level of identification was possible in 267 cases. The great white shark (Figure 5.8), called the white pointer or white death in Australia, topped the list with 32 (12 percent) attacks. The cosmopolitan tiger shark was a close second with 25 (9 percent) known attacks. The Australian gray nurse shark is also known to attack humans. (This species should not be confused with the rather sluggish nurse shark of American waters, *Ginglymostoma cirratum.*) The requiem sharks of the genus *Carcharhinus*, called whalers in Australia, have been responsible for many Australian attacks.

Australian waters harbor about 166 species of sharks, some of which have yet to receive scientific names. About 48 percent of the shark fauna is endemic to Australia. The whalers (family Carcharhinidae) comprise about 30 species, the catsharks (family Scyliorhinidae) about 32 species, the wobbegongs (family Orectolobidae) about six species; the dogfishes (family Squalidae) form the largest family with about 40 species.

Notwithstanding the facts given above, Australians are more likely to bite sharks than the other way round. The gummy shark, *Mustelus antarcticus,* and the school shark, *Galeorhinus australis,* are important commercial species. Most of the fish sold at fish and chip shops in Australia is shark, which also masquerades under the name 'flake' in some fish markets.

Australian Shark Attacks

Australia has the rather unpleasant reputation of being a center of shark attacks. Whitley (1940) listed 226 Australian attacks from 1803 to 1940, and Coppleson (1962) recorded 23 attacks from 1940 to 1961. Less than half (118, 47 percent) of these attacks were fatal. Green (1976) recorded 320 shark attacks in Australian waters from 1791 to 1975, of which 126 (39 percent) were fatal. West (1993) reported a total of 471 shark attacks in Australian waters as of February 1991, with a 39 percent fatality rate.

Of the 1165 cases documented in a worldwide shark attack file (Baldridge 1974), 319 were in Australian waters:

New South Wales	137	South Australia	12
Queensland (excluding		Tasmania	10
the Great Barrier Reef)	57	Great Barrier Reef	9
Torres Strait	43	Northern Territory	5
Western Australia	29	Non specific	2
Victoria	15		

An unusual ventral view of Port Jackson sharks, *Heterodontus portjacksoni*, in the acrylic viewing tunnel at Ocean World, Sydney. This stocky, benthic species of horned shark can be up to 1.7 m (5.6 ft) long.

The first record of a shark attack in Australian waters dates back to the early colonial period. Undoubtedly many Aboriginals have been attacked during their more than 40 000 year history in Australia, but there is no written record of such events. The Australian Shark Attack File at Sydney's Taronga Zoo lists 76 fatal attacks in New South Wales since 1791, 69 in Queensland, 14 in South Australia, eight in Tasmania, seven each in Victoria and Western Australia, and three in the Northern Territory (West, 1993). Since protective meshing of beaches was instituted most attacks have occurred in more remote, unprotected areas. From 1983 to 1992, six deaths occurred in Queensland, four in South Australia, and one each in New South Wales and Tasmania. No fatalities were reported from Victoria, Western Australia, or the Northern Territory.

The most famous Australian shark story is undoubtedly the 1935 'shark arm murder' (Ellis 1976). A 4.4 m (14.5 ft) tiger shark, while on display in a Sydney aquarium, regurgitated a human arm which had a rope tied around the wrist and a tattoo of two boxers on the shoulder. Australia's leading shark experts, Coppleson, Stead, and Whitley, agreed that the arm had been severed by a knife, not by a shark bite. The victim was identified by the tattoo and fingerprints. The Supreme Court decreed that an arm was not a corpse and that there could be no murder trial of the suspect alleged to have dismembered the body and thrown pieces into the ocean.

· · · · · · · · · ·
MEGAMOUTH

In 1988–89 I was Research Associate at the Western Australian Museum in Perth, on sabbatical from Ohio State University and working on the life history of the tiny salamanderfish. The phone rang early in the morning of 18 August 1988. It seemed that some strange large animal had washed up on the beach at Mandurah, 50 km south of Perth. So began the saga of Megamouth III.

Megamouth, *Megachasma pelagios,* so named because of its huge mouth, is a large shark species that, at the time, was known only from two other specimens. The first specimen became entangled in a US Navy sea anchor off Oahu, Hawaii, on 15 November 1976. It was a completely unknown fish, which belonged to a new family (Megachasmidae), genus, and species of shark. Eight years later, on 29 November 1984, a second specimen was caught in a gill net off Santa Catalina Island, California. Now we had the third specimen in Western Australia.

Our 1988 megamouth, similar to the other two, was a male 5.15 m (17 ft) long and weighing 690 kg (1518 lb). It was transported by truck from the beach at Mandurah and placed in a freezer. It attracted so much attention that we put the frozen animal on public display for a few hours several days later. Several thousand people viewed the beast.

The third known specimen of Megamouth, *Megachasma pelagios,* which washed up on the beach south of Perth on 18 August 1988. It is a 5.15 m (17 ft) long male. The Megamouth is now on permanent display at the Western Australian Museum in Perth.

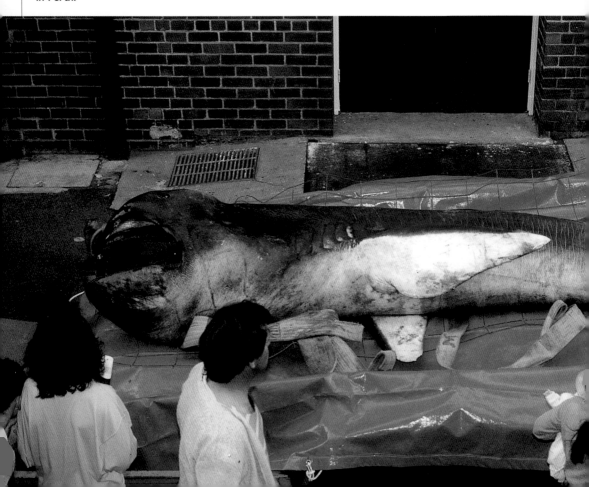

Eventually it was decided that the shark had to be preserved. As an interim measure, a coffin-shaped hole was excavated in sandy soil on museum property in Fremantle. The pit was lined with plastic swimming pool liner. A large hypodermic needle was fashioned from a meter length of stainless steel tubing and concentrated formalin was pumped into the defrosted shark. The hole was filled with 1600 liters of concentrated formalin and diluted with water to make a 10 percent formalin solution that covered the shark. Megamouth III now resides in a permanent display tank at the Western Australian Museum.

Since 1988, six other specimens of megamouth have been reported worldwide. Figure 5.9 shows the distribution of the nine cases. The stomach of Megamouth I contained a large quantity of euphausiid shrimp, indicating that megamouth is a filter feeder. Sonic transmitters were attached to Megamouth VI captured alive off California. It was videotaped, released, and then followed for two days. It descended to about 150 m (492 ft) during the daytime and rose to about 15 m (49 ft) at night. This vertical migration is probably in response to the movements of its food organisms.

Megamouth's teeth are tiny and harmless to humans. Like the other two species of very large filter-feeding sharks—the whale shark, *Rhiniodon typus,* and the basking shark, *Cetorhinus maximus*—megamouth most likely swims with its mouth agape, filtering food from large quantities of water that pass over its closely spaced gill rakers. Its flesh is very flaccid indicating that megamouth is not a powerful swimmer.

Megamouth I and our Megamouth III bore bite marks of the cookie cutter shark, *Isistius brasiliensis,* a small 50 cm (20 in) oceanic shark. This fish attacks whales, elephant seals, and even the rubber sonar domes of nuclear submarines. With a twist of its small body, the large saw-like teeth of the lower jaw of the cookie cutter shark remove a conical plug of flesh from the sides of its prey, leaving a crater-like wound. For more information on megamouth, see Berra (1997). As page proofs of this book were being corrected, I learned that a 10th specimen of megamouth has been obtained. It is a female from Toba, Japan. The specimen is about 5 m long and weighs 1000 kg.

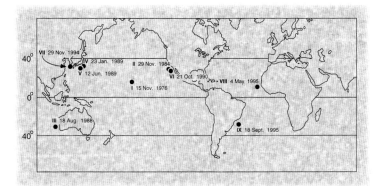

Figure 5.9
Distribution of reported specimens of megamouth shark, *Megachasma pelagios.*

Murray River crayfish, *Euastacus armatus,* is a food delicacy but it is also a formidable creature with large claws and spiked tail.

Some Interestin$g$ Invertebrate$s$

Australia is well known for its oddball mammals, and the plants are equally unusual. The uniqueness of the Australian biota resulting from its evolution in isolation extends even to the invertebrates.

GIANT EARTHWORMS

The southeastern region of Victoria known as Gippsland is home to the largest earthworm species on earth, the giant earthworm, *Megascolides australis* (Figure 6.1). These animals reach 3.7 m (12 ft) in length and 2.5 cm (1 in) in diameter. Their egg capsules are about 6.3 cm (2.5 in) long. They live in a cool moist climate, in burrows on the hillsides of the Bass River watershed and the southern slopes of the Strzelecki Ranges in soft volcanic soil with thick leaf litter. The worms make a gurgling sound that can be heard on the surface as they progress through the soil.

FRESHWATER CRUSTACEANS

Mountain Shrimp

Tasmania is the home of a group of 'living fossils' called mountain shrimp. These primitive crustaceans are members of the Syncarida. Very similar fossils date back to the Carboniferous and Permian times of Europe, North America, and Australia, 360–248 million years ago. The best known mountain shrimp is *Anaspides tasmaniae* (Figure 6.2). This animal may reach a length of 56 mm (2.2 in). It lives in cool pools and small streams at altitudes over 300 m (1000 ft) in central and southern Tasmania. Unlike true shrimp, which are decapods, the syncarida have no carapace, and, therefore, the thorax and abdomen bear obvious segments. Unlike decapods, which carry their eggs, mountain shrimp lay them singly on submerged objects. They feed on algae. There are five other species of Tasmanian syncarida and two species on the Australian mainland.

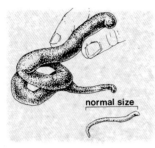

Figure 6.1
Giant earthworm,
Megascolides australis.

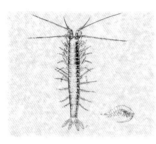

Figure 6.2
Tasmanian mountain shrimp,
Anaspides tasmaniae, lives in
cool pools and small streams of
central and southern Tasmania.

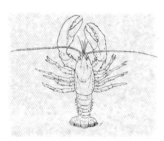

Figure 6.3
Murray River crayfish,
Euastacus armatus.

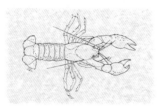

Figure 6.4
Yabby, *Cherax destructor,* is the
larger of the yabby species that
can be up to 250 mm long.

Freshwater Crayfish

The freshwater crayfish family of the southern hemisphere,
Parastacidae, is large and interesting. Most of the species are endemic
to Australia, with a few species being found in South America,
Madagascar, Papua New Guinea, and New Zealand. The northern
hemisphere group is the Astacoidea (families Astacidae and
Cambaridae) found in North America, Europe, and northern Asia.
There are no crayfish in Central America, Africa, India, or
Southeast Asia.

There are about 110 species in nine genera in Australia, all of
which are endemic. Australia has both the largest and the smallest
crayfish in the world. The Tasmanian *Astacopis gouldi* reaches 4 kg
(8.8 lb) and 60 cm (2 ft), while the adult *Tenuibranchiurus glypticus* of
southern Queensland reaches only 2.5 cm (1 in).

The Murray River crayfish, *Euastacus armatus,* is a formidable beast
with large white claws and numerous, big, white, thorn-like, spines
on its tail (Figure 6.3). It can reach over 35 cm (14 in) in length and
weigh up to 2.7 kg (6 lb). Found in the Murrumbidgee, Darling and
Murray rivers, it is fished by commercial fisherman.

Yabby

The yabby, *Cherax destructor* (Figure 6.4), is smooth-shelled and
widespread in inland southeastern Australia. It has been introduced
into Western Australia. It can withstand droughts by sealing itself
inside its deep, water-filled burrow for up to eight years. It is an
important item in the diet of freshwater fishes. Yabbies are fished
commercially from the wild, and are also raised commercially in
pond aquaculture; some details of their life history are known.
Large individuals of *C. destructor* may reach 250 mm (9 in) and 300 g
(0.66 lb), but most are under 100 g (0.2 lb), with males being larger
than females. Their usual habitats are streams, swamps, and
billabongs, but they can tolerate a wide variety of conditions,
including low oxygen levels. They dig burrows but not long complex
tunnel systems as some other species of crayfish do. Mating takes
place face to face from spring to autumn, and the females carry the
fertilized eggs on their swimmerets. Large females may carry as
many as 1000 eggs under their tails. Incubation takes about three
weeks in summer temperatures, after which time the young remain

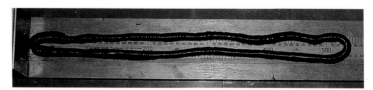

Giant earthworm, *Megascolides australis,* from Victorian Gippsland region,
can be up to 3.7 m (12.2 ft) long and 2.5 cm (6.3 in) in diameter. The one
in the photograph is over 1 m (3.3 ft) long.

with the female and undergo three molts during a 2–3 week period. They then become independent juveniles.

Yabbies usually feed at night on decomposing organic matter, often moving overland on damp vegetation to forage on plant matter. Feeding and growth slow down in cold months. Since crayfish are covered by a hard shell, they must shed this calcified skin to be able to grow. This process is called ecdysis or molting, and is controlled hormonally. Calcium carbonate is removed from the shell and stored in the stomach as small, paired stones, called gastroliths. The weakened shell splits open and the animal climbs out with its new skin that is soft and shiny. It immediately begins to resorb the calcium carbonate stored in the gastroliths and redeposit it in the new shell, which then hardens. This process may take from a few minutes to a few hours. The claws are the first part to harden, and eventually the gastroliths are resorbed. Sometimes fishermen catch fish with small flat stones in the stomach, which may be the gastroliths of crayfish they have eaten. Empty shells are usually devoured by their former occupant, which helps recycle calcium. As the yabbies get older they molt less frequently. Juveniles molt every 10 days or so, young adults 4–5 times per year, and older animals may molt once or twice a year.

Female Murray River crayfish with eggs ('in berry').

Marron

The marron, *Cherax tenuimanous,* of southwestern Western Australia can reach 380 mm (15 in) in length and weigh as much as 2.7 kg (5.9 lb), but most weigh less than 200 g (0.4 lb). They are highly prized for food and are raised in aquaculture. There is no commercial fishing of wild marron, but recreational fishing is permitted.

Land Crayfish

Land crayfish of the genus *Engaeus* live at some distance from streams in deep burrows. Some build mud chimneys above the entrance to their burrows.

Freshwater Crabs

Australia has a family of freshwater crabs, the Potamonidae. Five of the six Australian species occur on Cape York Peninsula, but one, *Parathelphusa transversa,* ranges deep inland to Lake Eyre and the Darling River (Figure 6.5). Their burrows may be up to 1 m (3.3 ft) deep. When the rains come these freshwater crabs appear in large numbers on the surface. The ancestors of this crab were from southeast Asia and made their way to Australia via the Indonesian islands and Papua New Guinea, which was connected to Australia, across the Torres Strait, during the Pleistocene. This family is also found in South America, southern Europe, Africa, and Asia.

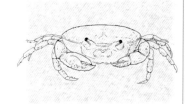

Figure 6.5
Inland freshwater crab, *Parathelphusa transversa.*

Top: Marron, *Cherax tenuimanous*, from southwest Western Australia, is able to survive up to eight years of drought by sealing itself in its water-filled burrow.
Middle: Southeastern marine crayfish, *Janus novaehollandiae*.
Bottom: A rock-hard termite mound near Darwin, Northern Territory. These complex structures can reach up to 7 m (23 ft) in height.

MARINE CRAYFISH

Technically, there are no lobsters in Australian waters. Lobsters are found in the North Atlantic and the Mediterranean. True lobsters (Homaridae) have large grasping claws on the first pair of legs (celipeds) and smooth shells. The North American species, *Homarus americanus,* can weight up to 20 kg (44.5 lb), but the average is closer to 1 kg (2.2 lb).

The large marine crustaceans fished commercially and by sport divers in Australia are crayfish of the family Palinuridae. This decapod family lacks celipeds and the carapace may be spiny in some species. Marine crayfish (spiny lobsters) are found worldwide in the tropics and subtropics. *Janus verreauxii* from temperate New South Wales can weigh 7.7 kg (17 lb) and reach 1 m (3.3 ft). *J. novaehollandiae* from the cooler waters of Australia's south coast can weigh up to 2.3 kg (5 lb).

The Western Australian crayfish, *Panulirus cygnus,* smaller at 1.5 kg (3.3 lb), nevertheless supports the richest Australian commercial fishery, especially from the reefs of Houtman Abrolhus, off Geraldton. Fishers set baited pots along an extensive area ranging from shallow coastal reefs to the 200 m depths of the continental shelf edge, from the Northwest Cape to Cape Leeuwin. Uncooked frozen tails are exported as 'rock lobster'. Western rock lobsters are nocturnal, live in dense groups, and feed on molluscs, small crustaceans, worms, algae, and seagrasses. They are prey for octopus and for predacious fishes such as sharks and snapper (Lutjanidae). Some of the larger specimens may be 30 years old. Rock lobsters take about five years to reach maturity, and a single female may carry as many as 500 000 tiny eggs on her swimmerets.

The other large southern hemisphere marine crayfish family is the Scyllaridae. They are clawless and have flattened bodies. Their antennae are enlarged and modified into rounded plates used for burrowing. A 25 cm (10 in) scyllarid from Queensland, called the Moreton Bay bug, *Thenus orientalis,* is taken by prawn trawls and considered a delicacy by gourmets. It is sometimes sold as 'sand lobster'.

No matter what you call them, all of the Australian 'crays' are a delicious treat in the Australian diet and some are significant export earners.

INSECTS

Insects are the most successful group of animals on earth in terms of number of species. The number of insect species that have been given scientific names is about 950 000, but the actual number of insect species may be closer to 8 000 000 (Groombridge 1992). Roughly 40 percent of all known insects are beetles. It can be said that about 80 percent of the described animals are insects, and that about one in every three animal species is a beetle. The reasons for the phenomenal success of insects are their high and rapid rate of reproduction, their protective armor, their ability to fly and to utilize many kinds of food, their wide temperature tolerance, and their ability to conserve water.

Australia has about 6 percent of the ice-free area of the earth and about 8.5 percent of the described insect species (Table 6.1). The 67 000 or so species of named Australian insects will probably grow to 100 000 when all species are named.

There are 661 families of insects in Australia; 29 of the world's 32 insect orders are represented (Table 6.1). The largest order is the beetles (Coleoptera) with about 20 000 described species in 113 families. Of these, about 6000 are weevils and their relatives.

Figure 6.6
Hercules moth, *Coscinocera hercules,* is one of the largest moths in the world with a wing span of 25 cm.

Butterflies and Moths

The Lepidoptera (butterflies and moths) are the second largest group of insect species in Australia, with about 10 000 described species, of which only about 400 are butterflies. One of the world's largest moths, the Hercules moth, *Coscinocera hercules* (Figure 6.6), of rainforests in north Queensland and Papua New Guinea, has a wingspread of 25 cm (10 in). The largest Australian butterfly is the Cairns birdwing, *Ornithoptera priamus* (Figure 6.7). The black and white females have a wingspan of 19 cm (7.5 in) and the black, green, and yellow males have a 16 cm (6.3 in) wingspan.

Figure 6.7
Male birdwing butterfly, *Ornithoptera priamus.* The males of this species are colored black, green and yellow, and the females are black and white.

Flies, Wasps, Bees, and Ants

The flies (Diptera) and wasps, bees, and ants (Hymenoptera) have about 8000 and 15 000 species respectively. The ants are very widespread, with 1000 or so species, and may be considered one of the dominant groups of insects in Australia. Some Australian termites (Isoptera) live in immense colonies numbering a million or more individuals. Their society is rigidly controlled by a system of chemical-signal substances. In northern Australia some termite mounds may be up to 7 m (23 ft) tall and 4 m in diameter at the base. The mounds are composed of cemented soil, plant, and fecal material and are honeycombed with chambers and tunnels. One species orients its mound on a north-south axis.

Dragonfly

The aridity of the Australian continent is legendary and much of the Australian biota, and especially the insects, are adapted to drought and high temperature. However, a rainforest region of northern Queensland is so continuously damp that the larval stages of a dragonfly, *Pseudocordulia,* in the family Corduliidae are completed in the leaf litter far from standing water. The larvae are very hairy and flat, which may increase the surface area for obtaining moisture. No other continent has produced a similar adaptation in this order of insects (Odonata), which normally have aquatic larvae.

TABLE 6.1 **INSECT ORDERS OF THE WORLD AND THE APPROXIMATE NUMBER OF SPECIES DESCRIBED FROM AUSTRALIA**

ORDER	COMMON NAME	NUMBER OF SPECIES WORLDWIDE*	NUMBER OF SPECIES DESCRIBED FROM AUSTRALIA *
Collembola	Springtails	6 000	1 630
Protura	Proturans	500	30
Diplura	Diplurans	800	31
Archaeognatha	Bristletails	350	7
Thysanura	Silverfish	370	28
Ephemeroptera	Mayflies	2 500	84
Odonata	Dragonflies	5 000	302
Plecoptera	Stoneflies	2 000	196
Blattodea	Cockroaches	4 000	428
Isoptera	Termites	2 300	258
Mantodea	Praying mantids	1 800	162
Grylloblattodea	Ice crawlers	25	0
Dermaptera	Earwigs	1 800	63
Orthoptera	Grasshoppers	20 000	2 827
Phasmatodea	Stick-insects	2 500	150
Embioptera	Webspinners	200	65
Zoraptera	Zorapterans	30	0
Psocoptera	Booklice	3 000	299
Phthiraptera	Lice	3 000	255
Hemiptera	Bugs	40 000	5 650
Thysanoptera	Thrips	4 500	422
Megaloptera	Dopsonflies	300	26
Raphidioptepa	Snakeflies	175	0
Neuroptera	Lacewings	5 000	623
Coleoptera	Beetles	300 000	20 000
Strepsiptera	Twisted-wing parasites	532	27
Mecoptera	Scorpionflies	500	27
Siphonaptera	Fleas	2 380	88
Diptera	Flies	150 000	7 786
Trichoptera	Caddisflies	7 000	478
Lepidoptera	Moths & butterflies	110 000	10 000
Hymenoptera	Wasps, bees, ants	110 000	14 781
Total		**786 562***	**66 723***

*Many more species remain to be described. See Groombridge (1992) for estimates.
Source: CSIRO (1991) and May (1988).

Aboriginal Insect Delicacies

Several insects provide food for tribal Aboriginals. The Bogong moth, *Agrotis infusa,* congregates by the millions in caves and mountaintop crevices in the high country south and west of Canberra. Aboriginals from the Snowy Mountains region used to feast on roasted moth bodies, the weight of which was about 60 percent fat.

Another highly sought after delicacy is the witchetty grub—
a name that is actually applied to several types of moth and beetle
larvae. In the South Australian deserts a moth species, *Xyleutes
leucomochla*, produces an 8 cm (3 in) larva that feeds on the roots of
Acacia ligulata. This species of grub is easy to locate because it tunnels
to the surface and can be collected without digging. The grubs are
eaten raw, or are cooked by rolling them in warm coals from a
campfire. Those whose palates soar above all prejudice report that
when raw the grubs taste like scalded cream and, when cooked, like
roast pork rind. Grasshoppers and locusts are also eaten, after being
skewered on sticks and grilled like a shish kebab.

If that does not seem appetizing, consider the following. The
distended bodies of honeypot ants, *Camponotus inflatus*, of Central
Australia store large amounts of sweet liquids. These ants live in large
colonies and hang from the roof of their underground nests. The
bodies of the worker ants are living storehouses. When food is scarce,
they regurgitate their stored nectar for other members of the colony.
Aboriginals bite off the enlarged abdomens, which resemble small
amber grapes, and allow the nectar to flow over their tongues.

Figure 6.8
Bushfly, *Musca vetustissima*,
plagues much of Australia. The
females are particularly
bothersome, and unfortunately
more numerous.

BUSHFLIES AND DUNG BEETLES

Much of Australia is plagued by bushflies, *Musca vetustissima*
(Figure 6.8). The females are especially bothersome, and are about
three times more numerous than the males. While they do not bite,
these irritating little beasts will cover your body, and if you are not
careful they will fly up your nose, climb in your mouth, and get
between your eyes and your spectacles, seeking both protein for egg
development and moisture. They are also vectors of eye disease in
livestock. They are responsible for the 'great Australian salute'—a
waving action of the hand near the head. This gesture is common to
anyone venturing outside in summer. However, bushflies do not
readily enter buildings, which might explain why the archetypical
Aussie 'cork' hat works: bushflies do not like going 'inside' the line of
corks hanging from strings around the rim of the hat.

The bushfly life cycle has four stages: egg, larval, pupal, and adult.
The female lays up to 50 eggs in wet dung, where the eggs hatch in
less than 24 hours—the higher the temperature the quicker they
hatch. The larvae (maggots) feed on the dung while passing through
three instars (stages between molts). They then enter the soil, where
pupation occurs. Adult flies emerge at dawn 3–18 days later,
depending on the temperature. They live a maximum of four weeks in
cool weather, and in hot weather they live only about one week.

The rapid build-up of bushflies in spring occurs when dung quality
is high due to periods of rapid pasture growth. Bushflies die out each
winter in the southern third of Australia, as they do not tolerate
temperatures below 12°C (53°F). This area is repopulated by flies that
overwinter north of the Queensland-New South Wales border area. In

Figure 6.9
African dung beetle, *Onthophagus gazella,* was introduced into Australia to control the bushfly population.

September and October, overwintering flies are carried southward by the wind, some reaching as far as Tasmania by December.

The native bushfly and the more northerly, introduced buffalo fly, *Hamematobia irritans,* both lay their eggs in dung. Before Europeans introduced cattle the only habitat for the flies was the small, relatively dry dung pellets of the large herbivorous marsupials such as kangaroos and wombats. This dung was readily decomposed by about 250 species of native scarab beetles called coprids. Thus the breeding grounds of the flies were limited by the dung beetles.

In 1788 the first English colonists disembarked with five cows, two bulls, seven horses, and 44 sheep. With the introduction and wide proliferation of domestic cattle, the habitat for bush flies changed. The large moist cow dung pads are unattractive to most of the native dung beetles. As the cow pads dry out they remain undecomposed for months or even years, with rough unusable grass growing at their edge. With 30 million cattle producing 300 million dung pads per day, Australia was losing up to 2.4 million ha (6 million acres) of pasture every year and providing an enormous nursery for bushflies. A single cow pad can give rise to 2000 bushflies.

The CSIRO began to investigate the possibility of introducing dung beetles from Africa that could control cattle dung. Africa, with its large herds of native bovids, has over 2000 species of dung beetles. Most coprid beetles dig tunnels under cow pads and carry the dung into the excavations. They roll the dung into balls, and the females lay their eggs in this food source. The larvae eat the surrounding dung ball. The actions of the dung beetle are beneficial to the soil by increasing fertility, improving permeability to water, increasing aeration, and helping seed germination. Experiments have shown that the African dung beetle, *Onthophagus gazella* (Figure 6.9), can break up and bury cow pads in 30–40 hours. This has resulted in an 80–100 percent decline in bushfly populations. In 1967 the first introduction of *O. gazella* took place in tropical Australia. It spread rapidly and has colonized the northern third of Australia. Two other African species, *Euonitiellus intermedius* and *E. africans,* have been successfully introduced into New South Wales and Queensland. Dung pads in these areas are now being decomposed rapidly and completely. There is no danger that the introduced dung beetles will get out of control and eat crops, as their mouth parts are adapted to sucking juices exclusively from dung. Only dung beetle eggs are imported and these are surface-sterilized to prevent introduction of diseases.

Much remains to be done, however. Beetles of different species that are active at the proper time of year to control all phases of fly reproduction are still being sought. Likewise, beetles that can tolerate the various degrees of drought and heat found in Australia are being investigated. An interesting (and unfortunate) problem occurs with the introduced toad, *Bufo marinus*—the so-called cane toad—which can eat 80 beetles per night. Day-flying coprids are being sought to cope with the nocturnal toad.

SPIDERS

There are some 15 000 species of spiders in Australia, representing about 38 percent of the world's total of about 40 000. Most of the species in Australia are endemic.

Funnelweb Spider

The Sydney funnelweb spider, *Atrax robustus* (Figure 6.10), is one of the most venomous spiders in the world. This large black spider is named for its habit of building a funnel-shaped burrow in the ground, usually beside a log or stone. Both sexes are venomous and aggressive. When annoyed they rear up and exude venom droplets from their fangs before striking. People have died as the result of a bite from this species. A related species, *A. formidabilis,* is also capable of causing human death. In the mid-1980s, after a period of research spanning almost 20 years, Dr Struan Sutherland of the Commonwealth Serum Laboratories produced an effective antivenom for funnelweb spider bites.

The Redback

The other Australian spider deadly to humans is the redback, *Latrodectus hasselti* (Figure 6.11). This species is related to the American black widow, *L. mactans.* It is jet black with a red stripe on the dorsal surface of the abdomen. An antivenom for the bite of this species was produced in 1956. Prior to that, at least 13 people had died from the redback's bite, but since 1956 no deaths have been recorded. It is the female that bites; the male is harmless. Its distribution is Australia-wide.

The life history of the redback spider is an interesting challenge in evolutionary thinking. The tiny male is eaten 65 percent of the time by the relatively enormous female. In fact, the male may be consumed while copulation is in progress. What is the survival value of such a mating system, one may ask. Given that the male weighs only about 2 percent of the female's weight, his sacrifice cannot be for the purpose of nutrition. Male-male competition may explain this bizarre situation. A male that allows himself to be eaten can copulate for a longer time, thus ensuring that more of his sperm is transmitted. In addition, a female that eats a male is more likely to reject another suitor. Thus the cannibalized male gets more of his genes into the next generation. The female lives for up to two years, but a male lives only 2–4 months after maturation. Even if he is not eaten, he is not likely to live long enough to find another mate. Therefore natural selection has programmed redback spider males to put all their resources into their one shot at reproduction.

Figure 6.10
Sydney funnelweb spider, *Atrax robustus,* so named because it builds funnel-shaped burrows in the ground. This large black species is very venomous, exuding venom from its fangs when provoked.

Figure 6.11
The deadly redback spider, *Latrodectus hasselti,* is related to the American black widow.

Golden Perch (rear)
Macquarie ambigua
and Silver Perch,
Bidyanus bidyanus.

The Freshwater Fishes

There are approximately 24 700 described species of fishes, of which 41 percent are found in fresh waters. Yet there are only about 200 species (c. 1 percent) in the fresh waters of Australia, as compared to the United States where the freshwater fish fauna numbers about 800 species (3.2 percent). The Australian freshwater ichthyofauna has three characteristics. It is:

1. depauperate;
2. derived from marine ancestors; and
3. highly endemic.

These characteristics are the result of two factors. Firstly, Australia separated from Antarctica and the rest of Gondwana about 50 MYA, before the dominant freshwater fish families of the world (Cyprinidae, Cyprinodontidae, Cichlidae, Percidae, and others) could reach Australia. Secondly, Australia is the driest continent; the lack of rivers discourages the development of a large freshwater fish fauna.

The continent's greatest river system, the Murray-Darling (Figure 7.1), has less than 30 native freshwater species, whereas the much shorter Ohio River in the United States has over 170 species. The Murray is about 2600 km long (1614 miles) and has a total catchment of 1 073 194 km^2 (414 253 mi^2), yet its annual discharge is only 1 500 000 hectare-meters (12 000 000 acre-feet). This is because the average annual rainfall over the Murray catchment is only 425 mm (17 in). Many of its tributaries cease to flow and even dry up periodically. Rainfall is variable, and there are no permanent snowfields to maintain flow during the dry summers. On the other hand, periodic flooding can be devastating. As an example of the extremes of the Australian aquatic environment, consider the Darling River, the largest tributary of the Murray. Its average annual flow is 269 125 ha-m (2 153 000 acre-feet), but this has varied from 125 000 to 1 375 000 ha-m (1000–11 000 000 acre-feet).

Most of the freshwater fishes of Australia have evolved from marine invaders that adapted to fresh water. This has resulted in a unique fish fauna with many species found nowhere else in the world. Table 7.1 (page 128) lists the families found in the fresh waters of Australia.

The mouth of a male pouched lamprey, *Geotria australis*.

• • • • • • • • • •
NOTABLE FAMILIES

Australian Lampreys

There are three species of lampreys (Petromyzontidae) in Australia's fresh waters. The pouched lamprey, *Geotria australis* (Figure 7.2), is found in southeastern and southwestern Australia as well as in New Zealand and southern South America. Adults may reach 510 mm (20 in) in length, and the males have a bizarre throat pouch, the function of which is unknown. *Geotria* feeds parasitically on the body fluids of other fishes by rasping their flesh with horny teeth.

Adult *Geotria* migrate to freshwater streams to spawn and die. The ammocoete (the larval stage) filter-feeds in fresh water until it reaches the large-eyed stage (macrophthalmia) and then moves downstream to the sea. After several years at sea, the adults return to fresh water. The two other species of Australian lampreys are *Mordacia mordax* (Figure 7.3) and *M. praecox*. The latter is free-living.

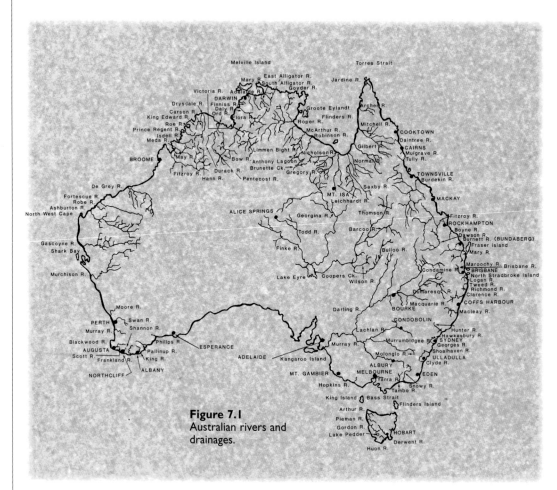

Figure 7.1
Australian rivers and drainages.

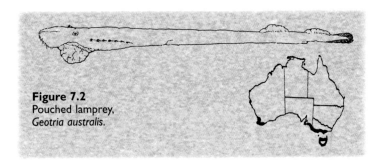

Figure 7.2
Pouched lamprey,
Geotria australis.

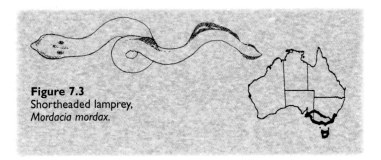

Figure 7.3
Shortheaded lamprey,
Mordacia mordax.

Australian Lungfish

The Australian lungfish (family Ceratodontidae) is the most primitive of the three lungfish families, all of which are confined to the southern hemisphere. The other two are the Lepidosirenidae of South America and the Protopteridae of Africa. Lungfishes are a very ancient group that evolved and dispersed before Australia became isolated from Gondwana. The Australian species, *Neoceratodus forsteri* (Figure 7.4), is the only living member of its family and is one of the two truly freshwater families found in Australia. (A truly freshwater group, also called a primary division freshwater group, is one that is confined to fresh water and has very little salt tolerance. Primary division freshwater fishes have not depended on the sea for dispersal.)

Neoceratodus is native to the Mary and Burnett river systems of Queensland, and has been successfully introduced to other Queensland rivers such as the Brisbane, Coomera, and Stanley. Its preferred habitat is slow-flowing deep pools. The Australian lungfish does not estivate in the mud at the bottom of dried-out lakes and streams as do the other lungfishes. It can utilize atmospheric oxygen in its single lung when the dissolved oxygen level of its habitat decreases. The body of this fish is covered with large scales, and the paired fins are paddle-like. The caudal fin is pointed and extends dorsally and ventrally. *Neoceratodus* lays large eggs among aquatic plants in shallow water from August to October, and it can grow to 180 cm (32 in) and 45 kg (99 lb). It is carnivorous.

An extinct genus, *Ceratodus,* was very similar and was widely distributed around the world from the lower Triassic to the upper Cretaceous, about 240–65 million years ago. *N. forsteri* is known from the Cretaceous of New South Wales and has apparently remained unchanged for over 100 million years, making it one of the oldest living species of any vertebrate.

TABLE 7.1 **FISH FAMILIES IN THE FRESH WATERS OF AUSTRALIA**

FAMILY	COMMON NAME	FAMILY DISTRIBUTION	SOME AUSTRALIAN GENERA OR SPECIES
Petromyzontidae	Lampreys	Worldwide temperate zone	*Geotria australis, Mordacia*
Carcharhinidae	Requiem sharks	Cosmopolitan	*Carcharchinus*
Pristidae	Sawfish	Pantropical	*Pristis*
Dasyatidae	Stingrays	Pantropical	*Dasyatis*
Ceratodontidae	Australian lungfish	S.E. Queensland	*Neoceratodus fosteri*
Osteoglossidae	Bony tongues	South America, Africa, Asia, N. Australia	*Scleropages*
Megalopidae	Tarpons	Pantropical	*Megalops cyprinoides*
Anguillidae	Freshwater eels	Cosmopolitan	*Anguilla*
Clupeidae	Herrings	Cosmopolitan	*Nematolosa, Pomatolosa*
Engraulidae	Anchovies	Cosmopolitan	*Thryssa*
Chanidae	Milkfish	Indo-Pacific	*Chanos chanos*
Ariidae	Forktailed catfish	Pantropical	*Arius, Cinetodus*
Plotosidae	Eeltailed catfish	Indo-Pacific	*Tandanus, Neosilurus, Copidoglanis*
Lepidogalaxiidae	Salamanderfish	S.W. Western Australia	*Lepidogalaxias salamandroides*
Retropinnidae	Southern hemisphere smelts and graylings	S.E. Australia, New Zealand	*Retropinna, Prototroctes*
Galaxiidae	Galaxiids, Tasmanian whitebait	South America, S. Africa, Australia, New Zealand	*Galaxias, Paragalaxias Galaxiella, Lovettia sealii*
Hemiramphidae	Halfbeaks	Cosmopolitan	*Arrhamphus*
Belonidae	Needlefish	Cosmopolitan, N. Australia	*Strongylura*
Atherinidae	Silversides	Cosmopolitan	*Craterocephalus, Atherinosoma, Quirichthys*
Melanotaeniidae	Rainbowfish	New Guinea, Australia	*Melanotanenia, Pseudomugil*
Syngnathidae	Pipefish	Cosmopolitan - E. Australia	*Syngnathus*
Synbranchidae	Swamp eels	South America, W. Africa, Asia, Australia	*Ophisternon, Monopterus*
Scorpaenidae	Scorpionfish	Cosmopolitan	*Notesthes*
Platycephalidae	Flatheads	Indo-Pacific	*Platycephalus*
Centropomidae	Giant perches	Pantropical	*Lates calcarifer*
Chandidae	Glass perches	Pantropical	*Ambassis, Parambassis*
Percichthyidae	Percichthyids	South America, Australia	*Maccullochella, Macquaria, Bostockia, Nannoperca, Edelia, Kuhlia, Gadopsis*
Teraponidae	Grunters	Indo-Pacific	*Bidyanus, Leipotherapon, Hephaestus, Scortum*
Apogonidae	Cardinalfish	Pantropical	*Glossamia*
Lutjanidae	Snappers	Cosmopolitan	*Lutjanus*
Gerridae	Mojarras	Pantropical	*Gerres*
Sparidae	Porgies	Cosmopolitan	*Acanthopagrus*
Sciaenidae	Drums	Cosmopolitan	*Sciaena, Johnius*
Monodactylidae	Batfish	West Africa, Indo-Pacific	*Monodactylus*

Continued next page

TABLE 7.1 **FISH FAMILIES IN THE FRESH WATERS OF AUSTRALIA**
(Continued)

FAMILY	COMMON NAME	FAMILY DISTRIBUTION	SOME AUSTRALIAN GENERA OR SPECIES
Toxotidae	Archerfish	Asia, New Guinea, Australia	*Toxotes*
Scatophagidae	Scats	Africa, Asia, New Guinea, Australia	*Scatophagus, Selenotoca*
Mugilidae	Mullets	Cosmopolitan	*Mugil, Liza, Myxus*
Polynemidae	Threadfins	Pantropical	*Polydactylus*
Bovichthyidae	Tupong	South America, S.E. Australia, New Zealand	*Pseudaphritis urvilli*
Eleotridae	Gudgeons	Cosmopolitan	*Philypnodon, Gobiomorphus, Hypseleotris, Mogurnda*
Gobiidae	Gobies	Cosmopolitan	*Arenigobius, Redigobius, Pseupogobius, Awaous*
Kurtidae	Nurseryfish	Asia, New Guinea, N. Australia	*Kurtus gulliveri*
Soleidae	Soles	Cosmopolitan	*Brachirus*
Tetraodontidae	Puffers	Cosmopolitan	*Sphaeroides*

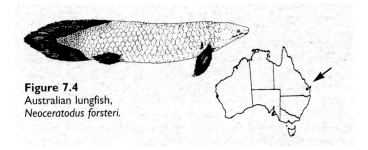

Figure 7.4
Australian lungfish,
Neoceratodus forsteri.

Australian lungfish,
Neoceratodus forsteri.
Note the large scales and
paddle-like fin.

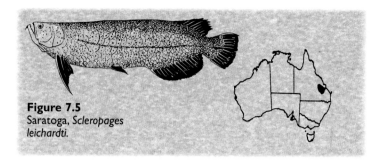

Figure 7.5
Saratoga, *Scleropages leichardti.*

Bonytongues

The family Osteoglossidae is the other primary division freshwater group native to Australia. The bony-tongued fishes are found in tropical South America, Africa, Malay Archipelago, Borneo, southern New Guinea and northern Australia. It is the only family of primary division freshwater fishes to cross Wallace's Line, that imaginary boundary between the Asian and Australian realms (Figure 2.15, page 48). The family's distribution pattern, along with that of the lungfish, may reflect the previous Gondwanan pattern with subsequent break-up and drifting of the ancient continent.

There are two Australian species: *Scleropages leichardti* (Figure 7.5), found only in the Fitzroy River system of eastern Queensland; and *S. jardini,* which occurs in rivers flowing into the Gulf of Carpentaria and the Timor Sea. Both species are buccal incubators—they carry their eggs and larvae in their mouth. They have upturned mouths and feed on insects, fish, and crustaceans. These long, primitive, compressed fishes have heavy scales, and posteriorly positioned dorsal and anal fins. *S. leichardti* may grow to about 1 m (3.3 ft) and weigh 4 kg (8.8lb), and *S. jardini* reaches 900 mm (3 ft) and can weigh up to 17.2 kg (37.8lb).

Australian Eels

Four species of *Anguilla* occur in Australia. Two species common in southeastern Australia are the longfinned eel, *A. reinhardtii,* which may reach 2 m (6.5 ft) and weigh 16 kg (35 lb), and the shortfinned eel, *A. australis* (Figure 7.6), which grows to 1.1 m and can weigh up to 6.8 kg. The shortfinned eel is fished commercially and raised in aquaculture. Eels spawn at sea at depths greater than 300 m (984 ft).

A glass eel, *Anguilla* sp., at the juvenile stage after the leptocephalus larva and before the pigmented elver.

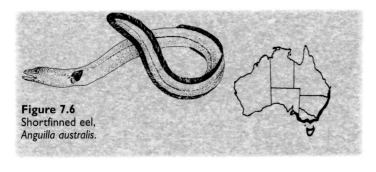

Figure 7.6
Shortfinned eel,
Anguilla australis.

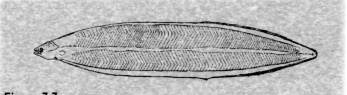

Figure 7.7
Leptocephalus larvae of shortfinned eel, *Anguilla australis.*

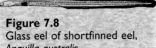

Figure 7.8
Glass eel of shortfinned eel,
Anguilla australis.

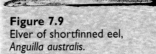

Figure 7.9
Elver of shortfinned eel,
Anguilla australis.

Longfinned eels, *Anguilla reinhardtii*, from the Tambo River in Victoria; the larger one is 1.1 m (3.6 ft) long. The trap is a fyke net used by eel fishermen.

For both, the spawning area is believed to be near New Caledonia. Each females produces 5–10 million eggs. The leaf-like leptocephalus larvae (Figure 7.7) are carried on currents to the coast, where they metamorphose into transparent glass eels (Figure 7.8). In fresh water they develop pigment and are called elvers (Figure 7.9). Females grow in fresh water for up to 20 years before beginning their downstream migration. Males tend to be smaller and spend more time near river mouths. Eels are carnivorous and feed at night on crustaceans, molluscs, insects, and fish.

Herrings

The Clupeidae is a worldwide marine family which has a few freshwater representatives. Clupeids have a sawtoothed ventral surface and easily-shed cycloid scales, and lack a lateral line. The bony bream, *Nematalosa erebi* (Figure 7.10), may reach 470 mm (19 in) and

Figure 7.10
Bony bream,
Nematalosa erebi.

weigh 2 kg (4.4 lb). It is widespread in Australia. It is a mud-feeder and an important forage species for Murray cod, cormorants, and pelicans. Bony bream die in large numbers following a rapid drop in water temperature. However, their fecundity is enormous. A large female can produce over 800 000 eggs.

The freshwater herring, *Pomatolosa richmondia,* is found along the southeast coast and is usually smaller than 300 mm (1 ft) and weighs 354 g (0.75 lb). It is carnivorous.

Catfish

The Ariidae is a worldwide marine family with freshwater representatives in Australia and New Guinea which belong to the forktailed catfish genus *Arius* (Figure 7.11). The males carry developing eggs in their mouth.

The family Plotosidae has some marine representatives in the Indo-Pacific, but most species live in fresh water in Australia and New Guinea. This family lacks an adipose fin, which is unusual for a catfish family. Its caudal and anal fins are confluent and extend dorsally, accounting for the name eeltailed catfish. *Tandanus tandanus* (Figure 7.12) is widespread throughout the Murray-Darling river system and in streams along the eastern coast. It can grow to about 900 mm (35 in) and weigh 6.5 kg (14.3 lb).

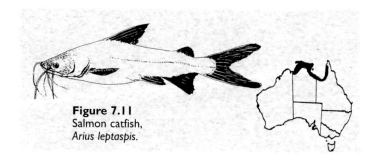

Figure 7.11
Salmon catfish,
Arius leptaspis.

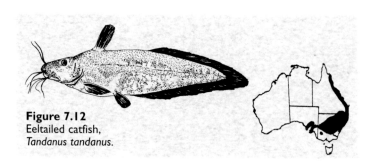

Figure 7.12
Eeltailed catfish,
Tandanus tandanus.

There are about seven species in the plotosid genus *Neosilurus,* and several more undescribed forms. This genus is distributed throughout the various northern drainages and in Lake Eyre. Its species are much smaller than *Tandanus.*

Southern Hemisphere Smelts and Graylings

The family Retropinnidae is found only in southeastern Australia and New Zealand. These small fishes have an adipose fin. There are two Australian smelts, *Retropinna semoni* (Figure 7.13) and *R. tasmanica,* which are small osmeriform fishes that serve as important forage for many of the larger game fishes. They rarely exceed 100 mm (4 in). *R. semoni* is one of the most widespread fishes in southeastern Australia, ranging from southern Queensland to eastern South Australia. They also occur in the internal Lake Eyre drainage system. *R. tasmanica* is smaller and occurs only in Tasmania. Smelts are carnivorous.

The southern hemisphere graylings were formerly considered a separate family but are now grouped with the smelts. The New Zealand grayling, *Prototroctes oxyrhynchus,* and the Australian grayling, *P. maraena* (Figure 7.14), are the only members of the genus. The New Zealand species is probably extinct. The Australian grayling occurs in coastal rivers from about the latitude of Sydney to several hundred kilometers west of Melbourne, and around the Tasmanian coast. It spawns in the coastal rivers during the fall and the young are swept downstream. Graylings grow for six months in brackish water, then return in November to fresh water where they spend the rest of their lives, which may be up to five years. Graylings feed on algae and insects. Their skin has a peculiar cucumber odor, as does the skin of smelts. This is due to the presence of a particular molecule which is also found in a cucumber. Australian graylings may reach 330 mm (1 ft) and weigh 450 g (1 lb).

Australian smelt, *Retropinna semoni,* is an important forage species widespread throughout southeastern Australia.

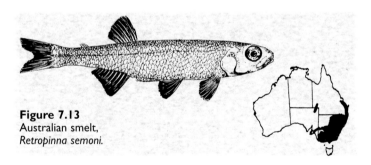

Figure 7.13
Australian smelt,
Retropinna semoni.

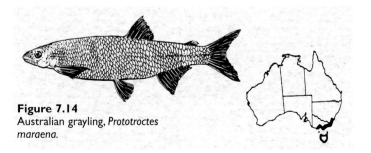

Figure 7.14
Australian grayling, *Prototroctes maraena.*

Lepidogalaxiidae

Male salamanderfish,
Lepidogalaxias salamandroides.
The bent neck position is
possible because of the large
gaps between vertebrae.

The salamanderfish, *Lepidogalaxias salamandroides* (Figure 7.15), is one of the strangest freshwater fishes in the world. There is only one species in this family, and it is confined to the southwest corner of Western Australia, and nowhere else in the universe. Its evolutionary relationships are still under investigation. It is most likely a branch of the osmeriform fishes of the southern hemisphere, such as the galaxiids. It does not get very big; 67 mm (2.6 in) from snout to base of tail is the world's record. The generic name, *Lepidogalaxias*, means scaled-galaxid and the specific name, *salamandroides*, refers to the salamander-like behavior and appearance of this unusual, elongated fish.

The isolated pools and streams in which it lives are very acidic (pH = 4.0) and ephemeral. The pools are about the size of a small room, and contain some 150 salamanderfish. The water is about 1 m deep. This habitat dries up during the austral summer. How these fish survive without water, and where they go were mysteries my colleague, Gerry Allen, and I tried to unravel.

As our study pools near Northcliffe, Western Australia, evaporated in the heat of December, we searched for the fish by digging in the bottom sand. Eventually, we uncovered specimens sitting on damp sand at the top of the water table, several centimeters below the dry surface of the pool. Apparently salamanderfish can pass the summer in this fashion as long as they stay moist. We noticed that they burrow deeper as the water table subsides. We found them as deep as 60 cm (2 ft) below the surface. We determined that they had no accessory respiratory structures such as lungs, and that they did not use their swimbladders for respiration. Experiments performed in the laboratory demonstrated that they can absorb oxygen and give off carbon dioxide through their skin. This is what the lungless salamanders of North America

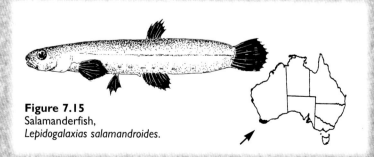

Figure 7.15
Salamanderfish,
Lepidogalaxias salamandroides.

do. Since salamanderfish are not feeding while they are underground, they utilize their fat reserves as an energy source and probably have a lower metabolic rate during their estivation.

After a brief overnight shower of 8 mm (0.3 in), we found fish were swimming in pools that were dry the day before. I wondered how long it takes for the fish to re-emerge after rainfall refills a dry pool. I decided to simulate rain by releasing 2700 liters (713 gal) of water from a fire truck into a dry pool. The result was dramatic: I caught fish within ten minutes. Instant fish — just add water! During a particularly dry summer they may remain underground for many months, but in a wet summer they go up and down like yoyos as the pools partially refill, then evaporate. This must be a very difficult life, but they are tough little fish.

Back in the laboratory, I noticed that salamanderfish in aquariums bent their heads at right angles to their bodies while tracking live food such as the crustacean *Daphnia*. This was a strange sight, for one does not normally think of fishes as having 'necks'. When I examined the skeleton of *Lepidogalaxias*, I noticed that their vertebrae were widely separated from one another and from the skull, and that their ribs were very reduced. These are anatomical adaptations that allow a great deal of flexibility to the spine and body and thus facilitate burrowing behavior. Likewise, their skull is very robust. It can withstand the shearing forces of the weight of damp sand over the fish and is studded with tiny fierce teeth. The fish feeds on insects.

The species is sexually dimorphic, and males can be differentiated from females when they reach about 25 mm (1 in). The males have a modified anal fin, which presumably functions in the transmission of sperm to the female, but very little is known of their reproductive biology. Population analysis indicates that spawning takes place in the rainy winter months, and that the fish can live for several years. Females tend to be larger than males, and the young develop rapidly so that they are ready to estivate when their pools dry up just five or six months after they are born (hatch). See Berra and Pusey (1997) for more details and references.

Ephemeral pool habitats such as this one of salamanderfish near Northcliffe in Western Australia dry up in summer. The fishes survive the dry summers by burrowing into damp sand on top of the water table. Note kangaroo paw flowers in foreground.

Galaxiids

The Galaxiidae have a very interesting distribution in the southern hemisphere, occurring in South America, the tip of South Africa, southern Australia, and New Zealand, as well as Lord Howe Island, New Caledonia, and the Falkland Islands. This is one of Australia's largest native freshwater fish families, with about 20 species—half the total number of species in the family. Most of the other species occur in New Zealand. The Tasmanian galaxiids are especially diverse, with 15 species in three genera. Galaxiids are small, elongate, soft-rayed fishes that lack scales and have a posterior dorsal and anal fin. They do not have an adipose fin or a lateral line. They are found in a variety of niches and can be considered the ecological equivalents of the northern hemisphere salmonids.

Galaxiella munda from southwest Western Australia.

Many galaxiids feed on insects and have been adversely affected by competition from introduced trout. Galaxiids are a unique part of the Australian ichthyofauna and are an important link in the aquatic food web.

The eastern little galaxias, *Galaxiella pusilla,* reaches only 29 mm (1.1 in) for males and 39 mm (1.5 in) for females, while the spotted mountain 'trout', *Galaxias truttaceus,* may grow to 200 mm (8 in) in total length.

G. maculatus (Figure 7.16) is the most widespread species, occurring from southern Queensland to Adelaide, Tasmania, and southern Western Australia. It is also found on Lord Howe Island, in New Zealand, on Chatham Island, in southern South America, and on the Falkland Islands. *G. maculatus* has one of the most widely disjunct distributions of any freshwater fish (see Figure 7.17).

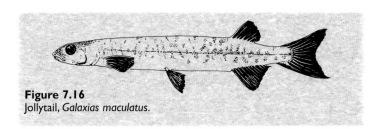

Figure 7.16
Jollytail, *Galaxias maculatus.*

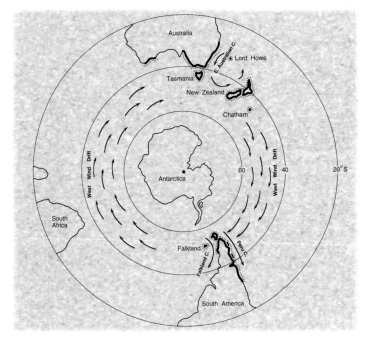

Figure 7.17
Distribution of *Galaxias maculatus* and ocean currents.

Two hypotheses have been proposed to explain the disjunct distribution of *G. maculatus:* dispersal (movement through the sea) and vicariance (continental drift). With Walter Ivantsoff at Macquarie University in Sydney, I recently carried out some experiments that addressed that problem. We looked at the genetic variation of populations of *G. maculatus* from New South Wales, Western Australia, New Zealand, and Chile. We found very little genetic difference among these populations, indicating that gene flow is occurring in spite of the geographical distances. The gene pool is kept more or less uniform by the dispersal of the salt-tolerant larval stage of *G. maculatus.* This is the only galaxiid to breed in brackish

Jollytail, *Galaxias maculatus.*

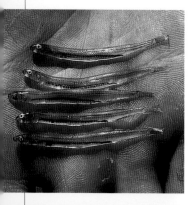

Top: Whitebait stage of *Galaxias maculatus*. Adults spawn in brackish waters and the hatchings spend about six months growing at sea, returning up river as whitebait.
Above: *Galaxias occidentalis* from Western Australia.

water, and the larvae spend about six months at sea before ascending streams as transparent juveniles, called whitebait. The direction of the ocean currents (Figure 7.17) is such that the larvae could be swept to South America. This infusion of Australian and New Zealand genes into the South American population prevents the fragmentation of this species into several species. Furthermore, it is unlikely that populations of *G. maculatus,* if isolated in South America since the break-up of Gondwana at the end of the Mesozoic, would be conspecific with populations from Australia and New Zealand.

The family Aplochitonidae, with two species in southern South America and the larva-like Tasmanian whitebait, *Lovettia sealii* (Figure 7.18), on the north and east coast of Tasmania has recently been assigned to the Galaxiidae. *Lovettia* is elongated, reaching about 77 mm (3 in) in length, is scaleless, and has a small adipose fin and a lateral line. It was at one time fished for canning during its spawning migration. It is carnivorous. Preliminary studies suggest that it may be the sister group of salamanderfish.

Figure 7.18
Tasmanian whitebait, *Lovettia sealii.*

Silversides

The Atherinidae is a large worldwide marine family (160 species) with many representatives in fresh water. Of the approximately 20 Australian species, nine are found in fresh water. These small fishes have two dorsal fins and no lateral line. *Craterocephalus stercusmuscarum,* the fly-specked hardyhead (Figure 7.19), is common in eastern Australia, and *C. eyresii* occurs in the internal Lake Eyre system as well as elsewhere in eastern Australia. The small-mouthed hardyhead, *Atherinosoma microstoma,* is extremely abundant in shallow brackish waters on the southeast coast. Most of the atherinids are under 90 mm (3.5 in) long.

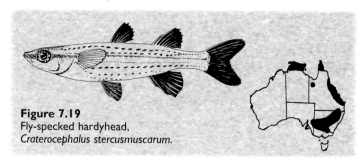

Figure 7.19
Fly-specked hardyhead, *Craterocephalus stercusmuscarum.*

Rainbowfishes

The family Melanotaeniidae occurs only in the fresh waters of
Australia and New Guinea. There are about 50 species in eight
genera, of which approximately one-third are Australian. They are
related to the silversides. Rainbowfishes are often very colorful and
are frequently kept in aquariums. They have two dorsal fins, the
second of which may be very long. They lack a lateral line and have
tiny, rather elaborate teeth. Most species are smaller than 180 mm
(7 in) and some do not reach 70 mm. The major Australian genus is
Melanotaenia (Figure 7.20). The male blue-eye, *Pseudomugil signifer*, has
pronounced filamentous extensions to its fins, especially the first
dorsal. The eye is surrounded by an iridescent blue ring.
Melanotaeniids fill many niches occupied by cyprinids, characins, and
cyprinodonts on other continents.

McCulloch's rainbowfish,
Melanotaenia maccullochi, from
northern Australia.

Figure 7.20
Crimson-spotted rainbowfish,
Melanotaenia splendida.

Banded rainbowfish,
Melanotaenia trifasciata,
from the Northern Territory
and Cape York.

Barramundi or giant perch,
Lates calcarifer.

Ken Griffiths

Giant Perch

The family Centropomidae occurs in the coastal waters of the New
World and the Indo-Pacific as well as in fresh water in Africa and
Australia. The Australian species, called barramundi or giant perch, is
Lates calcarifer (Figure 7.21), which occur in rivers and estuaries of the
Indo-Pacific from the Persian Gulf to China and around the north
coast of Australia. It can live to at least 20 years old and grow to
1.8 m (6 ft) and 60 kg (132 lb) in fresh water. The lateral line of *Lates*
extends to the posterior margin of the caudal fin. In Australia and
Papua New Guinea, barramundi are protandrous hermaphrodites.
Small specimens are males which mature sexually at three and four
years of age. They change sex as they grow larger, at about seven
years of age. Specimens larger than 730 mm (2.4 ft) are usually
females, the larger of which are extremely fecund, producing up to
15–45 million eggs per year. They spawn at the beginning of the wet
season, returning to brackish water in inshore coastal areas. *Lates*
feeds on fishes and crustaceans and is highly prized as a food fish. In
fact, it is the most important freshwater commercial species in
Australia. (This species should not be confused with the bony-
tongues which are sometimes also called barramundi.)

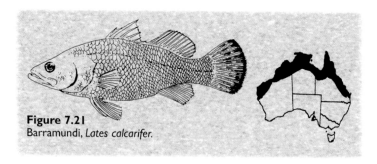

Figure 7.21
Barramundi, *Lates calcarifer*.

Glass Perches

This family, Chandidae or Ambassidae (both names are used in the recent literature), of small translucent perch-like fishes is distributed from the east coast of Africa to Australia. Eight species in three genera occur in Australian fresh waters. *Ambassis castelnaui* (Figure 7.22), the western chanda perch, is common in the Murray-Darling system, and other species of *Ambassis* and *Parambassis* occur in northern Australia and New Guinea. Most are under 150 mm (6 in) in total length and are important forage species.

Trout cod, *Maccullochella macquariensis*, an endangered species.

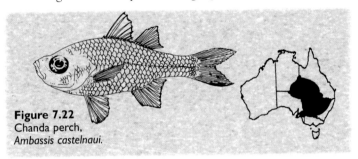

Figure 7.22
Chanda perch,
Ambassis castelnaui.

Freshwater Basses and Cods, Pygmy Perches, and River Blackfishes

The family Percichthyidae contains some of the most important game species in inland Australia. This family, which is not well understood taxonomically, also occurs in the fresh waters of South America.

The largest freshwater fish in Australia is the Murray cod, *Maccullochella peeli* (Figure 7.23). There is a report of a 114 kg (250 lb) specimen, and I have examined a 68 kg Murray cod in the Museum of Victoria. Anglers commonly catch 36 kg fish in the Murray-Darling system. The fish can reach 1.8 m (5.9 ft). The flesh is white

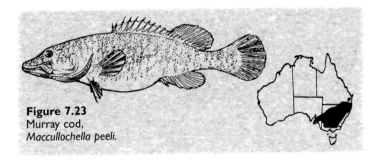

Figure 7.23
Murray cod,
Maccullochella peeli.

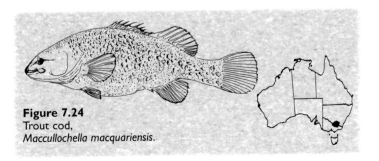

Figure 7.24
Trout cod,
Maccullochella macquariensis.

Figure 7.25
Golden perch,
Macquaria ambigua.

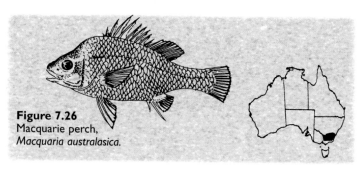

Figure 7.26
Macquarie perch,
Macquaria australasica.

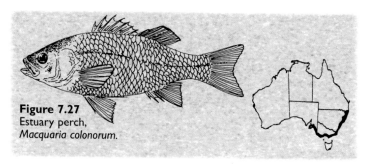

Figure 7.27
Estuary perch,
Macquaria colonorum.

and tasty. Naturally a species that reaches this size is highly sought after by anglers, although smaller specimens taste much better than the large ones. Spawning takes place in spring, around or in submerged hollow logs. Cod feed on crayfish, fish, and molluscs. Formerly, Murray cod were fished commercially, and today they are often stocked in farm ponds. They are sedentary fish, usually moving less than 10 km (6 miles) along the river. The smaller congener, the trout cod, *M. macquariensis* (Figure 7.24), is one of the most endangered fish species in Australia. Once broadly sympatric with the Murray cod, it is now known from only a few sites. There is an endangered cod species, *M. ikei,* known from the Clarence River in Queensland. The Mary River, also in Queensland, is the home of the subspecies *M. peeli mariensis.*

Other important inland Australian percichthyids are the golden perch, *Macquaria ambigua* (Figure 7.25), which may reach 760 mm (30 in) and weigh 23 kg (50 lb), and the smaller Macquarie perch, *M. australasica* (Figure 7.26). These species are typical of the Murray-Darling drainage. The golden perches undertake a longer migration, associated with flooding, than any other Australian freshwater fish.

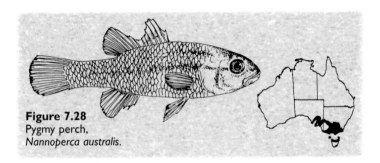

Figure 7.28
Pygmy perch,
Nannoperca australis.

They may travel 2000 km (1200 miles). Females produce up to 500 000 tiny eggs that float downstream. Spawning occurs at least 500 km from the river mouth, to prevent the pelagic juveniles from being swept into salt water. The juveniles disperse over the river's floodplain. The Macquarie perch, once widespread, is now seriously threatened. Coastal rivers of eastern Australia may contain the Australian bass, *Macquaria novemaculeata,* and the larger estuary perch, *M. colonorum* (Figure 7.27) which prefers more brackish waters. Both are game species. The nightfish, *Bostockia porosa,* is found in southwest Western Australia, reaches 160 mm (6.3 in), and is nocturnal. All of these species are found nowhere else in the world.

The Percichthyidae has recently been expanded to include six species in four genera that were formerly included in the Indo-Pacific Kuhliidae. *Nannatherina balstoni* is found in southwest Western Australia, and *Edelia* has one species in southwestern Australia and another species in southeastern Australia. *Kuhlia rupestris* is highly prized by anglers on the eastern Queensland coast. *Nannoperca australis* (Figure 7.28) is known from inland and coastal eastern Australia. These attractive little fishes make interesting aquarium specimens, seldom exceeding 70 mm (2.7 in) in total length, although *K. rupestris* may reach 450 mm (18 in) and weigh 3 kg (6.6 lb).

The two species of river blackfishes *(Gadopsis)* are exclusively

Blackfish, *Gadopsis marmoratus,* from a tributary of the Tambo River, Victoria.

Western pygmy perch,
Edelia vittata.

Australian. *G. marmoratus* (Figure 7.29) is found in the upper reaches
of coastal rivers and in the Murray-Darling system, and *G. bispinosus*
occurs in the King River and King Parrot Creek of northwestern
Victoria. *Gadopsis* feeds on insects, crustaceans, and molluscs. It has a
single long dorsal fin and a long anal fin, but the pelvic fins are
reduced to a few jugular rays. *G. marmoratus* may reach 5 kg (11 lb) and
600 mm (2 ft). The phylogenetic placement of this genus is under
discussion, with some ichthyologists considering it to be somewhere
near the blennies. A recent reassessment has included blackfishes and
pygmy perches within the Percichthyidae.

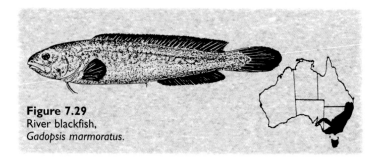

Figure 7.29
River blackfish,
Gadopsis marmoratus.

Freshwater Grunters

The Indo-Pacific Teraponidae has many species in the fresh waters of
Australia and New Guinea. Some species produce grunting sounds—
thus the common name. The silver perch, *Bidyanus bidyanus*
(Figure 7.30), occurs in most of the Murray-Darling system. A female
may produce up to 300 000 planktonic eggs. Adults reach 400 mm
(16 in) and weigh up to 1.5 kg (3.3 lb). The spangled perch,
Leiopotherapon unicolor, is found in warm water over much of
Queensland, Northern Territory, and Western Australia, while
Amniataba percoides and several species of *Hephaestus* occur in the
north and west of Australia.

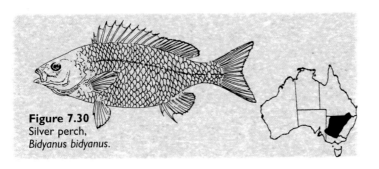

Figure 7.30
Silver perch,
Bidyanus bidyanus.

Archerfish, *Toxotes jaculator.*

Archerfishes

The Toxotidae includes six species, ranging from India to Australia. These fishes have the unusual ability to 'shoot down' insects with a droplet of water forcefully expelled from their mouth. A groove along the roof of the mouth, together with the tongue, forms a tube through which water drops are forcefully ejected by strong compression of the gill covers. Three species, *Toxotes chatareus,* *T. oligolepis,* and *T. lorentzi* occur in coastal rivers and estuaries of north central Australia. *T. jaculator* (Figure 7.31) is widespread in tropical areas of Australia and Papua New Guinea, but rarely invades fresh waters.

Figure 7.31
Archerfish,
Toxotes jaculator.

Bovichthyidae

This Antarctic family has a freshwater representative in coastal streams of southeastern Australia. The tupong or congolli, *Pseudophritis urvilli* (Figure 7.32), is a carnivorous, bottom dwelling, elongate species that reaches 360 mm (1.2 ft). Its reproductive biology is at present under investigation.

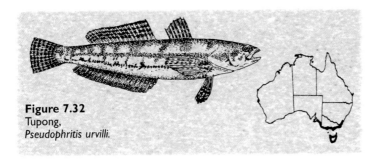

Figure 7.32
Tupong,
Pseudophritis urvilli.

Purple-spotted gudgeon,
Mogurnda adspersa.

Gudgeons (Sleepers)

The Eleotridae contributes about 50 fresh water species, more than any other family in Australia. Many others are from brackish waters, but some species such as *Hypseleotris klunzingeri* (Figure 7.33) and *Philypnodon grandiceps* are also found inland in the Murray-Darling system. Most of the Australian species are very small, 30–100 mm (1–4 in), although *Oxyeleotris lineolatus,* from northern Australia, can reach 480 mm (1.6 ft) and weigh up to 3 kg (6.6 lb). Colorful species such as *Mogurnda adspersa* from the upper Murray-Darling system and eastern coastal drainages of Queensland are attractive aquarium specimens.

Figure 7.33
Western carp gudgeon,
Hypseleotris klunzingeri.

Gobies

The Gobiidae is a huge worldwide marine family with about 1900 species, 400 of which are Australian. Most of these are brackish water or marine forms, except *Chlamydogobius eremius* (Figure 7.34), which occurs in small creeks and ponds associated with artesian springs in the internal Lake Eyre drainage. It is 60 mm (2.5 in) long.

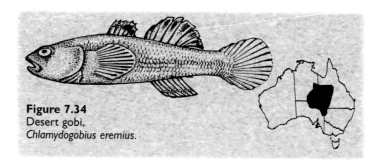

Figure 7.34
Desert gobi,
Chlamydogobius eremius.

Nurseryfishes

The Kurtidae is a remarkable family of two species: one is *Kurtus gulliveri* (Figure 7.35) found in northern Australia and in New Guinea, and the other is an Indo-Malaysian species. The males have a supra-occipital hook over the head, on which the eggs are hung like a cluster of grapes. It is not known how the eggs become attached to the hook, and from what antecedent this unusual system of parental care evolved. The fishes are carnivorous and may reach 600 mm (2 ft) in length.

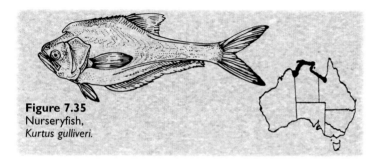

Figure 7.35
Nurseryfish,
Kurtus gulliveri.

• • • • • • • • • •

INTRODUCED FISHES

In addition to the characteristically Australian species discussed above, and the more complete listing given in Table 7.1 (page 128), 21 exotic species in five families are now established in Australian fresh waters. These are presented in Table 7.2. Many of these species have become pests and others compete with the native species, usually to the detriment of the latter. Anglers are grateful for the exotic brown and rainbow trout, but are less enthusiastic about the carp. Recently, plans to introduce the Nile perch, *Lates niloticus,* a large and voracious predator from Africa, were abandoned because of the fish's potentially disastrous impact on native fishes.

TABLE 7.2 **FISH INTRODUCED INTO THE FRESH WATERS OF AUSTRALIA**

FAMILY	COMMON NAME	SPECIES	DISTRIBUTION IN AUSTRALIA
Salmonidae	Atlantic salmon	*Salmo salar*	NSW
	Brown trout	*Salmo trutta*	WA, SA, Vic., Tas., NSW
	Brook trout	*Salvelinus fontinalis*	Tas., NSW
	Rainbow trout	*Oncorhynchus mykiss*	WA, SA, Vic., Tas., NSW (restocked)
	Chinook salmon	*Onchorynchus tshawytscha*	Vic. (restocked)
Cyprinidae	Carp	*Cypinus carpio*	WA, SA, Vic., NSW, Qld
	Goldfish	*Carassius auratus*	WA, SA, Vic., Tas., NSW, Qld
	Crucian carp	*Carassius carassius*	WA, NSW, Qld?
	Tench	*Tinca tinca*	SA, NSW, Tas.
	Roach	*Rutilus rutilus*	Vic.
Poeciliidae	Mosquitofish	*Gambusia holbrooki*	All states except Tasmania
	One-spot live bearer	*Phalloceros caudimaculatus*	WA
	Guppy	*Poecilia reticulata*	Qld
	Sailfin molly	*Poecilia latipinna*	Qld
	Swordtail	*Xiphophorus helleri*	Qld
	Platy	*Xiphophorus maculatus*	Qld
Percidae	European perch	*Perca fluviatilis*	WA, SA, NSW, Vic., Tas.
Cichlidae	Mozambique tilapia	*Oreochromis mossambica*	Qld
	Spotted tilapia	*Tilapia mariae*	Vic.
	Convict cichlid	*Cichlasoma nigrofasciatum*	Vic.
	Jack Dempsey	*Cichlasoma octofasciatum*	Vic.?

Source: McKay (1984).

Eastern water dragon,
Physignathus lesueuri,
from the Tambo River of Victoria.

Warts, Scales, Frills, and Thorns

8

AMPHIBIANS

Many vertebrate paleontologists think that amphibians evolved from an ancient group of fishes known as the rhipidistian crossopterygians about 400 MYA. The only surviving species of this ancestral lineage is the coelacanth, *Latimeria chalumnae*. Recently, molecular taxonomists have suggested that lungfish are the closest relatives of land vertebrates. See Meyer (1995) and references cited therein for a review of the coelacanth vs lungfish debate.

Fossil remains of the crossopterygian shoulder girdle and pectoral fin show the same basic limb pattern as that of ancient amphibians, as well as internal nares which provide a pathway for air flow to the lungs when the external nares are exposed above water. There are other similarities in tooth structure and vertebral and skull anatomy to support the suggested relationship. The ability to locomote on land and breathe atmospheric air was a selective advantage to the fish-amphibian transitional form during the periodic droughts of the Devonian Period. Thus the movement onto land by the early amphibians was paradoxically brought about by movement from drying pools in pursuit of additional aquatic habitats. A mechanism involving gene expression leading to the development of the tetrapod foot from a fish fin has recently been proposed by Sordino, van der Hoeven, and Duboule (1995).

Modern amphibians belong to three orders: Gymnophiona (caecilians); Caudata (salamanders); and Anura (frogs). The caecilians are limbless, worm-like burrowing animals. There are about 165 species in the tropics of Central and South America, Africa, India, and southeast Asia. Caecilians do not occur in Australia, but there are two species in Sabah and one species in Sarawak and the Philippines. Likewise, Australia has no salamanders, however, the exotic axolotl, *Ambystoma mexicanum,* can be bought from Australian pet shops. There are about 400 species of salamanders in North and South America, Europe, North Africa, and Asia.

Of the 4522 or so amphibian species of the world, about 3967 are frogs (in 23 families). In Australia there are only four native frog families, with about 200 species. This paucity of frogs is probably due to the arid environment. One exotic toad family, Bufonidae, has been introduced to Australia. Table 8.1 lists the families and genera of Australian frogs.

Corroboree frog,
Pseudophryne corroboree—
Australia's most colorful frog.

Myobatrachidae—The Southern Frogs

The largest amphibian family in Australia is the Myobatrachidae with about 109 species in 21 genera. This family was formerly grouped with the Leptodactylidae of South America, but the recent trend among herpetologists is to consider the Australian species as a discrete, endemic family, due to its long isolation and internal relationship. The Myobatrachidae most likely share a common origin with the leptodactylids. Many myobatrachids show adaptations for an arid existence, and most members of this family are terrestrial and lack expanded discs on fingers and toes as well as webbing between toes. Some species have aquatic egg and larval development, while others are strictly terrestrial. Myobatrachids are restricted to the Australian region, found from the coast to the arid parts of the continent.

TABLE 8.1 AUSTRALIAN FROG FAMILIES: THEIR DISTRIBUTION, GENERA, AND NUMBER OF SPECIES

FAMILY	AUSTRALIAN DISTRIBUTION	AUSTRALIAN GENERA AND NUMBER OF SPECIES
Myobatrachidae	All of Australia except Nullarbor Plain.	*Adelotus brevis, Arenophryne rotunda,*
		Assa darlingtoni, Crinia 14 sp (incl. *Ranidella*)
		Geocrinia 5 sp
		Heleioporus 6 sp
		Lechriodus fletcheri, Limnodynastes 12 sp
		Megistolotis lignarius, Metacrinia nichollsi,
		Mixophyes 5 sp
		Myobatrachus gouldii, Neobatrachus 9 sp
		Notaden 4 sp
		Paracrinia haswelli, Philoria 4 sp
		(incl. *Kyarranus*)
		Pseudophryne 10 sp
		Rheobatrachus 2 sp *Spicospinna flammocaerulea*
		Taudactylus 6 sp
		Uperoleia 23 sp
Hylidae	All of Australia except Nullarbor Plain.	*Cyclorana* 12 sp
		Litoria 57 sp
		Nyctimystes dayi
Microhylidae	Cape York Peninsula, Qld; peripheral NT.	*Cophixalus* 11 sp
		Sphenophryne 5 sp
Ranidae	Cape York Peninsula, Qld.	*Rana daemeli*
Bufonidae*	Eastern and northern Qld, western NT, northeast NSW.	*Bufo marinus*

*Introduced.
Source: Tyler (1989) and Cogger (1992).

Adelotus brevis is a small, flattened species called the tusked frog because the males have a pair of tooth-like tusks near the tip of the lower jaw. The function of these structures is unknown. This gray or brown frog may have a bright red patch in the groin or legs. It lives in moist forests or near water along the Great Dividing Range and the coast from southern New South Wales to southern Queensland.

The sandhill frog, *Arenophryne rotunda,* is found on dunes in a narrow zone along the coast of Western Australia at Shark Bay, and at Kalbarri, near the mouth of the Murchison River. This small (30 mm; 1.3 in) species burrows into sand to reach moisture about 10 cm down and thereby escapes the drying heat of the day. During the night it emerges to feed on ants. Its nocturnal forays leave tracks in the sand that terminate abruptly at a little crater where the frog has entered the sand head first. Mating occurs under the sand and the very large eggs (5.5 mm; 0.2 in) allow the embryos to develop within them, thereby omitting the need for freestanding water. Baby frogs hatch directly from the eggs.

The hip-pocket frog, *Assa darlingtoni* (Figure 8.1), from cool rainforests in the mountains of the Queensland-New South Wales border area, has an unusual feature. There are pouches in the groin of males where the tadpoles develop. After the eggs hatch on land, the tadpoles wriggle through the egg jelly into the male's pouches. Baby froglets emerge 48–69 days later.

Rheobatrachus silus (Figure 8.2), the gastric brooding frog, known only from a small area of the Conondale and Blackall Ranges in southeastern Queensland, has the most bizarre parental care of any frog. After swallowing fertilized eggs, the female incubates them in her stomach. The tadpoles secrete a prostaglandin that inhibits release of gastric acid from the stomach of the female. After about six weeks, as many as 21 fully formed juvenile frogs emerge from the female's mouth over a one week period. The inhibition of digestive juice and intestinal peristaltic movement is relieved when the babies leave the stomach. *R. silus* may now be extinct in the wild. Recent searches have failed to find it in fast-flowing rainforest streams. A second species, *R. vitellinus,* was recently discovered in central coastal Queensland, but it too is very rare.

In all of the 12 species of the widespread *Limnodynastes,* the breeding female develops a paddle-like flange on the inner finger which is used in a 'dog paddle' arm motion of 4–5 strokes. This creates tiny bubbles that are tossed backward and dispersed around the bodies of the amplexing pair. This paddling activity helps create a foam nest in which the eggs are laid.

Notaden bennetti is a colorful, globular, warty myobatrachid of arid New South Wales and Queensland. It exudes a viscous, smelly, white secretion when aroused. This substance is probably a defense against predators and dries to a strong elastic texture.

Pseudophryne corroboree (Figure 8.3) is one of Australia's most attractive frogs. This 30 mm (1.2 in) amphibian lives in the Australian

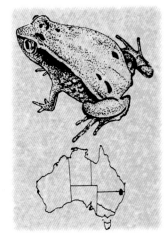

Figure 8.1
Hip-pocket frog,
Assa darlingtoni.

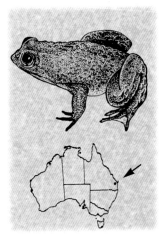

Figure 8.2
Gastric brooding frog,
Rheobatrachus silus.

Figure 8.3
Corroboree frog,
Pseudophryne corroboree.

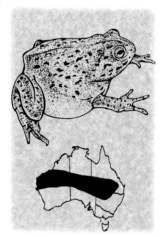

Figure 8.4
Water-holding frog,
Cyclorana platycephala.

Figure 8.5
Giant green tree frog,
Litoria infrafrenata.

Figure 8.6
Leaf green tree frog,
Litoria phyllochroa.

alps and surrounding high country near Mount Kosciuszko. Its dorsal surface is bright yellow and jet black, and its ventral surface is mottled black and yellow or white. It lives in sphagnum bogs above the 1500 m (4900 ft) tree line, and at lower altitudes it uses fallen logs and creek vegetation for shelter.

Hylidae—The Tree Frogs

The Hylidae is a large family of tree frogs found in North and South America, Europe, North Africa, Asia, and Australia. Hylids have expanded discs on their fingers and toes to facilitate climbing. The Australian-New Guinean hylids are considered by some herpetologists to be a separate family, Pelodryadidae. There are about 70 Australian species in the three genera, *Cyclorana, Litoria,* and *Nyctimystes.* Australian hylids are more common in wetter areas of the north and east, with their diversity decreasing in arid regions. Eggs are laid in water and hatch into tadpoles.

The water-holding frog, *Cyclorana platycephala* (Figure 8.4), from central arid Australia, stores large quantities of water in its bladder. Its loose outer layer of skin is shed as a complete cloak, and hardens to become a protective bag that lines its underground burrow and greatly reduces evaporative water loss. It may spend several years in this underground cocoon. Before it emerges during heavy rain to spawn in temporary pools, *Cyclorana* sloughs off its cocoon of dead skin cells and, on occasions, may stuff it into its mouth. This frog recycles itself!

Litoria, with 57 Australian species, was previously included in the *Hyla* complex. Most of these frogs are arboreal and have expanded digital discs that function as adhesive structures. They also have webbed toes. *Litoria infrafrenata* (Figure 8.5), the giant green tree frog, is the largest frog in Australia at 140 mm (5.5 in). It is present only around Cape York, but is widely distributed throughout New Guinea. Its congener, *L. caerulea,* is more widespread and commonly encountered as it frequents such structures as buildings, mailboxes, and toilets. It may reach 100 mm (4 in) in length. *L. rubella* is widespread over the northern half of Australia in habitats as diverse

Red-backed toadlet, *Pseudophryne coriacea,* is cryptically colored against fallen leaves.

as wet coastal forests and central deserts. *Litoria peronii, L. phyllochroa* (Figure 8.6), and *L. lesueurii* are common eastern frogs. *L. meiriana,* a tiny diurnal frog from northwest Western Australia and Arnhem Land, hops across water the way a flat stone can be skipped over a pond surface.

Nyctimystes dayi has vertical pupils compared to the horizontal or diamond-shaped pupils of *Litoria,* and the lower eyelid of *Nyctimystes* has a unique pattern of colorful venation. This species is found only in rainforest regions of Cape York Peninsula. Other species occur in New Guinea.

Microhylidae—The Narrow-mouthed Toads

The Microhylidae just barely reach northern Australia from New Guinea, where they are more common. They are found in North and South America, southern Africa, and southern Asia to New Guinea and northern Australia. The Australian members are minute, tiny-headed frogs, with stout bodies, unwebbed toes, and digital discs. Development of the eggs takes place on land. There is no aquatic tadpole stage, as fully formed frogs hatch from the eggs. There are two genera, *Sphenophryne* (5 species) and *Cophixalus* (11 species) in Australia, both of which are also found in New Guinea. Rainforests are their preferred habitat. *C. exiguus* is Australia's tiniest frog at 15 mm (0.6 in).

Ranidae—The 'True' Frogs

The dominant frog family of the world is the Ranidae. It is found on all continents, but only one species has reached Cape York Peninsula from New Guinea. *Rana daemeli* (Figure 8.7) lives in rainforests of northeast Queensland and reaches 80 mm (3.2 in). The only area of the world to lack ranids, other than most of Australia, is southern South America.

Litoria tyleri displaying vocal sac while calling.

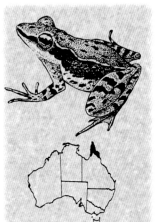

Figure 8.7
Wood frog, *Rana daemeli.*

The ventral surface of *Pseudophryne coriacea* is marbled black and white. Toadlets often remain motionless when placed on their backs.

Cane Toad, *Bufo marinus.*

Gerry Swan

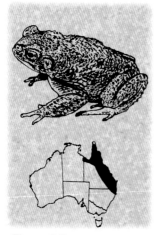

Figure 8.8
Cane toad, *Bufo marinus.*

Bufonidae—The 'True' Toads

The toads, Bufonidae, occur worldwide except in Australia and New Guinea. Unfortunately, the cane toad, *Bufo marinus* (Figure 8.8), the world's largest toad at about 200 mm (7.9 in), was introduced into eastern Queensland in 1935 from Hawaii. It had been introduced to Hawaii from its native Central and South America. The reason for its introduction into Australia was to control the pest, the sugar cane beetle, *Phyllophaga vandinei*. The toad was subsequently introduced to Florida as well as New Guinea, Fiji, Solomon Islands, Marianas Islands, Samoa, Palau, and throughout the Pacific. Now, however, the toad itself has become a pest, and has adversely effected indigenous animals. Its hardiness and tolerance of many conditions has guaranteed its rapid spread and prevented its eradication.

The transplantation of exotic fauna should be discouraged because of the history of negative impact of exotics on native animals and the environment. The introduction of the cane toad was considered a last resort by sugar cane growers who lacked insecticides in the 1930s. Today the cane toad is considered a 'noxious animal' or 'vermin' in most Australian states and very strict controls apply to its importation across state boundaries.

Bufo marinus is an opportunistic breeder and some individuals will be spawning at any given moment. In other words, there is no specific breeding season. Females produce jelly strings of eggs numbering 8000 to 25 000 in ponds, ditches, and sluggish streams. In spite of the specific name *marinus*, the cane toad is not very tolerant of salinity and needs fresh water. Captive specimens have lived 16 years, and the largest toad from Queensland weighed in at 1.36 kg (3 lb) and reached 241 mm (9.5 in) in length.

The 101 surviving toads originally imported from Hawaii in June 1935 reproduced to 3400 small toads that were released around

Cairns, Gordonvale, and Innisfail in July 1935. Their range has been expanding ever since and now the toad occupies eastern Queensland into northern New South Wales and the eastern Northern Territory. The cane toad does eat harmful insects, but it is an indiscriminate feeder and will eat anything that is smaller than its gape, including native lizards and other frogs. It seems especially fond of bees and the honey industry has suffered because of this. Toads congregate around hives and prey on bees in large numbers; bee stings to the mouth and stomach of the toads seem to have no effect. Elevating bee hives off the ground protects them from the toads, but adds to the cost each time hives are relocated to new flowering sites. Cane toads also eat dung beetles, which were introduced to clean pastures of cattle pads.

The milky venom from the pair of parotid glands on the back of the toad contains numerous toxic substances, including bufotenine, which has killed various animals that 'mouth' the toads. Victims include 'marsupial cats', kookaburras, goannas, and tiger snakes. The toads will brazenly eat pet food out of bowls. Domestic dogs and cats have died from holding the toad in their mouth, thereby apparently absorbing the venom through the mucous membranes of the mouth. Symptoms include excessive salivation, cardiac arryhthmia, hypertension, convulsions, and death. The venom can also be sprayed up to 1 meter (3.3 ft) in several directions when the toad is violently provoked. If the spray reaches the eyes it can be extremely irritating and a predator would surely release or avoid the toad. Apocryphal news stories had 'crazed hippies' in Australia licking toads in search of a psychedelic experience. Australian authorities denied that this was happening, but one boy was reported to have died after eating cane toad eggs.

It is not clear that cane toads have been effective in the control of insect pests, but nevertheless *Bufo marinus* has served some useful functions. For years it was used in pregnancy testing. About 10 ml of a woman's urine was injected under the skin of a male toad. A gonad-stimulating hormone is present in the urine of pregnant women, which causes the toad to release sperm. These sperm appear in the toad's urine about three hours later, and thus confirm pregnancy. Such tests are now old-fashioned and have been replaced by biochemical tests that yield an answer in a few minutes.

Cane toads have been used extensively for physiological experiments and for teaching anatomy by dissection in biology classes. A medical use may yet be found for bufotenine. Meanwhile, attempts at tanning toad skin for use in such items as leather wallets and belts are under way. In hindsight, however, I doubt if anyone would want to unleash these voracious, toxic pests upon Australia and its wildlife if they had the chance to do it all over again. That's the problem with introductions. Prediction of consequences is difficult and reversal is costly or impossible.

Top: Giant tree frog, *Litoria infrafrenata.* The white stripe along the lower lip is diagnostic for this species, which is Australia's largest frog.
Above: Verreaux's tree frog, *Litoria verreauxii.*

EVOLUTION OF REPTILES

Reptiles evolved from anthracosaurian labyrinthodont amphibians and first appeared in the Carboniferous Period some 310 MYA. By 230 MYA they had become the dominant form of life. The reptiles' greatest evolutionary contribution was the waterproof egg that freed them from the aquatic environment and allowed them to exploit the land. Their scaly skin helps prevent water loss and has allowed reptiles to thrive in arid environments. The Mesozoic Era is known as the Age of Reptiles due to the success of the dinosaurs and their relatives. The reptiles were replaced as the dominant vertebrates some 65 MYA by their descendants, the mammals. Some fossil reptiles such as *Lystrosaurus* (Figure 2.5, page 42)—found in South Africa, India, China, and Antarctica—and *Thrinaxodon*—found in Africa and Antarctica—confirm the land connection between the Gondwanan continents during the Triassic Period. The presence of *Lystrosaurus,* as well as the glossopterid flora in Antarctica, indicate that its climate was much more moderate than it is today, and that Antarctica could have served as a dispersal route for organisms before the complete separation of the southern continents.

Modern reptiles of the world are divided into the following groups:

Order	Rhynchocephalia	Tuatara	2 sp
Order	Crocodylia	Crocodiles	23 spp
Order	Testudines	Turtles	245 spp
Order	Squamata		

Suborder	Amphisbaenia	Worm lizards	140 spp
Suborder	Sauria	Lizards	3865 spp
Suborder	Serpentes	Snakes	2500 spp

TABLE 8.2 **AUSTRALIAN CROCODILES AND TURTLES**

ORDER (SUBORDER)	FAMILY	SCIENTIFIC NAME
Crocodylia	Crocodylidae	*Crocodylus johnstoni, C. porosus*
Testudines	Chelonidae	*Caretta caretta, Chelonia depressus, C. mydas,*
(Cryptodira)		*Eretmochelys imbricata,*
		Lepidochelys olivacea, Natator depressus
	Dermochelyidae	*Dermochelys coriacea*
	Carettocheylidae	*Carettochelys insculpta*
(Pleurodira)	Chelidae	*Chelodina expansa, C. longicollis, C. novaeguineae, C. oblonga,*
		C. rugosa, C. steindachneri, Elseya dentata, E. latisternum,
		Emydura krefftii, E. macquarii, E. signata,
		E. subglobosa, E. victoriae, Elusor macrurus, Pseudemydura umbrina, Rheodytes leukops

Source: Cogger (1992).

The tuatara survives as *Sphenodon punctatus*, in the family Sphenodontidae. This relict is found only on about 30 offshore islands of New Zealand, in close association with vast colonies of nesting shearwaters (muttonbirds). The birds and the lizard-like tuatara share burrows. The tuatara population of North Brother Island in Cook Strait may possibly warrant recognition as a distinct species, *S. gunther* (Daugherty et al. 1990). Australia has representatives of the other reptile orders (Tables 8.2 on page 156 and 8.3 on page 162), but lacks worm lizards (suborder Amphisbaenia). Of the 6774 or so species of living reptiles, about 718 or 11 percent are found in Australia.

●●●●●●●●●●

CROCODILES

The family Crocodylidae, with 23 species, is distributed throughout tropical areas. The freshwater crocodile, *Crocodylus johnstoni* (Figure 8.9), is found in freshwater rivers and swamps in northern Australia, usually above tidal influence. It can reach about 3 m (9.8 ft) and weigh 99 kg (218 lb). 'Freshies' feed on fish, frogs, birds, and small mammals. Females lay one clutch of 13 or so eggs in sandbanks towards the end of the dry season in August and September. High (33–34°C, 91–93°F;) and low (29–31°C; 84–87°F) nest temperatures yield females, and temperatures around 32°C (90°F) result in male hatchlings. Incubation time ranges from 65 to 95 days, depending on temperature. The mortality rate of hatchlings is high due to cannibalism and predation by goannas, birds, turtles, and fishes. Less than 1 percent survive to maturity. Freshwater crocodiles can gallop at speeds up to 18 km/h (11 mph) over short distances.

Freshwater crocodile, *Crocodylus johnstoni.* Note the long, slender snout compared to that of the saltwater crocodile.

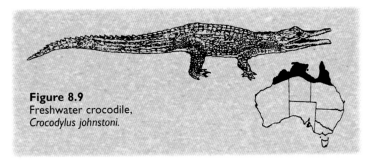

Figure 8.9
Freshwater crocodile, *Crocodylus johnstoni.*

Figure 8.10
Estuarine or saltwater crocodile, *Crocodylus porosus.*

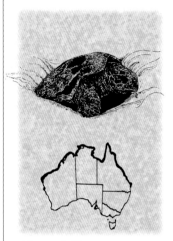

Figure 8.11
Loggerhead turtle,
Caretta caretta.

Figure 8.12
Flatback turtle,
Natator depressus.

Figure 8.13
Green turtle, *Chelonia mydas.*

Ken Griffiths

The estuary or saltwater crocodile, *Crocodylus porosus,* has been responsible for human deaths in tropical Australia.

This species is not considered dangerous to humans. It has a longer, more slender snout and finer, more needle-like teeth than the estuarine or saltwater crocodile, *C. porosus* (Figure 8.10), which occurs in coastal waters from India through the Malay Archipelago to northern Australia. Specimens of the saltwater species have been taken at sea, far from land.

In spite of their name, many saltwater crocodiles live in freshwater rivers and wetlands. It is the largest crocodile species, reaching about 7 m (23 ft) and over 1000 kg (2200 lb), which is a 10 000-fold increase over hatchling weight. They may live to at least 70 years. Males grow faster than females and reach maturity at about 16 years when they are about 3.3 m long; females reach maturity at 13 years when they are about 2.3 m long. Female *C. porosus* lay about 50 eggs high on river banks during the wet season (January–March). Incubation time ranges from 75 to 106 days at 33–29°C respectively. High temperatures yield females, but average nest temperatures of 31.6°C produce males. Flooding of nests is a major source of mortality. Females excavate their nests when the hatchlings emit a squeaking noise. This species feeds on vertebrates (such as dingos, wallabies, birds, and fish) and has been responsible for human deaths in Australia and throughout its range. Hunting for the valuable skin had previously reduced crocodile numbers in Australia, but they are now protected in the wild and their numbers are increasing rapidly. They are farmed commercially for both meat and their skin.

· · · · · · · · · ·

TURTLES

Cheloniidae—The Sea Turtles

All five of the circumtropical sea turtle genera (family Cheloniidae) are represented in coastal northern Australia. The loggerhead turtle,

Caretta caretta (Figure 8.11), can reach a massive 400 kg (880 lb) with a 1.5 m (5 ft) shell. It is a carnivorous deepwater forager and nests usually from October to May, when it lays about 50 eggs. Some individuals may nest at any time of the year. The species occurs along the northern Australian coast from northwest Australia to southeast Queensland.

The flatback turtle, *Natator depressus* (Figure 8.12), is the only sea turtle confined to Australian waters. This carnivorous species occurs around the northern and northeastern coast of the continent, where it reaches about 1.2 m (4 ft) and lays about 50 eggs. Its hatchlings are nearly twice the size of those of *Chelonia mydas*.

The green turtle, *Chelonia mydas* (Figure 8.13), is so called because of the color of its fat. It is found along the northern coast in a distribution similar to *Caretta's*, and it is common along the Great Barrier Reef. *C. mydas* reaches about 1 m (3.3 ft). The young are carnivorous and the adults are herbivorous. The green turtle nests every two to four years, with the female laying three to four clusters of eggs. She excavates beach sand—first with her foreflippers, then with her hindlegs—to form a deep nest. This is usually done at night above the maximum high tide line. She stops digging when her hindlegs can no longer reach the bottom of the nest chamber. The 100 or so eggs are dropped, and the nest is filled in with sand. The female returns to the sea and copulates with a male that fertilizes her next clutch of eggs to be laid in about two weeks. Upon hatching (6–8 weeks later) the young fight their way to the surface and scramble to the sea. They are vulnerable to predation by gulls and crabs as they race for the water, where they are preyed upon by fishes. Those that survive grow at sea and eventually return to their nesting beach 20–30 years later. Green turtles have declined around the world recently due to predation by human hunters who overturn nesting females on the beach, and to wild dogs and pigs that dig up the eggs.

The hawksbill, *Eretmochelys imbricata* (Figure 8.14), has an Australian distribution similar to those of the green and loggerhead turtles. Its beautiful shell has long been in demand for tortoiseshell jewelry. It is carnivorous, and reaches 1 m (3.3 ft) in length.

The Pacific ridley, *Lepidochelys olivacea* (Figure 8.15), has a restricted Australian distribution. It is known from Torres Strait and coastal Arnhem Land. It reaches 1.5 m (5 ft) and lays about 100 large eggs.

Dermochelyidae—The Leatherback Turtle

The leatherback turtle, *Dermochelys coriacea* (Figure 8.16), is the largest living turtle. In fact, it is one of the largest living reptiles, rivaling the saltwater crocodile in weight. An adult male found drowned in a fishing net in 1991 off the coast of Wales weighed 915 kg (2015 lb). Its clawless front flippers, which are longer than the body, were 2.4 m (8 ft) long. There is only one species in the family Dermochelyidae,

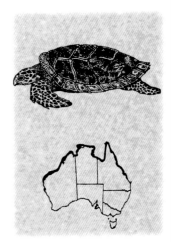

Figure 8.14
Hawksbill turtle,
Eretmochelys imbricata.

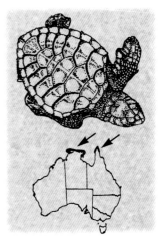

Figure 8.15
Pacific ridley turtle,
Lepidochelys olivacea.

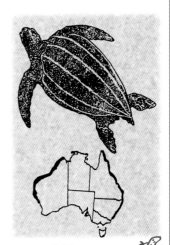

Figure 8.16
Leatherback turtle,
Dermochelys coriacea.

Figure 8.17
Pitted-shelled turtle,
Carettochelys insculpta.

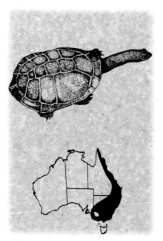

Figure 8.18
Long-necked turtle,
Chelodina longicollis.

Figure 8.19
Murray turtle,
Emydura macquarii.

but two sub-species, the Atlantic leatherback (*D. c. coriacea*) and the Pacific leatherback (*D. c. schlegelii*), are recognized. The species is distributed worldwide and wanders into cooler waters than do the other sea turtles.

The name leatherback is due to the fact that *Dermochelys* lacks epidermal scutes. In place of a typical carapace, it has a black leathery skin that contains a mosaic of small bony platelets called osteoderms. Adults have no scales on their back. There are seven narrow ridges of enlarged osteoderms that extend the length of its back and five longitudinal ridges on the plastron. These ridges are streamlining structures that help water flow over the animal. They function like the keel of a ship. Leatherbacks are powerful swimmers and deep divers. Dives as deep as 1200 m (4000 ft) have been recorded.

Most of what is known about leatherback biology comes from the study of nesting females and hatchlings. Very little is known about juveniles or adult males. Where hatchlings spend their time after they leave the nest is not known. Leatherbacks feed almost exclusively on jellyfish with the aid of flexible, backward pointing spines in the throat that prevent the escape of slippery prey. They may also eat crustaceans, mollusks, fishes, and marine plants. Leatherbacks make long migrations. A female was recaptured 5920 km (3700 miles) from her nesting beach less than one year after tagging. This species rarely breeds in Australia, but occasionally it will nest along a short stretch of the central Queensland coast.

Leatherbacks have survived for about 120 million years, 40–60 million years longer than other sea turtles, but they are now endangered from egg collecting on nesting beaches, discarded fishing lines and nets in which they become entangled, and lights from beachfront resorts that discourage females from coming ashore to nest, and disorient hatching young. Some individuals have even been found with plastic bags (which must look like a jellyfish to a leatherback) in their stomach.

Carettochelyidae—The Pitted-shelled Turtle

The pitted-shelled turtle, *Carettochelys insculpta* (Figure 8.17), is the only member of the family, Carettochelyidae. It is found in Papua New Guinea and in several river systems in the Northern Territory such as the Daly, Victoria, and Alligator. It lacks dermal scutes and has a thin layer of granulated soft skin over its pitted carapace. It has two-clawed, flipper-like forelimbs similar to those of sea turtles, and its nostrils are at the end of a fleshy proboscis. *Carettochelys* may be related to the soft-shelled turtles (Trionychidae) of North America, Africa, and Asia. Fossil carettochelyids are known from the Eocene Epoch of North America and Europe. It is the only Australian freshwater cryptodirous turtle (head and neck retract into the shell). It may reach 0.7 m (2.3 ft) and feeds on fish, mollusks, and fruits that fall into the water. About 15 eggs are laid in river sandbanks in the dry season. The embryos hatch out when the nest is flooded by rising waters at the beginning of the wet season.

Western swamp turtle, *Pseudemydura umbrina.*

Chelidae—The Sideneck Turtles

Aside from the sea turtles and *Carettochelys,* all other Australian turtles are pleurodires or sidenecked turtles of the family Chelidae. They are called this because, instead of withdrawing the head and neck vertically, as do cryptodires, they bend or fold their neck laterally under the front edge of the carapace. Pleurodires are found only in the southern hemisphere: South America, Africa, and Australia. The Chelidae is aquatic or semi-aquatic and is only found in South America, New Guinea, and Australia. Australia has 16 species in 6 genera (Table 8.2, page 156). The feet are clawed and webbed, not paddle-shaped as in the other Australian species. Australians call their turtles 'tortoises', a term herpetologists usually reserve for species with dome-like shells and stocky legs.

The long-necked turtle, *Chelodina longicollis* (Figure 8.18), the broad-shelled river turtle, *C. expansa,* and the Murray turtle, *Emydura macquarii* (Figure 8.19), are common in the Murray-Darling river system. *C. longicollis* is also found in coastal eastern Australia. Members of the genus *Chelodina* have exceptionally long necks and four claws on the forelimbs. *Emydura* have shorter necks and 5 claws. Both genera are aquatic and are widely distributed throughout Australia, except in the arid center. The two species of *Elseya* are confined to rivers in northern and eastern Australia, and the monotypic *Pseudemydura* is known only from two small, shallow, seasonal swamps at Bullsbrook northeast of Perth. There are only about 45 individuals of the western swamp turtle, *Pseudemydura umbrina* (Figure 8.20, page 162), of which about half are in the Perth zoo. The species is Australia's most endangered reptile, and its smallest turtle. Males, which are larger than females, reach 150 mm (6 in) in carapace length and weigh 550 g (1.2 lb).

Oblong turtle, *Chelodina oblonga,* from near Northcliffe, Western Australia.

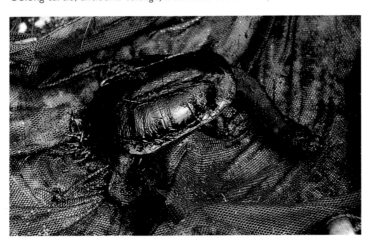

Top: Pitted-shelled turtle, *Carettochelys insculpta,* Australia's only freshwater cryptodire. Note the flippers and fleshy proboscis.
Middle: Murray turtle, *Emydura macquarii.*
Bottom: Saw-shelled turtle, *Elseya latisternum,* showing tubercles on the neck.

Figure 8.20
Western swamp turtle,
Pseudemydura umbrina.

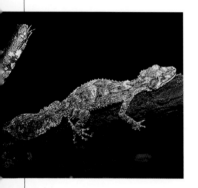

Southern leaf-tailed gecko,
Phyllurus platurus.

Although the western swamp turtle was first described in 1839, it was not reported again until 1953. *Pseudemydura* feeds on aquatic organisms such as crustaceans, insect lavae, and tadpoles. During the dry summer, when the swamps disappear, it estivates in holes in the ground or under deep leaf litter. Growth is slow, and it may live 15 or more years. Females lay only 3–5 eggs, in November or December in an underground nest. The hatchlings emerge about six months later and weigh only 6 gm (0.2 oz). This species is carnivorous, feeding on aquatic insects, crustaceans, and tadpoles. Habitat destruction in the form of clearing and draining of land for urban use, agriculture, and clay mining are major factors in the decline of the species. Predation by introduced foxes also exerts a deleterious effect on the numbers. An experimental breeding program at the Perth Zoo may eventually result in the release of animals into the wild.

A recently described new species, *Rheodytes leukops,* occurs only in the Fitzroy River system of eastern Queensland. It is apparently a derivative of *Elseya.* A new genus and species of short-necked turtle, from Queensland, was described in 1994. *Elusor macrurus* was known for 25 years only from specimens purchased in pet shops. After much detective work and searching, in 1990 specimens were finally obtained from the Mary River just north of Brisbane. As it turns out, *Elusor* may be the largest member of the family Chelidae with a carapace length of about 400 mm (16 in). The males of this species have a very long (up to 70 percent of carapace length) and thick tail.

· · · · · · · · · ·

THE LIZARDS OF OZ

Australia is the Land of Lizards. There are more species of lizards in Australia, especially in the desert regions, than anywhere else in the world. Australia's lizard fauna (Table 8.3) reflects an Oriental origin followed by an independent evolutionary history in Australia.

TABLE 8.3 **AUSTRALIAN SNAKES AND THE LIZARDS**

ORDER	SUBORDER	FAMILY	COMMON NAME	No. Aust. genera	No. Aust. species
Squamata	Sauria	Gekkonidae	Geckos	16	97
		Pygopodidae	Snake lizards	8	34
		Agamidae	Dragon lizards	13	63
		Varanidae	Goannas	1	25
		Scincidae	Skinks	32	306
	Serpentes	Typhlopidae	Blind snakes	1	31
		Boidae	Pythons	4	15
		Acrochordidae	File snakes	1	2
		Colubridae	Colubrids	9	11
		Elapidae	Elapids	20	77
		Hydrophiidae*	Sea snakes	13	33

*Includes *Laticauda.*
Sources: Wilson and Knowles (1988) and Cogger (1992).

Figure 8.21
Spiny-tailed gecko,
Diplodactylus ciliaris.

Gekkonidae—The Geckos

The geckos are found throughout the tropical and subtropical regions
of the world. These attractive lizards are mostly small, nocturnal, and
arboreal. They have large eyes with vertical slit-like pupils and
expanded friction pads on their toes that enable them to climb
vertical surfaces. They are frequently found in houses, where they eat
flies and other insect pests. Geckos are harmless and easily shed their
tail (autotomy) when attacked. A new tail is then regenerated. There
are about 97 Australian species in 16 genera. Geckos are absent from
Tasmania. *Diplodactylus* (Figure 8.21) is the largest Australian genus,
with about 36 species confined to Australia and distributed
throughout the continent. The western spiny-tailed gecko,
D. spinigerus, has a long tail with two rows of spines and glands that
secrete a repellent liquid which sometimes deters a hungry snake.
The clawless gecko, *Crenadactylus ocellatus,* Australia's smallest gecko—
35 mm (1.3 in) in snout-vent length (SVL)—is the only (but a highly
variable) species in this endemic genus, occuring in sandy deserts of
central and western Australia. The following are also endemic
Australian genera:

Carphodactylus	1 sp
Heteronotia	2 spp
Lucasium	1 sp
Nephrurus	9 spp
Oedura	13 spp
Phyllurus	4 spp
Pseudothecadactylus	2 spp
Rhynchoedura	1 sp

Northern leaf-tailed gecko,
Phyllurus cornutus, forages for
insects on tree trunks at night
in rainforests and wet
sclerophyll forests.

Common scaly-foot, *Pygopus
lepidopodus,* feeds primarily on
spiders in a variety of habitats
from semi-arid mallee in the
inland to coastal wet
sclerophyll forests.

Figure 8.22
Burton's snake-lizard,
Lialis burtonis.

Figure 8.23
Frilled lizard,
Chlamydosaurus kingii.

Figure 8.24
Thorny devil, *Moloch horridus.*

Pygopodidae—The Snake Lizards

The snake lizards evolved entirely within Australia and New Guinea and are the only endemic lizard family of the region. They are probably derived from a gekkonid ancestor, and some herpetologists place these within the Gekkonidae. There are 34 species in eight genera in Australia. The forelegs have been lost and the hindlegs are reduced to a pair of scaly flaps on each side of the vent. Many species are burrowers and can shed their tails. Some pygopodids can make squeaking noises like geckos, while others (*Pygopus* and *Paradelma*) mimic venomous snakes with their markings and posturing. The presence of external ear openings in some species separates the pygopodids from snakes, which lack such openings. Pygopodids have tails that are longer than their bodies, whereas the tails of snakes are shorter than body length. Snake lizards frequently clean their lidless eyes with a flick of their long fleshy tongues. (Snake tongues are narrow and forked.) Only two of the 34 Australian species occur in New Guinea and one New Guinean species is the only pygopodid that does not occur in Australia.

Delma is the largest genus, with 17 species. These lizards have well-developed hindlimb flaps and large external ear openings. *Lialis burtonis* (Figure 8.22) is the most widely distributed and highly variable snake lizard. It has sharp, recurved teeth and feeds on small lizards. It occurs throughout Australia except most of Victoria and Tasmania, and may reach 290 mm (11.5 in) SVL. It is also in New Guinea. *Pygopus nigriceps* and *P. lepidopodus* are also very widespread species, but the pygopodids do not reach Tasmania.

Agamidae—The Dragon Lizards

The dragon lizards are found in southern Europe, Africa, Asia, New Guinea, and Australia. They are the Old World ecological equivalent of the New World Iguanidae. There are many striking examples of convergent evolution between the dragons and the iguanas. In Australia there has been substantial adaptive radiation of dragons in the arid regions, but only one species reaches Tasmania. There are 63 species in 13 genera (Table 8.3, page 162), with eight of the 13 genera being endemic to Australia. All agamids lay eggs and most feed on insects. They are usually diurnal and very swift with good eyesight. Color change is possible in most species. Some species have greatly enlarged spines and tubercles. The largest agamids reach about 260 mm (10 in) SVL. Courtship displays include head bobbing and arm waving.

Chlamydosaurus kingii (Figure 8.23), the frilled lizard, is an impressive animal with a large loose collar that is folded around the neck and shoulder when at rest but is erected when the lizard opens its mouth. This is a formidable display by Australia's most distinctive lizard. *Chlamydosaurus* can run bipedally and is normally arboreal in

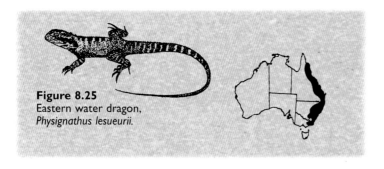

Figure 8.25
Eastern water dragon,
Physignathus lesueurii.

Painted dragon,
Ctenophorus pictus, from near
Kalgoorlie, Western Australia.

Figure 8.26
Bearded dragon,
Pogona barbata.

northern Australia. It also occurs in southern New Guinea. The genus is monotypic.

The small to moderately large dragons of the endemic genus *Ctenophorus* include about one-third of the Australian species in the Agamidae. This genus is distributed throughout the arid regions of Australia. Males are brightly colored during the breeding season.

Perhaps the most incredible-looking Australian lizard is the thorny devil, *Moloch horridus* (Figure 8.24). The slow-moving *Moloch* feeds only on black ants of the genus *Iridomyrmex.* A single meal may

Frilled lizards, *Chlamydosaurus kingii,* with neck frill folded.

Thorny devil, *Moloch horridus.*
This fearsome-looking beast
feeds only on ants in the sand
and spinifex deserts of central
Australia and the arid scrub
of Western Australia.

Forest or angle-headed dragon, *Hypsilurus spinipes*, from the rainforests of southeastern Queensland and northeastern New South Wales.

consist of 5000 ants eaten over a two-hour period. *Moloch* lives in the arid regions of west and central Australia. Its fat, depressed, reddish brown body is covered by large conical spines, and there is a prominent spiny hump on the back of the neck. Special scales between the spines direct rainwater and dew to the corners of the mouth. When walking, it resembles a wind-up toy with erratic movements. This endemic lizard may reach 90 mm (3.6 in) SVL. Females lay an average of eight eggs from November to December. In spite of its fearsome appearance, *Moloch* is harmless.

The eastern water dragon, *Physignathus lesueurii* (Figure 8.25), is found along rivers in eastern Australia. A ridge of spines extends along its back on to a long compressed tail. Its SVL is about 200 mm (8 in), but its tail can be 2.5 times that length. This species will drop from overhanging tree branches into the water when disturbed. It can remain submerged for at least 30 minutes. It is omnivorous.

The bearded dragon, *Pogona barbata* (Figure 8.26), has a well-developed dark beard under its chin with spiny scales that it erects when disturbed. It is common over the eastern third of Australia in dry forests, and feeds on flowers and insects. This familiar flattened lizard engages in a variety of courtship displays and is frequently observed basking on roads.

Other endemic agamids are:

Caimanops	1 sp
Chelosania	1 sp
Diporiphora	14 spp (one species is found in Papua New Guinea)
Lophognathus	3 spp (also in New Guinea and Tanimbar Island)
Hypsilurus	2 spp (widespread in Papua New Guinea, Moluccas, and Solomon Islands)
Tympanocryptis	8 spp, including *T. diemensis,* the only agamid to reach Tasmania

Varanidae—The Goannas

About 40 species of monitor lizards are found from Africa through to India and southeast Asia to Australia. This family includes the world's largest lizard, the Komodo dragon, *Varanus komodoensis,* from the lesser Sunda Islands of Indonesia. It may reach 3 m (10 ft) in length.

A fossil Australian species from 30 000 years ago, *Megalania prisca,* reached 7m and weighed 650 kg (1430 lb). The closest relatives of varanids are the earlier monitor lizard, *Lanthanotus borneensis* (in the monotypic family Lanthanotidae of Borneo), and the Helodermatidae of southwestern North America. The two species in this family, the gila monster, *Heloderma suspectum,* and the Mexican beaded lizard, *H. horridum,* are the only venomous lizards in the world. However, their bite is rarely fatal to humans. About 63 percent of the total number of varanid species are Australian. There is only one genus, *Varanus* (although taxonomic revision may change this), which includes 25 Australian species. There are no varanids in Tasmania.

These large lizards are called goannas in Australia. This name is a corruption of iguana, a family of lizards found mainly in South America. Goannas are swift-moving predators and scavengers that relish carrion. For this reason their bite is said to be unusually prone to bacterial infection. See Auffenberg (1981) for a description of several aseptic bites as well as cases of human death from massive sepsis. Varanid teeth are large, sharp, and recurved for holding prey. Broken teeth are replaced by new ones.

Goannas are the only Australian lizards with a long, slender, forked tongue similar to that of a snake. The tongue is used to 'taste' the air by bringing molecules to Jacobson's organ in the roof of the mouth. When searching for food, their snout is held close to the ground and the tongue darts in and out. Their long powerful tail, accompanied by loud hissing, can be used to knock predators off their feet and as a brace to stand erect for long periods. An erect posture provides an unobstructed view and may be the source of the name 'monitor' lizard. This bipedal posture and grappling with forelimbs is common in male-male combat over females.

Males tend to be larger than females and have two lateral swellings at the base of the tail, where the hemipenes are stored. These dual copulatory structures are inserted, one at a time, into the female for sperm transmission. The male uses his tail to elevate the female during mating, thus facilitating insertion of the hemipenis. After about 10 minutes, the female leaves the male and enters her burrow, sometimes re-emerging for further matings with the same male. This activity may last several days.

The *Varanus rosenbergi* female lays about 10–17 leathery-shelled eggs in a tunnel she excavates in an active termite mound. Soon the termites repair the mound, and the eggs are sealed in the tunnel. The termites control the mound's temperature at abut 30°C (86°F), and the humidity is near saturation. This creates a perfect incubator. The eggs hatch in spring about eight months after they were laid. The active young disappear quickly after hatching, probably moving up into trees in order to avoid the cannibalism of all varanid species. In captivity, several varanid species have lived 5–15 years.

During winter, varanids may remain in their burrows, but otherwise they are active during the day. Desert species may retreat to their burrows during the midday heat and re-emerge in late afternoon as a means of regulating their body temperature. Varanids usually dig their own burrows, but warrens of the introduced rabbit (*Oryctolagus cuniculus*) may also be used. If goanas get overheated, they open their mouth and vibrate their throat. This gular fluttering increases evaporative cooling and decreases high head temperatures that could be fatal. It also increases loss of water that must be replaced. Goannas get most of their water from their food. They are

Pogona vitticeps, one of seven species of bearded dragons.

very efficient at conserving water by resorbing it from feces and urine before excretion. Many varanids also have salt-excreting glands in the nasal capsules.

Varanids are unusual in the way they breathe. Mammals have a sheet of muscle (a diaphragm) that divides the chest cavity from the abdominal cavity. Reptiles lack a diaphragm. However, varanids, *Lanthanotus* and *Heloderma* have a diaphragm-like septum between the thoracic and abdominal cavities which increases the respiratory efficiency of their large lungs. Likewise, the heart of a varanid approaches a mammalian heart in that there is relatively little of the mixing of oxygenated and deoxygenated blood that commonly happens in typical lizard hearts. Amphibious goannas can remain submerged for as long as one hour, due to their unusual ability to tolerate lactic acid in their tissues.

Some varanid species are arboreal and some are amphibious, but most are terrestrial. The smallest varanid is *V. brevicauda* of the arid regions of Western Australia and the western Northern Territory. It reaches only 200 mm (8 in) in total length. Australia's largest lizard is the perentie, *V. giganteus,* which can reach 2.5 m (8.2 ft) in total length. It occurs in the arid interior from the coast of Western Australia to western Queensland. It feeds on insects, reptiles, birds, small mammals, and carrion. Juvenile perentie are especially attractive, with yellowish speckles on a dark background. *V. gouldii* is widely distributed over most of Australia except the southeast coast. *V. varius* (Figure 8.27), the lace monitor of eastern Australia, is an

Top: A young specimen of perenti, *Varanus giganteus,* Australia's largest lizard. See page 63 for Aboriginal painting and legend.
Middle: Sand monitor, *Varanus gouldii.*
Bottom: Ventral surface of the skink, *Lerista punctatovittata,* showing the limb reduction common in this genus of 76 fossorial species.
Right: Spiny desert or gidgee skink, *Egernia stokesii.*

Figure 8.27
Lace monitor,
Varanus varius.

arboreal species with a very long slender tail. It feeds on eggs and nesting birds. This agile goanna can reach 2 m (6.5 ft) and has the habit of spiraling upwards around a tree trunk in order to remain on the opposite side from an observer. Females lay a clutch of 8–12 parchment-shelled eggs in termite nests in a tree. Aboriginals consume several species of goannas, especially in the desert.

Scincidae—The Skinks

Figure 8.28
Eastern blue-tongued skink, *Tiliqua scincoides*.

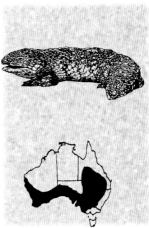

Figure 8.29
Shingle back, *Trachydosaurus rugosus*.

Skinks are found throughout the world except the far north and the tip of South America. They are exceedingly diverse in Australia, with 32 genera and 306 species. More species remain to be described. At least eight species of skinks have colonized Tasmania. Most skinks are small, diurnal insect eaters. Many burrowing forms have small limbs. The genera *Ctenotus* (81 species) and *Lerista* (76 species) are the most speciose, and *Carlia, Egernia,* and *Pseudemois* have 20 or more species each. The blue-tongued skinks, five species of the genus *Tiliqua* (Figure 8.28), are relatively large and reach about 30 cm (1 ft) SVL. These robust skinks are widespread throughout Australia, and are a common feature of the Australian fauna, familiar even to urban dwellers. They give birth to live young. When disturbed they inflate their body and hiss while displaying a pink wide-open mouth with the namesake blue tongue. The shingle back, *Trachydosaurus rugosus* (Figure 8.29), is often found on roads in the southern half of Australia. It moves very slowly. Its short stumpy tail is similar to its head, making it difficult to tell if it is coming or going.

Left: Eastern blue-tongued skink, *Tiliqua scincoides*, from Bombala, NSW, showing how it gets its name.
Below: Shingle back, *Trachydosaurus rugosus*, from Cocklebiddy, Western Australia.

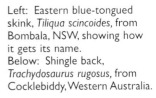

Amethystine python, *Morelia amethistina*, swallowing its prey.

SNAKES

Of the 2500 or so species of snakes in the world, Australia has about 169 (7 percent) in 48 genera. Most of these are venomous. Snakes most likely evolved from burrowing lizards more than 135 MYA, in the late Jurassic Period. However, Caldwell and Lee (1997) recently proposed that snakes evolved from marine varanoid lizards. The first clearly defined snake fossil dates back 120 MYA to the Cretaceous Period of Algeria. Snakes lack external ear openings and limbs, and they have a deeply forked, long, slender tongue that captures scent molecules and transmits them to paired Jacobson's (vomeronasal) organs in the snout. The widely separated tips of the tongue may allow a snake to detect a gradient in a scent trail and thereby find prey or a mate. Snakes are unable to blink because they lack eyelids. Each eye is covered by a transparent scale, called a spectacle, which is shed with the skin. Unlike lizards, snake tails are shorter than body length. The belly (ventral) scales of most snakes are enlarged into tranverse plates which help the snake locomote by providing friction and traction on the ground.

Typhlopidae—The Blind Snakes

The blind snakes are small, oviparous, burrowing, harmless, worm-like creatures found on all tropical continents. They are primitive snakes with a vestigial pelvic girdle, and they lack enlarged ventral scales. The 31 Australian species feed on ants and termites and their eggs. Highly polished, armor-like scales protect them from vicious ant bites. As the common name indicates, the eyes are vestigial in keeping with a fossorial existence. Anal glands emit a foul-smelling odor that may deter some enemies but not bandy bandys (fossorial elapids of the genus *Vermicella*)—their major predators. The only genus in Australia is *Ramphotyphlops* (Figure 8.30). Their maximum

Figure 8.30
Blind snake,
Ramphotyphlops australis.

size is approximately 750 mm (30 in). The genus is widely distributed throughout continental Australia, but no species reaches Tasmania. Because of their burrowing lifestyle, they are not commonly seen.

One member of this family is unique in the snake world. *R. braminus* reproduces by parthenogenesis, that is, females lay unfertilized eggs that hatch into clones of themselves. All of the young are identical, and there are no males of this species. This unusual mode of reproduction is also known in some lizards. It apparently arises from chromosomal changes produced by the hybridization of related species. *R. braminus* is Australia's smallest snake. It reaches a maximum size of 170 mm (7 in) and is found in the Darwin area of the Northern Territory. It was probably accidentally introduced into Australia by humans. It occurs in Asia, Indo-Malaysia, and New Guinea. The common name 'flowerpot snake' refers to the fact that it can live in small containers of soil. This may be how it arrived in Australia. The unique 'virgin birth' would enable a single colonist to develop into a colony upon arrival in a new location.

Diamond python, *Morelia spilota spilota,* from coastal New South Wales.

Boidae—The Pythons

Members of this family occur throughout the tropics and subtropics of the world and include the largest snakes on earth. The boas, which give birth to live young, are confined to the New World tropics, with the exception of two genera from Madagascar and one from the Western Pacific. The pythons lay eggs and are the Old World counterpart of the boas. They are found in Africa, Asia, and Australia. Pythons probably entered Australia from Asia relatively recently after Australia collided with Asia less than 25 MYA. This family is non-venomous and kills by constriction, suffocating the prey. Most

Carpet python, *Morelia spilota variegata.*

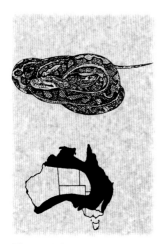

Figure 8.31
Carpet python,
Morelia spilota.

pythons have heat-sensitive pits in the lower lip scales used for detecting warm-blooded mammals and birds. Their recurved teeth hold prey securely while it is swallowed. Moveable bones in the skull and a distensible skin enable pythons to eat food items larger than their own girth. They then fast for long periods.

The Boidae is a primitive family of snakes with vestiges of a pelvic girdle. This reflects the evolution of snakes from lizards. Short spurs (vestigial hindlimbs) are present on each side of the cloaca. The males use these structures to stimulate the females during breeding. Females can generate enough heat by muscle contraction to elevate body temperature and thereby brood their eggs during a two month incubation period. Pythons are the only Australian snakes known to remain with their eggs. Pythons are excellent climbers and are usually nocturnal.

None of the 15 species of Australian pythons reaches Tasmania. One of the four genera, *Aspidites*, is endemic to Australia. It lacks the heat-sensing pits in the lip scales, and its two species are more likely to feed on other reptiles.

The amethystine python, *Morelia amethistina*, of Cape York and New Guinea, is Australia's largest snake; one specimen measured 8.5 m (28 ft), but most average less than 4 m. Its iridescent sheen accounts for its name. The most widely distributed Australian species is *Morelia spilota* (Figure 8.31), the carpet python. It is found throughout continental Australia in habitats from rainforest to semi-arid areas, but it is absent from most of Victoria. It grows to 4 m. The diamond python found in coastal New South Wales is a subspecies of *M. spilota* with a very different color pattern.

The green python, *Chondropython viridis*, is a beautiful arboreal bird-eating inhabitant of rainforests in Cape York. It begins life in a yellow phase and gradually becomes a brilliant emerald green at maturity about three years later. It is convergent with the emerald tree boa, *Corallus caninus*, of South America. The Children's python of northern Australia, *Liasis childreni*, is not so named because of its dietary preferences or because it makes a good pet. It is named in honor of a Mr Children of the British Museum.

Acrochoridae—The File Snakes

The file snakes are so called because of the coarse texture of their small, finely keeled and pointed scales. They have unusually stout bodies and loose-fitting skin that compresses into a paddle-like swimming aid. The family occurs from coastal India through southeast Asia to coastal northern Australia and does not appear to be closely related to other snake families. They are aquatic snakes and almost helpless out of water as they lack ventral shields. Their nostrils can be shut by valves. They feed on fishes, usually at night, and readily enter the sea. They are ovoviviparous and have no need to come ashore to lay eggs. They are non-venomous.

Figure 8.32
File snake,
Acrochordus granulatus.

The two Australian species are *Acrochordus granulatus* (Figure 8.32) and *A. arafurae.* Both are usually less than 1.5 m (5 ft) long and are shared with New Guinea. *A. granulatus* ranges into southeast Asia. File snakes are included in the diet of Aboriginals. *A. granulatus* has a very low metabolic rate and a very large blood volume. These adaptations enable it to remain submerged for long periods. According to Richard Shine (1991) of the University of Sydney, *A. arafurae* reproduces only about once each decade. This is the slowest rate of reproduction of any vertebrate species. Apparently it takes the female many years to store enough energy to fill her body cavity with eggs and then give birth to an average of 17 young. The young mature slowly and are not ready to breed for 6–9 years.

Colubridae—The Colubrids

The colubrids are the dominant snake family of the world with about 1600 species in 300 genera. They are common in North America, but there are only 11 species in nine genera in Australia. These are predominantly northern (tropical) and indicate a relatively recent

Ken Griffiths

Eastern brown snake,
Pseudonaja textilis.

Figure 8.33
Brown tree snake, *Boiga irregularis.*

invasion from Asia via New Guinea. There are no endemic Australian colubrids, probably due to their relatively recent arrival. This family includes both non-venomous solid-toothed snakes and the mildly venomous rear-fanged snakes. In Australia the non-venomous common tree snake, *Dendrelaphis punctulata,* occurs along the north and east coast. This swift, slender arboreal snake can reach 2 m (6.5 ft) and feeds predominantly on frogs. It is common in suburban areas.

The slender, rear-fanged brown tree snake, *Boiga irregularis* (Figure 8.33), has a distribution similar to that of the common tree snake. It is venomous and aggressive and will strike savagely with an open mouth, but it is not regarded as a threat to humans. This nocturnal predator feeds on small mammals, birds, eggs, and lizards. The brown tree snake was accidently introduced to Guam around the time of World War II and has devastated the bird fauna of that island. It is an egg-layer and lives in trees or rock outcrops.

Cerberus, Enhydris, Fodonia, and *Myron* are aquatic, weakly venomous, rear-fanged colubrids that feed on fishes and crabs in northern Australia. The live-bearing *Enhydris polylepis* is found in creeks and rivers, and the other genera live in the brackish or salt water of mangrove flats. The aquatic colubrids have round tails that distinguish them from the compressed tailed, highly-venomous sea snakes (Hydrophiidae). Some aquatic colubrids have upward-directed bulging eyes that enable them to see above the water surface when their body is submerged. The aquatic species are somewhat more robust than the tree-dwelling colubrids.

Elapidae—The Elapids

Australia is the only major biogeographic region in which the venomous snakes greatly outnumber the non-venomous ones. About 70 percent of Australia's snakes are either elapids or their derivative, the hydrophiids. The elapids contain some of the world's deadliest snakes, such as the African and Asian cobras, African mambas, and Asian kraits. Elapids reached Australia within the last 25 million years, most likely from Asia. North America's only elapids are the coral snakes. The family of 200 species in 50 genera is circumtropical, with great diversity in Australia.

Australian elapids are most diverse on the central eastern coast and in the south and west of the continent in well-watered areas near the coasts. This habitat is also favored by people, and the resulting interaction has been most harmful to the snakes. Elapids have two rather short, immovable, deeply grooved or hollow fangs on the upper jaw, each connected to a venom gland. The venom of Australian species is mostly neurotoxic—it affects the nerves controlling respiration and heart action, thereby resulting in suffocation. Vertebrates constitute the exclusive prey of elaphids. Most of the 77 Australian species are only mildly venomous and pose no hazard to humans. They eat mostly skinks. However, several widespread species are highly venomous and have caused human deaths. Australian elapids have exploited most habitats, but very few are arboreal or aquatic. Some elapids are egg-layers and have divided subcaudal scales. Young are born alive to species with undivided subcaudal scales (except blacksnakes). Live-bearing species tend to be in cooler climates. Giving birth to live young is a relatively recent (5 MYA) innovation in Australian elapid evolution, and is probably an adaptation that allowed expansion into cooler regions where embryos could be protected inside the mother.

Snake venom is modified saliva, and it evolved for the function of capturing and digesting food. Animals that chew their food begin the digestive process by working saliva into the prey. This lubricates the food, and the enzymes present initiate a predigestion. Snakes swallow their food whole. Their sharp, recurved teeth function to capture and hold struggling prey, but are not suitable for chewing. Natural selection has favored the enlargement of maxillary teeth followed by the evolution of grooved and eventually hollow hypodermic-like fangs that facilitate the delivery of venom.

The injection of salivary enzymes into the prey not only subdues it but also initiates digestion. The proteolytic and other enzymes

Yellow-faced whip snake, *Demansia psammophis psammophis*. This slender swift-moving, highly variable elepid is found throughout most of Australia except the tropical north-central region and Tasmania.

Ken Griffiths

TABLE 8.4   **TOXICITY OF VARIOUS SNAKE VENOMS
TO MICE**

SCIENTIFIC NAME	COMMON NAME	FAMILY	DISTRIBUTION	TOXICITY TO MICE (LD50)* (mg/kg)
Oxyuranus microlepidotus	Small-scaled snake	Elapidae	Australia	0.02
Pseudonaja textilis	Brown snake	Elapidae	Australia to New Guinea	0.05
Oxyuranus scutellatus	Taipan	Elapidae	Australia to New Guinea	0.08
Notechis scutatus	Eastern tiger snake	Elapidae	Southeastern Australia	0.12
Enhydrina schistosa	Beaked sea snake	Hydrophiidae	Persian Gulf to Australia	0.14
Hydrophis ornatus	Ornate sea snake	Hydrophiidae	Persian Gulf to Australia	0.16
Laticauda laticaudata	Sea krait	Hydrophiidae	India to Australia	0.17
Disteira majora	Sea snake	Hydrophiidae	Australia to New Guinea	0.19
Aipysurus laevis	Olive sea snake	Hydrophiidae	Australia to New Guinea	0.22
Crotalus scutulatus	Mojave rattlesnake	Viperidae	Southwestern United States to Mexico	0.25
Hydrophis elegans	Elegant sea snake	Hydrophiidae	Australia to New Guinea	0.27
Bungarus caeruleus	Indian krait	Elapidae	India to Pakistan	0.27
Pelamis platurus	Yellow-bellied sea snake	Hydrophiidae	East Africa to western Americas	0.28
Naja naja	Asian cobra	Elapidae	Southern Asia	0.29
Crotalus durissus	Brazilian rattlesnake	Viperidae	Southeastern Brazil to Argentina	0.31
Astrotia stokesii	Stoke's sea snake	Hydrophiidae	Southeast Asia to Australia	0.35
Acanthopis antarcticus	Death adder	Elapidae	Australia to New Guinea	0.37
Laticauda colubrina	Sea krait	Hydrophiidae	India to Australia	0.42
Laticauda semifasciata	Erabu sea krait	Hydrophiidae	China to Samoa	0.50
Lapemis hardwickii	Hardwick's sea snake	Hydrophiidae	Persian Gulf to Australia	0.62
Hydrophis cyanocinctus	Annulated sea snake	Hydrophiidae	Persian Gulf to Java	1.24
Ophiophagus hannah	King cobra	Elapidae	Southeast Asia	1.86
Dendroaspis angusticeps	Green mamba	Elapidae	East Africa	2.18
Vipera russelli	Russel's viper	Viperidae	West Pakistan to Taiwan	2.42
Echis carinatus	Saw-scaled viper	Viperidae	Africa to India	3.88
Crotalus horridus	Timber rattlesnake	Viperidae	Eastern United States	c. 4.50
Dispholidus typus	Boomslang	Colubridae (rear-fanged)	Africa	6.28
Bitis gabonica	Gaboon viper	Viperidae	Africa	6.60
Crotalus adamanteus	Eastern diamondback rattlesnake	Viperidae	Southeastern United States	9.55
Brothrops atrox	Fer-de-lance	Viperidae	Mexico to Argentina	10.02
Agkistrodon piscivorus	Cottonmouth moccasin	Viperidae	Southeastern United States	11.60
Aipysurus eydouxii	Sea snake	Hydrophiidae	Southeast Asia to Australia	11.70
Lachesis mutus	Bushmaster	Viperidae	Nicaragua to Brazil	36.90

* LD50 is the dose in milligrams of dry venom per kilogram of mice required to kill 50 percent of the test animals within a given period.
Sources: Modified from Heatwole (1987), Minton and Minton (1969), and Sutherland (1983).

interfere with various biochemical processes at the cellular and membrane level and break down the prey's tissues. Snake venom is pharmacologically extremely complex. In addition to several types of toxins, other substances are present. For example, hyaluronidase is found in the salivary secretions of most venomous snakes. This enzyme assures rapid diffusion of the venom, and the prey's circulatory system helps distribute the venom before death, when digestion begins. Because snakes have no pectoral girdles and very mobile skull bones, they can swallow large prey that stretch their bodies. Digestion can then proceed both externally from the gastric juices and internally from the injected venom. Eating large meals means that a snake may go for a long time without feeding. This reduces energy expended searching for prey, and allows a wider selection of prey items.

It is difficult to compare the toxicity to humans of various snake venoms because there are so many variables to take into consideration, such as size of the snake, dose delivered, chemical composition of venom, time of year, individual variation of snakes and humans, location and depth of bite (muscle, blood vessel, body cavity), and health and age of victim. One standard of comparison is to measure how much venom must be injected into mice to kill half the test animals within a given time. The lethal dose (LD50), measured in milligrams of dry venom per kilogram of mice, is shown in Table 8.4 for a variety of venomous snakes from around the world. The reaction of mice and humans to a given snake venom is not identical, but the data given in the table are probably a reliable indicator of relative toxicity. Note that many of the most deadly species are Australian.

Ken Griffiths

Red-bellied black snake,
Pseudechis porphyriacus.

Snake
Bites

In over half the instances of venomous snakebites little or no venom is actually injected. This fact should be used to calm a snakebite victim while getting him or her to a medical facility. Nevertheless, first aid should be administered because symptoms may not manifest themselves for several hours. The great majority of bites are on the extremities of the body. These bites are less dangerous than bites to the head, neck, chest, or abdomen. The old Boy Scout first aid treatment of 'cut and suck' is not advisable, as it can do more harm than good. The best first aid treatment for Australian snakebite is to wrap the bitten limb with a snug elastic bandage. A broad strip of cloth, firmly wrapped, can substitute for a crepe bandage. For example, if the victim is bitten on the ankle, wrap a bandage around the ankle, the foot, and up to the knee. The idea is to slow the movement of venom and keep it localized in the limb. This allows the toxin to be released slowly without overwhelming the body.
Do not apply a tourniquet. These are very dangerous because they stop the flow of blood into the limb and can result in gangrene and subsequent loss of the limb. The snug elastic bandage allows arterial blood to flow into the limb, but impedes the return of venus blood and lymphatic fluid containing the venom to the heart. Do not loosen the bandage once it is in place. This could release a 'slug' of venom. Let the doctor do this in a hospital. When hiking in the bush always carry an elastic bandage roll *with* you—don't leave it in the car.

It should be pointed out that the Australian pressure-immobilization method is not appropriate for pitviper bites (American rattlesnakes, etc). Keeping the venom localized with a pressure bandage may result in increased local necrosis due to the hemolytic nature of pitviper venom as opposed to elapid venom.

Antivenoms (also known as antivenenes or antivenins) are produced by the Commonwealth Serum Laboratories for all deadly Australian snakes. If possible, try to identify the snake so the proper antivenom can be used to neutralize the toxin, but do not risk being bitten. If you cannot make a positive identification, traces of venom remaining around the wound or in the bloodstream or urine can be quickly tested at the hospital with a venom ID kit. If the species cannot be identified, a polyvalent Australian antivenom can be used.

Antivenoms are made by injecting horses with snake venom. The animal builds antibodies to the toxin. Blood is eventually collected from the hyperimmunized horse and processed. The blood cells are removed and the remaining serum contains the antibodies to the toxin. This is injected intravenously into a bite victim to neutralize the venom. This should be done in a hospital under medical supervison because many people have an allergic reaction to the foreign serum and may go into anaphylactic shock. This is especially likely in persons who have previously been injected with horse serum. Shock symptoms can be life-threatening and must be treated. Because of this serious risk, antivenoms are usually administered only when the symptoms of envenomation appear.

If massive envenomation has occurred the victim may collapse and go into shock soon after the bite. This is relatively rare. However, some people faint after a bite, but soon revive. They will probably experience the more typical symptoms, including muscle ache and stiffness, within 30 minutes. Drooping eyelids and pain in the groin or armpits may be experienced, and severe limb pain may occur within two hours. Headache, drowsiness, nausea, vomiting, and double vision may follow, along with shallow breathing and low blood pressure. After several hours the urine may become dark reddish brown in severe cases, as muscle fibers decompose and myoglobin is released. Creatine kinase levels become elevated in the blood. If treatment is not given, death from respiratory paralysis may result in 12–24 hours, but it may take up to 48 hours. Some survivors of severe envenomation may have permanent muscular weakness and kidney damage, but most promptly and properly treated victims recover completely. On average, less than five people die annually from snakebite in Australia, out of an estimated 3000 snakebite cases, compared to 20 000 deaths each year in much more densely populated India.

Death adder, *Acanthophis antarcticus*.

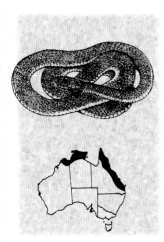

Figure 8.34
Taipan,
Oxyuranus scutellatus.

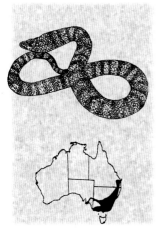

Figure 8.35
Eastern tiger snake,
Notechis scutatus.

Taipan, *Oxyuranus scutellatus*,
one of the world's most
deadly snakes.

The taipan, *Oxyuranus scutellatus* (Figure 8.34), of coastal, tropical northern Australia and New Guinea, and the small-scaled (or fierce) snake, *O. microlepidotus*, of eastern inland arid regions, are without a doubt two of the most dangerous snakes in Australia. The venom dose per bite of the 2m (6.5 ft) *O. microlepidotus* is nearly 100 times more potent than that of the Asian king cobra, *Ophiophagus hannah*, and 500 times that of the American eastern diamondback rattlesnake, *Crotalus adamanteus*. It is four times as potent as the taipan. *Oxyuranus microlepidotus* is the most venomous terrestrial snake known (Table 8.4, page 176), but, fortunately, there are only three confirmed snakebite cases for this species. All three victims survived.

Oxyuranus, an egg-laying genus, feed exclusively upon warm-blooded prey, which they bite and release. This prevents injury to the snake by the larger struggling prey such as rats and bandicoots. Taipans can reach 4 m (13 ft) but average half that size. They are the largest venomous snakes in Australia. Taipans are common in sugarcane fields. This may be due to the presence of the introduced toad, *Bufo marinus*, whose skin secretions are toxic to other snakes, thereby reducing competition. Taipans, with their preference for mammalian prey, leave *Bufo* alone.

The eastern tiger snake, *Notechis scutatus* (Figure 8.35), is responsible for a large proportion of snakebite deaths in Australia. Its venom is extremely potent and an antivenom to it was the first to be developed in Australia. The species' color can be highly variable, and many individuals are banded. Tiger snakes occur in southeastern and southwestern Australia in a broad range of habitats. They can be aggressive when disturbed and, like many Australian elapids, will assume a flattened neck, cobra-like posture. Tiger snakes are often found near fresh water and normally feed on frogs. The black tiger snake, *N. ater*, is one of the few snakes to have reached Tasmania and Kangaroo Island off South Australian coast. This species holds the record for the largest litter ever recorded for an Australian snake—109 babies. Most tiger snake births are more modest at 20–50 young.

Another deadly elapid of southeast Australia, including Tasmania, is the copperhead, *Austrelaps superbus* (Figure 8.36), which is no relation to the American copperhead, *Agkistrodon contortrix*, a crotalid. Its venom is not only neurotoxic but hemolytic as well. This species is active at lower temperatures than most snakes and is found above the snowline on Mount Kosciuszko. Frogs are the preferred food of adults, and lizards are preferred by juveniles.

The mulga, *Pseudechis australis* (Figure 8.37), is a dangerous 3 m (9.8 ft) snake found throughout most of continental Australia except the humid eastern and southern regions. The red-bellied black snake, *P. porphyriacus* (Figure 8.38), is beautiful iridescent black with a vermilion belly. It is usually associated with streams or standing water in eastern Australia, but it also frequents suburban areas. It is diurnal and feeds on frogs; however, its numbers have declined where *Bufo marinus* is common. Feeding on this poisonous toad is known to kill snakes. The bite of a red-bellied black snake is serious but not usually lethal. Unlike other elapids, it gives birth to live young. This has enabled the red-bellied black snake to expand its geographic range into the cooler southeastern area of Australia.

The eastern brown snake, *Pseudonaja textilis* (Figure 8.39), is very deadly, swift-moving and willing to strike if annoyed. It is found in the eastern half of continental Australia, especially in agricultural areas with an abundant supply of house mice, *Mus musculus.* It is probably the most commonly encountered dangerous snake in Australia. Its congener, *P. nuchalis,* occurs over the western two-thirds of the continent, except in the south. The dugite, *P. affinis,* from the southwest corner, and *P. nuchalis* have mostly mutually exclusive ranges. Both are deadly but respond to *P. textilis* antivenom treatment.

The death adders, *Acanthophis antarcticus* (Figure 8.40), *A. praelongus,* and *A. pyrrhus,* are found in all parts of Australia except the southeast corner and Tasmania. They appear viper-like, with a broad triangular head and cryptically colored heavy body. Their short stocky body enables them to strike rapidly. Death adders are ambush predators, and their tails end in a curved spine that is wriggled convulsively to lure mice or lizards within striking distance. Their fangs are long and their venom extremely potent. They reproduce, on average, once every two years and give birth to about eight live young.

The bandy-bandy, *Vermicella annulata* (Figure 8.41), is a strikingly colored black snake with white bands. This burrowing species occurs over most of eastern Australia with disjunct populations in the Northern Territory and Western Australia. They are nocturnal, reach 1 m (3.3 ft), and prey only on blind snakes (Typhlopidae).

Some pretty little striped or banded members of the desert genus *Simoselaps* feed exclusively on reptile eggs. Several specializations have evolved that enable these snakes to cope with such an extreme diet. Most of their teeth are reduced, thus enabling an egg to be swallowed more easily. They have an enlarged tooth on each side of the lower jaw that slits the egg shell as it is forced past the teeth during swallowing. As elapids, these snakes are venomous, but they usually do not attempt to bite and are considered harmless to humans and domestic animals. They are burrowing snakes, less then 60 cm (2 ft) long.

Many of the species mentioned above engage in male combat. Males may entwine and push and shove each other for the right to mate with a female. Such bouts may last from a few seconds to a few hours. Biting usually does not occur. In such wrestling species (taipans, copperheads, black snakes, brown snakes) the males tend to be larger than the females. Large size is of selective value in securing more matings and, therefore, in producing more offspring. In noncombatant species, such as blind snakes, file snakes, death adders and bandy-bandies, the female is the larger sex.

The sex of a snake can usually be determined by examining the tail. Males have a thick tail base that houses the dual sex organs, the

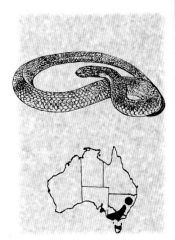

Figure 8.36
Copperhead,
Austrelaps superbus.

Figure 8.37
Mulga snake,
Pseudechis australis.

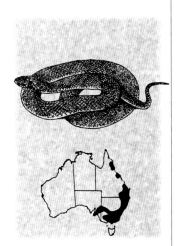

Figure 8.38
Red-bellied black snake,
Pseudechis porphyriacus.

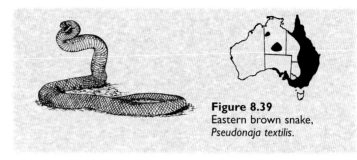

Figure 8.39
Eastern brown snake,
Pseudonaja textilis.

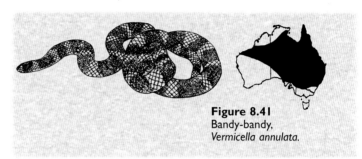

Figure 8.41
Bandy-bandy,
Vermicella annulata.

Top: Mulga, *Pseudechis australis*, feeds on small mammals and reptiles including other snakes. Above: Short fangs of *Pseudechis porphyiacus*, typical of most elapids.

hemipenes, while the tail of the females is more tapered. Only one hemipenes is inserted into a female at mating. The hemipenes are very ornate and studded with spines and hooks that grip the female's vagina. Copulation may last several hours.

Hydrophiidae—The Sea Snakes

The higher level classification of sea snakes is controversial. All herpetologists agree that sea snakes evolved from elapids. While some consider that sea snakes should be placed within the Elapidae, others feel that sea snakes have diverged sufficiently to occupy a family of their own, the Hydrophiidae. Still others classify the sea snakes within two families: the Laticaudidae or sea kraits of Asian elapid origin, and the Hydrophiidae of Australian elapid origin. For our purposes I shall adopt the middle ground and consider the family Hydrophiidae to be composed of two sub-families, the Laticaudinae and the Hydrophiinae. Recent biochemical studies suggest that even this may be excessive splitting and that generic distinction is sufficient.

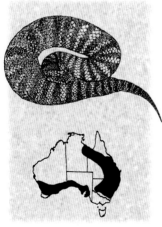

Figure 8.40
Death adder,
Acanthophis antarcticus.

The four species of laticaudids are all in the genus *Laticauda*, which is restricted to the waters of southeastern Asia and the southwest Pacific. Two of the these sea kraits reach the tropical coast of Australia—*L. colubrina* (Figure 8.42) and *L. laticaudata*. These snakes have highly toxic venom but an inoffensive disposition. They usually do not attempt to bite, even when newly caught. The sea kraits are much more dependent upon land than are the other sea snakes. *Laticauda* must come ashore to lay eggs, whereas the Hydrophiinae give birth at sea to live young. *Laticauda* has a flattened tail for propulsion in water, yet it also has enlarged ventral scales like terrestrial snakes. This provides traction on land and facilitates movement. The hydrophiids, which rarely leave the sea, have reduced belly scales. Sea kraits dine exclusively on eels, with the males feeding near shore and the females fishing in deeper waters.

In addition to the two species of *Laticauda*, Australia has 12 genera and about 30 species of Hydrophiinae. The genus *Hydrophis* is the largest with at least 12 species, and *Aipysurus* has about seven Australian species. The hydrophiids are found in the Indian and the Western Pacific Oceans but one very wide-ranging species, the yellow-bellied sea snake, *Pelamis platurus*, occurs from the east coast of Africa to the west coast of the Americas. There are no sea snakes in the Atlantic Ocean, probably because the cold waters of southeastern Africa and southern South America form a barrier to their passage. *Pelamis* ceases to feed at 18°C (66°F). This is a good reason for not building a sea level canal across Central America. Such a vast ditch would almost certainly allow the passage of venomous sea snakes into the Atlantic and Caribbean. *Pelamis* most likely reached the Eastern Pacific less than 3 MYA after the Isthmus of Panama had emerged. Sea snakes occur around the northern coast of Australia, and *P. platurus* more or less surrounds the Australian coast except for the colder south coast from Sydney to Perth.

All members of the Hydrophiidae have compressed tails. The paddle-like tail is an obvious adaptation to aquatic life. Sea snakes also possess vavular nostrils that contain erectile tissue. When this tissue swells with blood the nostrils seal, and water cannot enter. Snug-fitting mouth scales also provide a watertight seal.

Most sea snakes feed on eels and other fishes. Other sea snakes specialize in gobies or catfishes. Fish prey is detected by olfaction and vibration. Prey is bitten and swallowed, usually underwater. Of the Australian species, *Aipysurus eydouxi* and *Emydocephalus annulatus* dine only on caviar. They eat fish eggs foraged from burrows and crevices.

As derivatives of elapids, sea snakes have venom that is highly neurotoxic (Table 8.4, page 176). *Laticauda*, although venomous, is reluctant to bite humans and is usually not considered dangerous. On the other hand, some species of *Hydrophis* and *Aipysurus laevis* may attempt to bite humans with a minimum of provocation. There are even reports of *A. laevis* chasing divers for 100 meters or so before breaking off an attack. Their curiosity can be unnerving. A study at Mystery Reef on the southern end of the Great Barrier Reef estimated that 2000–3000 adult *A. laevis* live in the 1 sq. km reef lagoon.

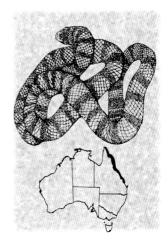

Figure 8.42
Sea krait,
Laticauda colubrina.

Sea krait, *Laticauda* sp. (family Hydrophiidae). Bites by these species can be fatal.

Bob Halstead

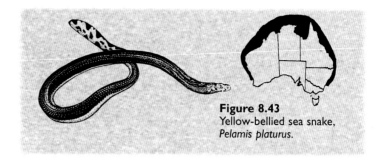

Figure 8.43
Yellow-bellied sea snake,
Pelamis platurus.

Figure 8.44
Sea snake,
Hydrophis elegans.

Enhydrina schistosa is rather ill-tempered and can deliver a large, lethal dose of venom. It has been responsible for human deaths in Malaysia. Sea snake antivenom is based on this species and the tiger snake *Notechis scutatus*. Symptoms as well as first aid and medical treatment for sea snake bite are similar to those for elapid envenomation. There are no authenticated cases of human death by sea snake bite in Australia.

Sea snakes have only a few enemies. Tiger sharks, *Galeocerdo cuvieri*, regularly prey on sea snakes. Saltwater crocodiles, moray eels, groupers, and sweetlips (Haemulidae) also occasionally eat sea snakes, as do white-breasted sea eagles, *Haliaetus leucogaster*. *Pelamis platurus* (Figure 8.43), with its strikingly colored bright yellow belly, black back, and black and yellow tail, is relatively undisturbed by predators. Its warning coloration probably deters potential attackers. Even hungry sharks refuse to eat it under experimental conditions. Experiments showed that naive predatory fish (those from the Atlantic side of Panama that had never been exposed to *Pelamis*) did ingest it, but some died after being bitten.

Sea snakes shed their skin several times more frequently than terrestrial snakes do. This helps remove fouling organisms that grow on their skin, such as algae, barnacles, and byrozoans. Indeed, one species of barnacle, *Platylepas ophiophilus*, is known only from sea snakes. Sloughing the skin every 2–6 weeks helps keep weight and hydrodynamic drag to a minimum.

Adult sea snakes range in size from 0.5 m (*Ephalophis greyi*) to 2 m (*Hydrophis elegans*—Figure 8.44). Most are slender and the sexes tend to be of different sizes in the various species.

The sea is saltier than the body fluids of sea snakes, so we might think that water would move out of the snake by diffusion, resulting in dehydration. However, the skin of sea snakes is impermeable to water, thereby preventing water loss to the environment. The sea snake's osmoregulatory problem is to get rid of extra salt. It does this by means of a sublingual salt gland in the floor of the mouth that excretes a concentrated brine, thus ridding the body of excess salt ingested with prey. Sea snakes get their fresh water from the tissue of their prey which is less salty than sea water. The kidneys of sea snakes cannot excrete a urine that is more concentrated than their body fluids.

Sea snakes dive to a maximum depth of about 100 m (328 ft) and remain submerged usually for about 30 minutes. However, dives lasting two hours have been recorded. They do not utilize anaerobic respiration or reduce their metabolism to achieve these remarkable bottom times. The unusual anatomy of the snake lung is responsible for this ability. The left lung is vestigial and the right lung extends the entire length of the body cavity. It is divided into an anterior tracheal lung, a central bronchial lung, and a posterior saccular lung. The tracheal lung is really a greatly modified windpipe, and is highly vascularized to function in gas exchange as a supplement to the bronchial lung, where the bulk of respiration takes place. The saccular lung is poorly vascularized and functions as an air storage space. In terrestrial snakes this arrangement allows the animal to hold its breath while swallowing very large prey. As a huge meal passes down the throat, the bronchial lung can extract oxygen from the air released from storage in the saccular lung, while the tracheal lung is free to exchange gases with the outside air. Such a system is preadapted to diving with prolonged breath holding. In addition, sea snakes can extract about 20 percent of their oxygen needs through their skin from oxygen dissolved in sea water.

A rather remarkable recent discovery demonstrated that the tail of the olive sea snake, *Aipysurus laevis,* is light-sensitive. This ability probably helps the snake keep its entire body concealed in a coral crevice during the day, and prevents a tempting piece of tail from dangling in the face of a predator.

It is not clear whether sea snakes are subject to the bends. Many species usually swim only briefly at the surface, so there is little time for nitrogen bubbles to form in the blood. However, some species do rest at the surface and seem unaffected by the bends. It may be that excess nitrogen is eliminated cutaneously, or that nitrogen is not picked up in the lungs because the blood is shunted to the muscles. More studies are needed on this issue.

Trial efforts to develop a sea snake leather industry are being made in Australia in an attempt to exploit snakes collected as a by-product of shrimp trawling. The Philippines is a leader in this craft industry.

Budgerigar,
Melopsittacus undulatus.

Aerial Australians: The Birds

....
A BIRD'S EYE VIEW

There are some 9700 bird species in the world. Australia has approximately 740 of them as native species or naturally occurring migrants. The United States has a similar total. About 350 of Australia's bird species (47 percent) are endemic to the Australian zoogeographical realm, which includes New Guinea and New Zealand. The non-endemic species are mostly widespread birds such as waders and oceanic wanderers. The molecular phylogeny produced by Sibley and Ahlquist (1990) has had a major impact on the classification of birds. Table 9.1 lists the bird orders as recognized by Sibley and Ahlquist. Family names and the sequence of their listing in the table largely reflect Sibley and Ahlquist, but with some modification by Christidis and Boles (1994) and Simpson, Day, and Trusler (1996).

Some groups are conspicuously absent from Australia. There are no *native* woodpeckers, hummingbirds, skimmers, sandgrouse, trogons, flamingoes, Old World vultures, bulbuls, hornbills, titmice, true wrens, nuthatches, shrikes, true finches, or buntings—to name just a few of the more widespread groups.

The consensus among ornithologists has been that the Australian avifauna is of Asian origin. However, if this is true, it must date from a very long time ago, as about 90 percent of Australia's land birds are confined to the Australian region. Recent anatomical and biochemical studies have shown that the relationship between certain Australian groups and their presumed Asian counterparts is doubtful, and that the Australian groups are more closely related to one another than to any non-Australian group. One large group, the Corvi (ancestors of crows and others), originated in Australia about 55–60 million years ago and has since spread to the rest of the world. For the time being, it must be said that the origin of Australia's avifauna is obscure.

It is interesting to note that only the southern hemisphere landmasses have large flightless birds. South America has the rheas, Africa the ostrich, Australia/New Guinea the emu and cassowaries, and New Zealand the extinct moas. This may reflect a Gondwanan origin.

Australia has many interesting and characteristic birds, a selection of which is described in this chapter. Except for the flightless birds (emus, cassowaries, and penguins), size measurements are from bill tip to tail tip.

TABLE 9.1 **BIR**

ORDER	FAMILY	COMMON NAME
Struthioniformes	Dromaiidae	Emu
	Casuariidae	Cassowaries
Craciformes	Megapodiidae	Mound-builders
Galliformes	Phasianidae	Quails, pheasants, turkeys
Anseriformes	Anseranatidae	Magpie geese
	Anatidae	Geese, swans, ducks
Turniciformes	Turnicidae	Button quails
Coraciiformes	Alcedinidae	Kingfishers
	Meropidae	Bee-eaters
	Coraciidae	Rollers, dollarbirds
Cuculiformes	Cuculidae	Cuckoos, koels, coucals
Psittaciformes	Cacatuidae	Cockatoos
	Psittacidae	Parrots, lorikeets
Apodiformes	Apodidae	Swifts
Strigiformes	Strigidae	Owls
	Tytonidae	Barn owls
	Podargidae	Frogmouths
	Aegothelidae	Owlet nightjars
	Caprimulgidae	Nightjars
Columbiformes	Columbidae	Pigeons, doves
Gruiformes	Rallidae	Rails and allies
	Otididae	Bustards
	Gruidae	Cranes
Ciconiiformes	Pedionomidae	Plains wanderers
	Jacanidae	Jacanas
	Rostratulidae	Painted snipe
	Burhinidae	Stone curlews
	Haematopodidae	Oystercatchers
	Charadriidae	Plovers, dotterels
	Recurvirostridae	Stilts, avocets
	Scolopacidae	Turnstones, sandpipers, etc.
	Phalaropodidae	Phalaropes
	Glareolidae	Pratincoles
	Stercorariidae	Skuas
	Laridae	Gulls, terns, etc.
	Pandionidae	Ospreys
	Accipitridae	Hawks, eagles, etc.
	Falconidae	Falcons
	Podicipedidae	Grebes
	Phaethontidae	Tropicbirds
	Sulidae	Gannets, boobies
	Anhingidae	Darters
	Phalacrocoracidae	Cormorants
	Ardeidae	Herons, bitterns
	Threskiornithidae	Ibises, spoonbills

AUSTRALIA*

WORLD DISTRIBUTION	TOTAL SPECIES (APPROX.)	AUST. SPECIES** (APPROX.)
Australia	1	1
New Guinea, Australia	3	1
Australasia	12	3
Worldwide	211	3
Australia, New Guinea	1	1
Worldwide	145	20
Eurasia, Melanesia, Australia	14	7
Worldwide	91	11
Africa, Eurasia, Melanesia, Australia	24	1
Africa, Eurasia, Melanesia, Australia	11	1
Worldwide	127	13
Indonesia, Melanesia, Australia	18	13
Pantropical	326	42
Worldwide	84	6
Worldwide	135	5
Worldwide	12	5
India, SE Asia, Australia	13	3
Melanesia, Australia	8	1
Worldwide	75	3
Worldwide	297	23
Worldwide	123	16
Africa, Eurasia, Australia	24	1
Worldwide	16	2
Australia	1	1
Pantropical	8	1
South America, Africa, Asia, Australia	2	1
Worldwide	9	2
Worldwide	7	2
Worldwide	64	16
Worldwide	13	3
Worldwide	85	43
Arctic, southern hemisphere	3	3
Eurasia, Australia	16	2
Oceans worldwide	6	5
Worldwide	88	28
Worldwide	1	1
Worldwide	219	18
Worldwide	61	6
Worldwide	20	3
Tropical oceans	3	2
Oceans worldwide	9	5
Worldwide	4	1
Worldwide	38	5
Worldwide	61	14
Worldwide	32	5

Continued next page

TABLE 9.1 **B I R D**

ORDER	FAMILY	COMMON NAME
Ciconiiformes	Pelecanidae	Pelicans
(cont.)	Ciconiidae	Storks
	Fregatidae	Frigatebirds
	Spheniscidae	Penguins
	Diomedeidae	Albatrosses
	Procellariidae	Fulmars, petrels, etc.
	Hydrobalidae	Storm petrels
Passeriformes	Pittidae	Pittas
	Climacteridae	Australian treecreepers
	Menuridae	Lyrebirds
	Atrichornithidae	Scrub birds
	Ptilinorhynchidae	Bowerbirds
	Maluridae	Fairy wrens
	Meliphagidae	Honeyeaters
	Ephthianuridae	Australian chats
	Pardalotidae	Pardalotes, scrub wrens, thornbills, etc.
	Eopsaltriidae	Australo-Papuan robins
	Pomatostomidae	Australo-Papuan babblers
	Cinclosomatidae	Quail-thrushes, whipbirds
	Neosittidae	Sittellas
	Pachycephalidae	Whistlers, shrike-thrushes
	Dicruridae	Flycatchers, monarchs, magpie larks, dror
	Campephagidae	Cuckoo-shrikes
	Oriolidae	Orioles and figbirds
	Paradisaeidae	Birds of paradise
	Artamidae	Wood swallows, butcherbirds, currawong
	Corcoracidae	Australian mud-nesters
	Corvidae	Crows
	Muscicapidae	Old World thrushes
	Hirundinidae	Swallows, martins
	Sylviidae	Old World warblers
	Alaudidae	Larks
	Nectariniidae	Sunbirds
	Dicaeidae	Flowerpeckers
	Passeridae	Weavers, pipits, grass-finches
	Alaudidae	Larks
	Nectariniidae	Sunbirds
	Dicaeidae	Flowerpeckers
	Passeridae	Weavers, pipits, grass-finches
	Zosteropidae	White-eyes

*The above list is loosely based on Simpson, Day, and Trusler's 1996 interpretation, but some of the classical arrangemen
**Introduced species are excluded.

F AUSTRALIA* (Continued)

WORLD DISTRIBUTION	TOTAL SPECIES (APPROX.)	AUST. SPECIES** (APPROX.)
Oceans worldwide	8	1
Worldwide	17	1
Tropical oceans	5	3
Southern oceans	18	11
Southern oceans	13	10
Oceans worldwide	70	43
Oceans worldwide	20	7
Africa, Asia, Melanesia, Australia	26	3
Australia, New Guinea	6	6
Australia	2	2
Australia	2	2
Australia, New Guinea	18	8
Australia, New Guinea	23	18
Australasia, Micronesia, Bali to Bonin Islands, Polynesia	151	73
Australia	5	5
Australasia	67	49
Australasia	46	20
Australia, New Guinea, Africa, Eurasia	5	4
Australia, New Guinea	15	8
Australia, New Guinea	2	1
Australasia	47	14
Africa, SE Asia, Australia, Melanesia	139	19
Africa, Asia, Melanesia, Australia	72	7
Africa, Eurasia, Australia, Melanesia	25	3
Australia, New Guinea	43	4
India, SE Asia, Australia, Melanesia	20	14
Australia	2	2
Worldwide	112	6
Worldwide	496	3
Worldwide	78	6
North America, Africa, Eurasia, Australia	400	8
Worldwide	82	2
Africa, Asia, Australia, Melanesia	117	1
India, SE Asia, Australia, Melanesia	58	1
Worldwide	294	26
Worldwide	82	2
Africa, Asia, Australia, Melanesia	117	1
India, SE Asia, Australia, Melanesia	58	1
Worldwide	294	26
Africa, Asia, Australia, Melanesia	84	3

...es have been retained in this table.

Figure 9.1
Emu, *Dromaius novaehollandiae*.

Emu, *Dromaius novaehollandiae*, with chicks. Up to 2m tall, this flightless bird runs on its long legs with a bouncy and swaying motion.

NATIVE SPECIES

Emu

The emu, *Dromaius novaehollandiae* (Figure 9.1), is a large flightless bird found throughout Australia, except in heavily forested or populated areas of the east coast and in Tasmania. It may stand 2 m (6.5 ft) tall, has large feet with three toes, and its body is covered with dark, shaggy feathers. In May or June the female lays up to 20 but usually 7–11 dark green eggs that measure 134 x 89 mm (5.3 x 3.5 in) and may weigh 700–900 g (1.5–2 lb). She takes no further part in care of the young. The male incubates the eggs for eight weeks, leaving rarely to eat. He may lose 4–8 kg (8.8–17.6 lb) during this time. After hatching, the striped chicks follow the male around and seek refuge beneath his feathers at night. Emus are very curious birds and can be attracted by waving a piece of cloth or flashing a mirror. Aboriginals have used this curiosity to lure the giant birds within spear range.

Emus feed on insects and a variety of seeds, fruits, flowers, and shoots of herbs and shrubs. They have done considerable damage to wheat crops in Western Australia, and have been hunted by farmers as a result. In 1932 an army detachment with two machine guns tried to exterminate the emus from an area of Western Australia, but fortunately, the 'Emu War' failed. Today many farmers in the east enjoy having a few emus on their property as an attraction. Emus are also farmed in Australia for their leather, oil, meat, eggs, and feathers.

Ken Griffiths

Recently, American farmers have begun raising emus for their lean, low-cholesterol meat. There is even a plan to produce an expensive moisturizer from emu oil.

Cassowary

The cassowary, *Casurius casuarius* (Figure 9.2), is native to the tropical rainforests of northeastern Queensland and New Guinea. Because of habitat encroachment by humans it is now confined in Australia to remote Cape York Peninsula and a narrow, 1100 km (660 mile) strip of rainforest in north Queensland. Cassowary numbers have declined to about 1000. *C. bennetti* occurs only in New Guinea.

The flightless cassowaries are the world's second heaviest bird after the ostrich, weighing up to 75 kg (165 lb) and measuring nearly 2 m (6.5 ft). Both males and females are covered with shiny black hair-like plumage. The neck is blue and a horny casque covers the head. Cassowaries can be aggressive and are armed with a 120 mm (4.7 in) long claw on the inside toe of each foot. They have been known to disembowel dogs.

The female lays four light green eggs measuring 140 x 95 mm (5.5 x 3.7 in) that the male incubates for two months. The male also broods the chicks for about nine months.

They feed on fruit that falls from the rainforest as well as practically any other edible matter, including dead birds. Seeds pass unharmed by their digestive juices to the forest floor, thus ensuring regeneration of plant life.

Mallee Fowl and Brush Turkey

Mallee fowl, *Leipoa ocellate* (Figure 9.3), is found in dry scrub of the semi-arid zone of southern Australia from western New South Wales to the west coast. This mound-builder is one of Australia's most fascinating endemic species. Both sexes are gray and brown, measuring up to 600 mm (2 ft).

Ken Griffiths

Top: Head of male emu showing blue skin on the neck, which is usually darker in females. Males tend to have darker feathers around the head during the breeding season.
Above: Powerful three-toed feet of an emu.
Left: Emu eggs (center) are large, as can be seen when compared with a jumbo hen egg (right) and an ostrich egg (left).

Figure 9.2
Cassowary,
Casuarius casuarius.

Figure 9.3
Mallee fowl,
Leipoa ocellate.

Mallee fowl mate for life, but are seldom together. (Perhaps that is the secret of a successful marriage!) The male excavates a depression in the ground about 1 m (3.3 ft) deep and 3–4 m (9.8–13 ft) in diameter. He fills the hole with leaves from the immediate vicinity, then digs an egg chamber about 0.5 m wide and 0.5 m deep in the top of the leaf pile. After rain has wet the leaf litter, the male covers the mound with sand to as high as 1.5 m (4.9 ft) and some 5 m (16 ft) in diameter. The female lays her eggs, one at a time, in the egg chamber that the male opens for each egg. She usually lays 15–24 eggs but can produce as many as 33. The interval between each episode of egg-laying is related to rainfall. It can be two days in wet weather or up to 17 days in dry weather. Incubation takes about 49 days but this varies with the temperature of the mound, which is determined by the heat of decomposition of the vegetative litter. The male regulates the temperature at about 33°C (91°F) by removing the covering to let heat escape or to allow the sun to warm the chamber, or by adding to the covering to insulate the nest. When a chick hatches it must climb through 1 m (3.3 ft) of leaves and sand. Within a few hours the chick can run, and it can fly within 24 hours. The parents do not brood or feed their precocious young.

These extraordinary birds are losing their habitat due to sheep grazing that consumes most of the acacias, the seeds of which comprise an important part of mallee fowl's diet. They are also vulnerable to introduced foxes that dig up their eggs.

The brush turkey, *Alectura lathami,* is another species of mound-builder or megapod, so called because of its huge feet. It occurs along the eastern edge of Australia from Cape York to the middle of New South Wales.

Recently it has been the subject of studies by Roger Seymour of the University of Adelaide. His work has shed light on the remarkable adaptations of the eggs and hatchlings. In winter, the male brush turkey kicks leaves, twigs, moss, and other litter backwards into a heap about 1 meter (3.3 ft) high and 5 m (16 ft) in diameter. This pile may weigh up to 6.8 metric tonnes and be 12 cubic meters in volume. The male repeatedly digs holes into this mound and mixes in new material to create over two weeks a large mound that consists of fine litter covered with twigs and sticks. Fungi are the chief decomposers in this compost heap, the heat that is released incubates the eggs. The temperature some 60 cm (2 ft) into the nest eventually stabilizes at about 33°C (91°F) and remains within a degree or two of that level throughout the breeding season, because of the large thermal inertia of the nest. The female lays an egg in a trench dug into the mound every few days for five to seven months. As many as 16 eggs, each at a different stage of development, have been found in a single mound.

Like the mallee fowl, the brush turkey probes the mound with its 2.5 cm (1 in) beak to sense the temperature. If the temperature is too low, fresh litter is added. This increases nutrients to the decomposers, which results in higher heat production. Moisture content is also controlled by adjusting the mound's shape. Humidity is kept near

99 percent. A depression in the top can serve to collect rain, or the top can be rounded to shed water. The actual water content must be kept low enough for air to circulate into the spaces surrounding the eggs, thereby enabling the developing embryo to absorb oxygen through the shell and give off carbon dioxide. Diffusion of gases through the egg shell is facilitated by a thin shell—about half as thick as that of a regular bird egg, relative to size. The shell becomes even thinner during development, making it easier for the chick to punch through the egg during hatching. The increased permeability of the shell to gases offsets the relatively low oxygen levels and high carbon dioxide levels around the egg. The structure of the mound protects the thin-shelled egg from being crushed. Incubation lasts about 49 days, which is almost two and a half times the normal 20 days of other birds of the same size.

As water evaporates from the egg during incubation, an air space is created under the blunt end of the egg shell. In most bird eggs, the chick forces its beak through the egg membranes into this air pocket and begins to breathe air into its lungs. A few hours later, the chick cracks the shell and breathes outside air. Brush turkeys, however, do it differently. There is no air pocket in a brush turkey egg, and the chick does not breathe air until it is hatched. Instead of the brush turkey using its vestigial egg tooth to break the shell, the chick uses its large feet to force its back against the shell, which, because it is so thin, cracks easily. The hatchling lies on its back and scratches the roof of the cavity surrounding it, and the litter that falls on it is compressed by its body. The chick takes about 2.5 days to emerge from the mound. Brush turkeys hatch with full primary feathers and some actually fly on the first day out of the mound. No parental care is given to these super precocious chicks.

Given that the chicks of mound-building birds have a vestigial egg tooth and hatching muscles on the back of the neck, mound building, with its attendant thin eggs, is considered a derived activity or character. Normal eggs and parental brooding are considered ancestral, that is, the megapods evolved from a group of birds that produced thick egg shells and sat on their eggs. Mound building freed the parents from incubation and allowed them to produce large numbers of young that could not possibly be incubated in the usual way. This high fecundity is of great selective advantage to these birds, but it is often matched by high first-year mortality.

Black Swan

The stately black swan, *Cygnus atratus* (Figure 9.4), is the emblem of Western Australia and the trademark of a Perth brewer. The 1.2 m (4 ft) body of the adult is black with white wing tips and a red bill. Black swans occur on lakes and estuaries throughout the southern third of the continent and extend up the east coast. Older birds usually form permanent pairs. They feed on aquatic plants and nearby pastures.

Figure 9.4
Black swan,
Cygnus atratus.

Below: Cape Barren goose,
Cereopsis novaehollandiae, one of
the world's rarest water birds.
Right: Black swan,
Cygnus atratus.

Figure 9.5
Cape Barren goose,
Cereopsis novaehollandiae.

Cape Barren Goose

The Cape Barren goose, *Cereopsis novaehollandiae* (Figure 9.5), is endemic to Australia. It is one of the world's rarest water birds and is in a subfamily by itself. It nests on islands off the south coast of Australia. It is a large bird, measuring up to 1 m (3.3 ft) when adult. It is gray with a white crown and yellow and black bill. Cape Barren geese probably mate for life, are extremely aggressive, and will attack anything that threatens their eggs or young. Their numbers have suffered in the past at the hands of sealers, who ate the birds and their eggs, and farmers who hunted them as agricultural pests. The birds are now protected by law.

Shelduck

The shelduck, *Tadorna tadornoides,* is a beautiful chestnut and black duck found in large flocks of 1000 or more birds in Western and southeastern Australia. Shelducks probably mate for life, and mated pairs often return to the same nesting site each year. Like many monogamous species, the sexes are similar. The female may be distinguished by the white ring around the eye. While the female is incubating the 5–14 eggs in a hollow tree limb 2–25 m (6.5–82 ft) off the ground, the male may be 2 km (1.6 miles) or more away, establishing a territory. After an incubation period of 30–35 days the

Sheld or mountain duck,
Tadorna tadornoides. Males and
females of the species are
similar in appearance, but the
female has a white eye ring.

young and the female are led overland by the male to his territory where the chicks are reared. On land they graze on grasses, and in the water they forage on aquatic plants. They may also eat grain crops, which sometimes earns them the wrath of farmers.

Kookaburra

The kookaburra, *Dacelo novaeguineae* (Figure 9.6), is the largest member of the kingfisher family. This archetypical Australian lives in family groups in woodlands and open forests along the eastern third of Australia and in the southwest corner of the continent, where it was introduced in 1897. It was introduced to Tasmania in 1905. Kookaburras have a large beak, up to 65 mm (2.6 in) long, and feed on insects, snakes, lizards, rodents, and small birds. Their rolling laugh of *koo-koo-hoo-hoo-hoo-haa-haa-haa-haa* is a delightful alarm clock that goes off in chorus each morning at dawn. (An old common name for the kookaburra was 'bushman's clock'.)

The social system of kookaburras is unusual. They form permanent pairs, and the female lays two or three eggs in early spring in a hole in a tree trunk. The young hatch in 24 days, are fledged in 36, and are then fed by the parents for up to 13 weeks. Instead of leaving the parents' territory, the young remain with the parents for up to four years, and help defend the boundaries and care for their parents' new offspring. They spend about one-third of the time required incubating their parents' eggs and brooding the young, and they provide up to 60 percent of the nestlings' food. This long juvenile period reduces breeding potential, but this does not pose a problem as kookaburras live up to 20 years. Their social system has survival advantage for all members of the group.

Kookaburra, *Dacelo novaeguineae*, chick eating a yellow-naped snake, *Furina barnadi*.

Figure 9.6
Kookaburra,
Dacelo novaeguineae.

Kookaburra, *Dacelo novaeguineae*. Its large beak can be up to 65mm long.

Ken Griffiths

Glen Threlfo, AUSCAPE International

Figure 9.7
Palm cockatoo,
Probosciger aterrimus.

Figure 9.8
Red-tailed black cockatoo,
Calyptorhynchus banksii.

Figure 9.9
Gang-gang cockatoo,
Callocephalon fimbriatum.

Cockatoos and Parrots

Australia is well known for its varied assortment of colorful birds belonging to the order Psittaciformes. About one-sixth of all the world's parrots are endemic to Australia. The palm cockatoo, *Probosciger aterrimus* (Figure 9.7), of Cape York Peninsula is Australia's largest parrot, reaching a length of 600 mm (2 ft). The red-tailed black cockatoo, *Calyptorhynchus banksii* (Figure 9.8), is similar to the palm cockatoo but has a broad red band on the tail. It is much more common, ranging throughout Australia except the center and the southeast corner. The red-tailed black cockatoo was the first parrot from Australia to be illustrated by Joseph Bank's artist, Sydney Parkinson, in 1770.

The gang-gang cockatoo, *Callocephalon fimbriatum* (Figure 9.9), inhabits the mountainous areas of the southeast corner of Australia. It shows little fear of humans and can readily be approached while feeding. The males have a gray body with a bright red head, and the females are gray. The gang-gang cockatoo can frequently be seen in the suburbs of Canberra.

The long-billed corella, *Cacatua tenuirostris*, uses its beak to excavate bulbs and roots. It nests in the hollows of trees, and is found in woodlands and grasslands of southeastern South Australia, Victoria, and Western Australia. Its range is expanding as more land is cleared for agriculture.

The galah, *Cacatua roseicapilla* (Figure 9.10), is one of the most characteristic features of the Australian landscape. This gorgeous bird with a pink body, gray wings, and a white crown occurs in large flocks throughout Australia. This bird is so common that Australians take it for granted, and even use the word 'galah' to mean a stupid person—possibly because of the noise, wing flapping, crest raising, and playful aerial acrobatics the bird performs at sunset. Like most

Left: Gang-gang cockatoo, *Callocephalon fimbriatum.*
Right: Long-billed corella, *Cacatua tenuirostris.*

Ken Griffiths

parrots, it is a seed-eater. The establishment of stock-watering facilities has allowed this species to spread into arid scrub where wattle seed is available.

Galahs usually breed in spring. The females have pink-red eyes and the males (and immatures) have dark brown eyes. Every 2–3 days an oval white egg is laid in the hollow of a tree until a clutch of 2–7 eggs has been deposited. Males and females take turns incubating the eggs for 22–26 days. Since the eggs are laid on different days, hatching is asynchronous. Fledglings are fed by the parents for about two months, with most pairs fledging two young per year. Mortality is high among nestlings and young birds, especially in the winter when grain shortages occur. Galahs form permanent pair bonds and live a long life in large flocks of hundreds of birds. They spend a large portion of each day sitting in trees digesting their grain meal, or sitting on television and radio antennae and power lines, which they can damage by their collective weight. Galahs have been hunted by farmers because of the damage they cause to their grain crops, and they are still trapped for the illegal overseas bird trade. Nevertheless, they thrive.

The pink, or Major Mitchell, cockatoo, *Cacatua leadbeateri* (Figure 9.11), is another beautiful bird. It has white wings and a deep pink crest. It is found in the arid and semi-arid interior.

The sulphur-crested cockatoo, *Cacatua galerita* (Figure 9.12), occurs by the thousands in northern and eastern Australia. This all-white bird with yellow crest is well known to bird enthusiasts around the world. It frequently flocks with galahs, and can make an enormous racket while roosting for the evening.

The eclectus parrot, *Eclectus roratus* (Figure 9.13), is widespread in New Guinea but it only occurs in the patch of rainforest on eastern Cape York. The bird is remarkable for its striking sexual dimorphism.

Left: Rainbow lorikeet, *Trichoglossus haematodus.*
Right: Galah, *Cacatua roseicapilla.*

Figure 9.10
Galah,
Cacatua roseicapilla.

Figure 9.11
Major Mitchell cockatoo,
Cacatua leadbeateri.

Figure 9.12
Sulphur-crested cockatoo,
Cacatua galerita.

Ken Griffiths

The males are bright green with blue and red on the wings, and the females are red with blue on the chest and wings.

The seven species of lorikeet are small, brightly colored members of the parrot family. The rainbow lorikeet, *Trichoglossus haematodus* (Figure 9.14), lives up to its common name, having a violet head, yellow-green neck, green dorsum, yellow-orange breast, and blue belly. It is widespread in forested areas of eastern Australia and New Guinea, where it visits the flowers of trees and Proteaceae, extracting nectar with its brush-tipped tongue and pollinating the flowers.

The king parrot, *Alisterus scapularis* (Figure 9.15), of eastern Australia, has green wings and back, a red head and ventral surface, and a blue tail. King parrots sometimes gather in large numbers to feed on fallen seeds and fruits. They also raid orchards and do significant damage to corn crops. The red-winged parrot, *Aprosmictus erythropterus,* is especially fond of eucalypt and acacia seeds. It inhabits open woodlands and grasslands of northern and northeastern Australia. The males of this species are bright green with red wing coverts.

Figure 9.13
Eclectus parrot,
Eclectus roratus.

Top: The all-white sulphur-crested cockatoos, *Cacatua galerita.*
Below: A male (left) and female (right) king parrots, *Alisterus scapularis.*

Figure 9.14
Rainbow lorikeet, *Trichoglossus haematodus.*

Ken Griffiths

Ken Griffiths

Figure 9.15
King parrot,
Alisterus scapularis.

The princes parrot, *Polytelis alexandrae,* is a delicate blend of pink, yellow, olive, and blue. This beautiful parrot is found in open arid regions, acacia scrubland, and along watercourses in dry areas. This rare and highly nomadic species travels in small flocks of 15–20 birds. It feeds on seeds of grasses, especially spinifex, and herbaceous plants, as well as acacia flowers and mistletoe berries.

The cockatiel, *Nymphicus hollandicus,* is a crested, gray bird with a yellow head, orange ear patch, and white wing patch. It occurs in large numbers in the dry interior of the continent where it feeds on the seeds of grasses, including grains, and on shrubs and trees, especially acacia. It nests in hollow limbs of trees, and it is a noisy, fast flier. It is an extremely popular aviary bird, second only to budgerigars. Hand-reared cockatiels usually become excellent pets and show affection to their caretaker. They can learn to whistle or even mimic a limited vocabulary.

Probably the best known parrot in the world is the budgerigar or parakeet, *Melopsittacus undulatus* (Figure 9.16). It is kept as a cage bird

Figure 9.16
Budgerigar,
Melopsittacus undulatus.

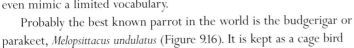

Top: Eastern rosella, *Platycerus eximius,* presents a brilliance of color as it flits from trees to the ground.
Below: Major Mitchell's cockatoo, *Cacatua leadbeateri,* showing off its crest with bright yellow and red bands.

Ken Griffiths

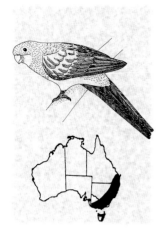

Figure 9.17
Eastern rosella,
Platycerus eximius.

around the world. In Australia, it lives in huge flocks that can darken the sky as they pass. This small green bird, barred with black and yellow on the back and upper surface of the wings, is highly nomadic and widely distributed throughout Australia, especially the interior.

One of the best known Australian parrots is the multicolored eastern rosella, *Platycerus eximius* (Figure 9.17), found in the southeastern corner of the continent. This bird occurs in lightly forested country and presents a whirl of brilliant colors as it flits from trees to the ground where it forages for seeds. The crimson rosella, *P. elegans,* can be stunningly crimson and cobalt-blue in the southeastern part of its range.

Unfortunately, Australia's wonderfully diverse parrot fauna is the object of illegal trade in Australian wildlife. Many birds die while being smuggled, which further inflates the prices. Customs officials in Australia and the United States have carried out elaborate 'sting' operations in their attempts to crack down on this crime, but the problem still exists today, and now some smugglers have turned their attention to eggs rather than adult birds.

Fork-tailed Swift

Fork-tailed swift, *Apus pacificus,* nests in northern Asia, but in the non-breeding season, if air currents are suitable, it may enter Australian airspace in huge numbers. However, it literally never sets foot in Australia. It drinks on the wing by scooping up water with its beak, and, apparently, it even sleeps while flying at high altitudes. To sustain its energy, it engages in aerial refuelling by consuming great quantities of flying insects.

Tawny Frogmouth

The tawny frogmouth, *Podargus strigoides* (Figure 9.18), is common in open woodlands throughout Australia. The softly mottled gray bird resembles tree bark so perfectly that it is very difficult to locate. It forages at dusk for insects, snails, and small vertebrates, and swoops silently from its vantage point in a nearby tree to its moving prey on the ground below. It has a huge gape and large eyes.

Crested Pigeon

The crested pigeon, *Ocyphaps lophotes,* is common in lightly wooded grasslands throughout Australia except for the far north, southwest Western Australia, and southeastern Australia. The range is actually expanding due to the clearing of closed woodlands for pasture. It is a swift flier and feeds mostly on seeds. The crested pigeon is brownish gray with an upright crest, and it has an iridescent bronze patch on the wing feathers. The males use this colorful feature in their courtship display by holding the wings open and twisting them forward to show off the color.

Figure 9.18
Tawny frogmouth,
Podargus strigoides.

Far left: Crested pigeon, *Ocyphaps lophotes,* has an iridescent bronze patch on the wing feathers, which the males display attractively during courtship.
Left: Tawny frogmouth, *Podarqus strigoides,* camouflages itself extremely well against tree bark.

Ken Griffiths

Wedge-tailed Eagle

The wedge-tailed eagle, *Aquila audax* (Figure 9.19), is a majestic bird. It is the largest bird of prey in Australia, measuring up to 1 m (3.3 ft) in length with a 2.5 m (8.2 ft) wingspread. It is found throughout the continent. Carrion forms a significant part of its diet. Other food items include the introduced rabbit, and kangaroos, wallabies, and reptiles. The unsupported belief that wedge-tailed eagles kill lambs has led to its destruction by graziers, but recent studies show that eagles have a negligible effect on the sheep industry, taking only sick or dead lambs. The eagle is now protected.

Figure 9.19
Wedge-tailed eagle, *Aquila audax.*

Wedge-tailed eagle, *Aquila audax,* is the largest bird of prey in Australia, with a wing span of 2.5m.

Sacred Ibis

The white or sacred ibis, *Threskiornis aethiopica,* has a black, naked neck and head, and black plumes near the tail on an otherwise white body. The bill is long, recurved, and black. The sexes are similar in appearance. This large bird is common over northern and eastern Australia. It can be found roaming around in pastures and parklands, even at the Royal Botanic Gardens near the Sydney Opera House. In Australia, if you look up and see a V-formation of birds with outstretched necks, it is probably a flock of sacred ibises. The birds flap their wings simultaneously, then glide together.

Sacred ibises are colonial nesters and build large, cup-shaped stick nests in trees or bushes over water. They lay 2–4 eggs, and both parents take turns at incubating them. The changing of the guard is accompanied by deep bows between parents. The young leave the nest at about three weeks but remain nearby to be fed by the parents for several more weeks.

Ibises feed mainly in the water, which probably accounts for their naked head and neck. They eat crustaceans, fish, snails, frogs, and snakes. A blow from their large beak can open a mussel shell held under their foot on a rock surface.

Australian Pelican

The Australian pelican, *Pelecanus conspicillatus* (Figure 9.20), occurs all over Australia, wherever there are large bodies of water such as lakes, rivers, swamps, estuaries, or lagoons. It can be found in large numbers, on both fresh and salt water and as far away as Indonesia and New Zealand. It frequently soars on air currents high above the water. The Australian pelican is graceful when in the air but clumsy, almost humorously so, on take-off. It can measure up to 1.8 m (5.9 ft) from tip of bill to end of tail, with a bill up to 455 mm (18 in). The body is covered with white plumage and the wing edges, shoulders, and rump are black. During courtship the front two-thirds of the pink, fleshy pouch becomes bright red. The female lays 2–4 white eggs at any time of the year when there is sufficient water. Incubation lasts 32–35 days, and both male and female take turns on the nest. After feeding, the chicks seem to go berserk, displaying violent convulsions which last about one minute before they collapse on the ground.

Figure 9.20
Australian pelican, *Pelecanus conspicillatus.*

Little
Penguin

The little penguin, *Eudyptula minor* (Figure 9.21, page 207), more frequently called the fairy penguin in Australia, is the smallest of the 18 living species of penguins (family Spheniscidae) and the only one to breed in Australia. It occurs along the southern coast of Australia from around Port Stephens in New South Wales to Fremantle in Western Australia. It can be easily observed at Phillip Island, southeast of Melbourne, where each year over 300 000 tourists watch the nightly penguin parade. The fairy penguin is also found around the New Zealand coast.

Fairy or little penguin, *Eudyptula minor*, is the only penguin species to breed in Australia and the only nocturnal penguin species on land.

This small bird stands only about 330 mm (13 in) tall and weighs 1 kg (2.2 lb). It has beautiful dark blue plumage on the dorsal surface and a white ventral surface. It lives in highly vocal colonies that may number in the thousands, usually associated with sand dunes along the coast and on nearby islands. This delightful little bird pops out of the water at dusk (it is the only nocturnal penguin species on land) and parades in groups up the sand dunes to its plant-lined burrows. Its tunnels are about 60–80 cm (24–32 in) long, tall enough for the bird to stand up in, and end in a nest chamber. Burrows are usually spaced at least 2 m (6.5 ft) apart.

The fairy penguin's short wings function much like paddles. It is propelled by powerful pectoral muscles attached to the large keel on the sternum, and the penguin actually 'flies' under water. The back muscles (supracoracoideus) are also well developed to pull the flippers against water resistance on the up stroke. Unlike other birds, which have hollow bones, penguin bones are solid. Nevertheless, the penguin is positively buoyant due to the air trapped between its feathers. When it stops swimming during a dive, it bobs back to the surface. Its feathers are waterproofed with a waxy secretion from an oil gland located above the base of the tail. The waxes are spread by the beak during grooming.

Males and females are practically indistinguishable, but studies have shown that males have a stouter beak with a more pronounced hook than females. The cornea of the penguin eye is flattened, allowing sharp vision in both air and water. The very dense plumage with its trapped air insulates the penguin's body from cold seas down to about 12°C (52°F). A human would die within a few hours if immersed in such waters, but the penguin can spend weeks at sea. However, below 12°C temperatures, the fairy penguin cannot maintain its body heat, which explains why it is not found in subantarctic waters. Likewise, it does poorly above 35°C (95°F), but the penguin rarely encounter such temperatures as it is nocturnal. During the day, the temperatures in the burrows are cooler than those on the surface.

The penguin extracts oxygen from its lungs with great efficiency, which allows it to resupply oxygen quickly between dives. It makes short, relatively shallow dives, 10–30 m (32–98 ft) deep for an average duration of 23 seconds, and apparently does not get the bends. Like all birds, it lacks a diaphragm and its lungs are inflexible. Structurally, however, its lungs are similar to those of the emu and are unlike the lungs of other birds. Why this is so is uncertain.

It is a noisy bird. Perhaps this is related to nocturnal activity, where vocal behavior is important. Mating birds may sing duets during mutual displays as well as sounding off for territorial defence. If a vocal threat fails, an intruder will be nipped with the sharp-edged beak and thrashed with the flippers. There are also calls that signify nest relief and announce arrival from the sea.

Fairy penguins have only one mate during a given breeding season, but about 25 percent change partners each year due to death or 'divorce'. Less than 10 percent keep the same mate for five years. The breeding season reflects the seasonal abundance of food items. Spring and summer are times of high marine productivity due to changes in ocean currents. This triggers breeding. In the mating act, a male grips the female's neck in his beak, stands on her lower back, and brings his cloaca against hers, thus transmitting the sperm. Two white eggs of similar weight and volume are laid in spring 1–4 days apart. Each parent has a brood patch, a vertical area of highly vascularized skin that develops on the abdomen between feather tracts at each breeding season. The heat of the blood supply transfers warmth to the eggs arranged vertically along the gap in the feathers as the penguins lie on the eggs. Both males and females take turns incubating the eggs in 1–2 day shifts over a 36 day incubation period. The off-duty parent feeds in the sea and then relieves its mate the next day.

The embryo has an egg tooth on the beak with which it chips its way out of the egg over a 1–3 day period. The first laid egg is the first to hatch. Hatching success is about 60 percent. In a good year, both chicks may survive, but in poor years the chick that is hatched second may not survive, because the parents feed the stronger chick first. The number two chick is fed only when the number one chick is satisfied —and then only if there is any food left in the adult's stomach. This brood reduction strategy is an adaptation to unpredictable food resources and ensures that at least one chick has a good chance of survival. If the first clutch of eggs is destroyed, a second clutch may be laid. Death of one parent usually means failure of the clutch because a single parent cannot adequately feed the growing chick.

The young are covered with down. Their eyes, closed at hatching, partially open around the second or third day, and are completely opened by the end of the first week. The helpless chicks are brooded for 2–3 weeks. After this period the chicks require so much food that both parents must fish daily and return to the nest to regurgitate fish into the chicks, who insert their open beaks into the parent's mouth. At four weeks the feathers of the adult plumage emerge. Parents continue to feed the chick until about the eighth week. After that the young leave the land and disperse at sea. The mortality rate for the newly fledged fairy penguin is about 20 percent.

Figure 9.21
Little penguin, *Eudyptula minor*.

After the breeding season, adults fatten at sea for about six weeks, then molt, and they cannot go back to the sea to feed until their new waterproof plumage appears. During this two week starvation period their weight drops about 50 percent, from a fat 2 kg to a normal 1 kg (4.4 lb to 2.2 lb). Stored fat is metabolized during this period. The birds emerge with a new, very dense multiple layer of waterproof plumage. They then return to sea to feed within about 10 km (6 miles) of the shore. Around June, the males, followed shortly by the females, return to land to begin the cycle again.

While swimming, the fairy penguin can reach speeds of up to 6 km/h. In Victorian waters its diet consists of 76 percent fish (anchovies and pilchard) and 24 percent squid. In some places and during certain seasons it will also eat krill. Its average life expectancy is about 6.5 years, and the average annual mortality rate is 14 percent. Some banded fairy penguins are known to have lived to 21 years.

In the past, coastal Aboriginals ate penguins and their eggs. Early European seal hunters actually used dried penguins as 'fire wood' (the corpses burned well because of their oils) for the purpose of boiling the oil from seal blubber. However, the fairy penguin was spared systematic exploitation, probably because of its relatively low density, small size, and burrowing lifestyle. Fleas and ticks are common external parasites and researchers who stick their arms into a burrow to tag penguins may emerge covered with fleas, not to mention being bitten and vomited upon with a semi-digested fish meal.

Natural predators include sharks and seals at sea, while lizards and snakes on land may take eggs and chicks. Feral dogs and cats can also do much damage to a breeding population. The biggest threat to the penguin colonies, however, is human activity. Automobiles and habitat destruction have reduced penguin numbers dramatically over the years. Today there are about 20 000 fairy penguins on Phillip Island; in 1918 the number was estimated at 200 000.

Figure 9.22
Superb lyrebird,
Menura novaehollandiae.

Superb Lyrebird

Lyrebirds, so called because the tail feathers of the male resemble the musical instrument (lyre), are endemic to the forested southeastern coast of Australia between Brisbane and Melbourne, on both sides of the Great Dividing Range. The superb lyrebird, *Menura novaehollandiae* (Figure 9.22), occupies most of this range. Another species, Albert's lyrebird, *Menura alberti,* occurs at the northern end near the New South Wales-Queensland border. The evolutionary relationships of lyrebirds have been a point of contention since their discovery, but recent DNA-DNA hybridization studies suggest they are related to the scrub birds of the family Atrichornithidae and to bowerbirds. The following information is about the more common species, the superb lyrebird .

Lyrebirds are rather secretive, and their forested habitat makes them difficult to observe. However, they are often heard. Males sing to announce their territories to other males and to advertise their availability to females. Songbirds (passerines) form their songs in the syrinx, where the lower tracheal rings fuse to form a resonating drum at the junction of the trachea and bronchi. Lyrebirds lack complete fusion of these rings and have only three pairs of intrinsic muscles, whereas most songbirds have four or more pairs. However, lyrebirds are excellent mimics and can copy the songs of many other birds. They often accurately imitate other large species with loud calls, such as black cockatoos, rosellas, kookaburras and currawongs. In fact, lyrebirds have been shown to imitate several bird calls simultaneously by producing separate sounds from each bronchus. Many Australians say they have heard lyrebirds mimic the sounds of chainsaws, trains, and other mechanical noises, but this is more 'bush lore' than fact. However, there is good evidence of a lyrebird copying tunes from a flute while it was in captivity and incorporating the tunes into its song, which were subsequently learned by other generations of lyrebirds after the bird's release from captivity. The tunes were evident in the population 50 years later.

The superb lyrebird is roughly the size of a domestic fowl. Males weigh up to 1.5 kg (3.3 lb) and females 900 g (2 lb). The only larger songbirds are some of the New Guinea birds of paradise. The lyrebirds' plumage, which is rather drab, dark brown above and gray below, helps them blend into the dappled light of the forest understorey. They are weak fliers with short, rounded wings. Lyrebirds use their powerful legs for running and leaping as well as for scratching earthworms, crustaceans, beetles, spiders, and centipedes from the soil. These items make up the bulk of their diet.

The tail display of the male lyrebird is spectacular. The long tail is raised and brought forward over the head like an open fan, and then vibrated. The lateral tail feathers, called lyrates, are spotted with silver, chestnut, and black and are elegantly S-shaped. The male's tail may be 50–70 cm (20–28 in) long. Females also have a long tail that

Ken Griffiths

they can raise, but it is less showy and is not used for courtship or territorial defence. Males attain mature tail plumage between six and eight years of age.

Males occupy territories that vary in size from one to 25 ha (2.5–87 acres), depending on the richness of the habitat and the population density. Male territories may overlap as many as six female territories. The males scratch up mounds of vegetation nearly a meter in diameter and 10 cm (25 in) high. They use these platforms to announce their presence by song and tail display. They also sing from perches in trees near the mounds. The male entices a female to the mound where copulation takes place. A single egg is laid in winter (June–August) in a nest constructed of sticks on the ground, by the female during April and May, usually on a bank or at the base of a tree. The egg is the largest of any passerine bird, weighing about 62 g (2.2 oz). It is roughly the size and shape of a chicken egg. Incubation

Male lyrebird, *Menura novaehollandiae*, displaying its magnificent tail during courtship.

Figure 9.23
Satin bowerbird,
Ptilinorhynchus violaceus.

Male satin bowerbird,
Ptilonorhynchus violaceus,
in his bower.

Ken Griffiths

takes a relatively long time (6–7 weeks) because the female does not sit continuously on the egg. This causes the embryo to cool and delays its development. A lyrebird egg incubated continuously by a chicken can hatch in 28 days. Males play no part in nest building, incubation, or care of the young. Females vigorously defend their nest and the chick with alarm calls and, if necessary, their powerful feet. The egg is not usually replaced if it is lost. The breeding success rate averages only about 0.1–0.2 fledglings per female. With a life span of about 20 years and a maturation age of about six years, the female typically produces about 1.4–2.8 offspring during her life.

Feral cats and dogs, as well as kookaburras, ravens, and goshawks are all possible nest predators. Male lyrebirds were slaughtered for their tail feathers well into the 20th century, but now the species is protected.

Bowerbirds

The bowerbirds and their relatives, the catbirds and riflebirds, traditionally are placed in the same family, Paradisaeidae, as those most magnificent of all birds, the birds of paradise of New Guinea. However, recent DNA studies suggest that bowerbirds are not so closely related to birds of paradise, but are closely allied to the lyrebirds. Based on the DNA data, bowerbirds are placed in separate family, Ptilinorhynchidae. There are eight species of this family in Australia.

The satin bowerbird, *Ptilonorhynchus violaceus* (Figure 9.23), occurs in forests of the east coast. The males are black with a purple sheen and have luminescent sapphire-blue eyes. The females are greenish gray.

The male constructs an elaborate bower, which is not a nest but rather a display. He erects 20–30 cm (7.9–11.8 in) high, parallel walls of sticks on a circular mat of twigs. The walls are placed in a north-south direction, and 'painted' by the with a mixture of saliva and charcoal dust, which he dabs on with his bill. The bird has an unusual penchant for the color blue and will decorate the 'stage' at the end of the bower with blue objects found far and wide. Items might include snakeskin, cicada cases, feathers, berries, flowers, and a host of objects such as scraps of blue ribbon, glass, paper, and cloth. When a female is attracted to the bower, the male struts and offers the female various accumulated blue objects. Mating takes place in the bower. The male mates with several females.

Superb Fairy Wren

The Australian wrens (Maluridae), also called fairy wrens, are small birds with iridescent blue plumage on most males, black bills, and very long, stiff tails that they carry at a characteristic angle to the body. This family is endemic to Australia and New Guinea. One of its members, the splendid wren, *Malurus splendens* (Figure 9.24), is among the most beautiful tiny birds (140 mm; 5.5 in) of any continent. It is found in the southern half of Australia.

The closely related superb fairy wren, *M. cyaneus,* is also a tiny (9 g) and colorful species, occurring in southeastern Australia. Its life history has been studied in some detail. The males have a blue-black throat and chest with grey-white underparts. The head cap, cheeks, and back of the neck are bright blue. The females (and the males when in eclipse plumage) are brown. Superb fairy wrens mate for life, but they are classic philanderers. After doing his duty at the home nest, the male visits every female's nest in the vicinity. He frequently comes courting with an offering of flowers in his beak. Every so often he gets lucky and mates with a female other than his own.

These wrens are highly social. In most bird species, the young leave the nest area before becoming adults, but the young males of superb fairy wrens remain in their parents' territory for several years and help feed and guard their parents' new chicks. The juvenile females disperse. The species is not migratory.

The breeding season is about six months long, from September to March. The adult female builds her dome-shaped nest from dry grasses in dense vegetation without help from a male. The nest may be lined with feathers. Three or four eggs are laid, and the female incubates them for two weeks. In another two weeks the young are fledged. Within a few days the female builds another nest, leaving her mate and 'sons' to care for the fledglings. About half of all nests are destroyed by predators such as the pied currawong *Strepera graculina,* feral cats, foxes, rodents, and snakes. As many as seven nests may be built and 20 or more eggs produced in a single season.

While the female is busy building another nest, her mate visits neighboring females and makes a displays. A given female may receive the attention of seven or more males. By repeated visits, with colorful flower petals in his beak and with his superb coloration, a male may eventually win a copulation, but this is under the control of the female.

At the beginning of the mating season, males have huge testes, at least twice as large as those of other similar-sized birds. This is probably an adaptation that permits frequent copulations and the transfer of a large volume of sperm. After the breeding season the testes decrease to 1/300 or 1/400 of their previous size. Studies using DNA fingerprinting techniques have shown that almost every nest has at least one chick fathered by an interloper. In one study done by ornithologist Raoul Mulder, 152 of 200 nestlings were not the offspring of the resident males.

Since the females are the gatekeepers of this system, via their selection of males, we might ask what benefits the females receive from this extra-pair copulation. The answer is that males of high color and persistence live longer and are healthy, and they are well adapted to their environment and lifestyle. This success reflects 'good

Figure 9.24
Splendid wren, *Malurus splendens.*

Ken Griffiths

Male superb fairy wren, *Malurus cyaneus.*

genes'. Also, the older and the healthiest males begin their molt earlier, and therefore are at peak color longer and have a head start at impressing the females by repeated visits. Choosing an early molter is one way for a female to select a superior male with proven ability to survive. The females will eventually succumb to a colorful, persistent male, and thereby gain those favorable genetic traits for their offspring, who in turn, will also be successful. Of course, half of the female's genes are also passed on to the offspring; and that is the name of the game of evolution—to get one's genes into the next generation.

Further DNA fingerprinting tests support this theory. Of 68 adult males examined in a study area, just nine fathered all of the extra-pair young, and most of these successful males were related! The sons inherited their father's genes for the traits that females viewed as high quality, and the female offspring presumably inherited their mother's genetic predisposition for the traits she selected in the male.

The cuckolded male continues to feed the nestlings even though he did not father all of them. He has no way of knowing which are his. The female continues to mate with the extra-pair males, but occasionally will allow her mate to copulate with her, thereby keeping him as a caregiver for her nestlings. When helper sons are at the nest, the female can reduce her copulations with her mate. This, in turn, reduces his visits to the nest, and he may father only one in every ten nestlings. However, by attacking and chasing his sons he can keep them on the nest, feeding what may or may not be his genetic offspring, while he visits other females. As an evolutionary strategy for the Australian wrens, rampant infidelity works!

Figure 9.25
Willie wagtail,
Rhipidura leucophrys.

Willie Wagtail

The ubiquitous little black and white flycatcher, *Rhipidura leucophrys* (Figure 9.25), is common in towns and cities throughout continental Australia. During the breeding season, from June to February, it is very aggressive and will chase much larger birds such as kookaburras, magpies, and even wedge-tailed eagles. In addition to the willie wagtail, Australia has many small, colorful flycatchers (Eopsaltriidae) called robins. Most are insect eaters, with bristles on each side of their large mouth.

Butcherbirds

The butcherbirds (Artamidae) are another endemic Australian-New Guinean family. Many of these birds are black and white. The Australian magpie, *Gymnorhina tibicen* (Figure 9.26), is found throughout Australia and will nest in trees in parks and in cities. It reach 440 mm (17 in). The magpie lives in small groups, and all

Figure 9.26
Australian magpie,
Gymnorhina tibicen.

Ken Griffiths

members defend the group's territory, which can range from 2 to 18 ha (5–44 acres). During the breeding season, the magpie will swoop down on unsuspecting humans who pass under its nest, which is usually in a tree 6–16 m (20–52 ft) above the ground. Magpies have been known to draw blood from a nip on the ear or neck. There is even one recorded death (in 1946) of a 13-year-old boy from northern New South Wales after he was repeatedly pecked on the head by a nesting magpie. He died from tetanus.

Australian magpie, *Gymnorhina tibicen.*

Other Perching Birds

Other interesting Passeriformes include the chowchillas, quail-thrushes, and whipbirds (Cinclosomatidae), which are endemic to Australia and New Guinea. The treecreepers (Climacteridae) are a group of six Australian species and one New Guinean species, in the genus *Climacteris.* They have the ability to climb tree trunks without using their short tail for support.

The honeyeaters (Meliphagidae) have shown great adaptive radiation in Australia, which has 73 of the 151 species. These blossom-visitors are characteristic features of the Australian scene. Honeyeaters have a distinctive brush-like tongue which is used to obtain nectar from flowers. They are the main pollinators of *Banksia,*

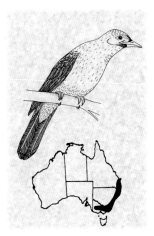

Figure 9.27
Bell miner,
Manorina melanophrys.

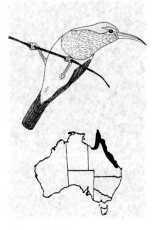

Figure 9.28
Yellow-bellied sunbird,
Nectarinia jugularis.

Figure 9.29
Zebra finch,
Taeniopygia guttata.

Grevillea, and other proteaceae. They also feed on *Eucalyptus* flowers. The family is highly diversified, *Lichenostomus* being the largest genus with about 18 species in Australia.

My personal favorite is the bell miner, *Manorina melanophrys* (Figure 9.27), of southeastern Australia. This small, olive green bird nests in colonies in sclerophyll forests. Its metallic call of 'tink-tink', which identifies the colony, is carried a long way and provokes other bell miners to answer. The result is a cacophony of 'tink-tinks' that proves you are in Australia. It defends communal territories in the forest canopy from other birds, thus reducing competition for its favorite food, the nymphs and secretions of psyllids (insects of the order Homoptera). Wherever there are bell miners the foliage is infected with psyllids and the trees appear unhealthy. If the bell miners leave the area, other birds invade the forest and control the psyllids. The trees recover from the insect damage. One reason why the bell miners do not control psyllids is that the birds frequently remove only the protective carbohydrate covers (lerps) of psyllid nymphs, leaving each nymph alive to regenerate a new lerp.

The blue-faced honeyeater, *Entomyzon cyanotis,* is one of the larger and more visible honeyeaters, and is so called because of the turquoise blue wattle (large bare skin patch) around the eye. This species occurs in small flocks in eucalypt woodlands, rainforest edges, and swamps in an arc from Broome in Western Australia to southeastern Australia. It feeds on insects, nectar, pollen, and fruits. In coastal Queensland it is called banana bird because of the damage it does to this important cash crop. The blue-faced honeyeater often utilizes the readymade nest of the gray-crowned babbler, *Pomatostomus temporalis* (family Pomatostomidae), by lining the old nest with new bark and grass. This is not a problem for the babbler because it constructs many roosting nests or dormitories other than the breeding nest in which it lays eggs.

The colorful endemic Australian chats (Ephthianuridae) have tongues similar to those of the honeyeaters. The yellow-bellied sunbird, *Nectarinia jugularis* (Figure 9.28), also feeds on nectar but lacks the brushtip tongue. Its bill is long and curved, and its tongue is tubular but curled in the reverse direction from that of the honeyeaters. Australia is also home to about 26 species of weavers and finches of the family Passeridae. The tiny (100 mm; 3.9 in) zebra finch, *Taeniopygia guttata* (Figure 9.29) occurs throughout Australia and is very popular with bird fanciers around the world. The exquisitely colored Gouldian finch, *Erythrura gouldiae,* must be seen to be believed.

Blue-faced honeyeater, *Entomyzon cyanotis*, so called because of the blue patch around its eye.

INTRODUCED SPECIES

About 40 species from other countries have been introduced to Australia, and at least 23 of these are well established. Some of the exotics are:

starlings, *Sturnus vulgaris*

mynas, *Acridotheres tristis*

house sparrows, *Passer domesticus*

true sparrows, *P. montanus*

European goldfinches, *Carduelis carduelis*

greenfinches, *C. chloris*

red-whiskered bulbuls, *Pycnonotus jocosus*

domestic pigeons, *Columba livia*

peafowl, *Pavo cristatus*

blackbirds, *Turdus nerula*

song thrushes, *T. philomelos*

cattle egrets, *Ardeola ibis*

mallards, *Anas platyrhynchos*

The starlings, house sparrows, and blackbirds have all become agricultural pests, causing damage in wheat and fruit growing areas.

The platypus,
Ornithorhynchus anatinus.

Mostly Marsupials: The Mammals

10

A WORD ON MAMMALS

Mammals evolved from mammal-like reptiles called therapsids about 220 million years ago (Triassic Period), but they didn't really 'inherit the earth' until the extinction of the dinosaurs about 65 MYA (Cretaceous). The therapsids were transitional between primitive reptiles and mammals, and they had their heyday between 260 and 190 MYA, before the ascendancy of the dinosaurs. The therapsids originated from an earlier reptilian group, the Pelycosaurs, of the Carboniferous, about 300 MYA. Recently a mammal-like reptile similar to *Kannemeyeria* of South America, Africa, and India was found in Triassic deposits (230 MYA) of southeast Queensland.

The characteristics of mammals include the presence of hair and the production of milk from mammary glands of the female. Mammals are homeothermic—they have a high and relatively uniform body temperature. The joint between the lower jaw and the skull is formed by the dentary and squamosal bones, respectively. Mammals also have three middle ear bones, whereas reptiles have only one, the stapes. The extra two bones, the malleus and incus, are derived from the reptilian articular and quadrate bones which are the articulating elements between the lower jaw and skull in reptiles. There are other skeletal differences as well.

Today there are about 4630 living or recently extinct (i.e. possibly alive in the preceding 500 years) species of mammals, of which about 6 percent are Australian. Traditionally, these are classified into two subclasses: the Prototheria, which includes the monotremes; and the Theria, which includes the marsupials and the placental mammals. However, following the discovery of an early Cretaceous monotreme fossil, *Steropodon*, with therian dentition, monotremes are now considered to be therians (Archer *et al.* 1985). While this book was in press, two studies utilizing nucleotide sequence data from mitochondrial DNA were published by Janke *et al.* (1996, 1997). These data show that monotremes and marsupials are sister groups that diverged from each other about 115 MYA, and that the monotreme/marsupial lineage separated from the more ancient placental mammals about 130 MYA. These studies refute the traditionally held view that marsupials and placentals form a lineage to the exclusion of the monotremes. There are 26 orders of mammals, two of which (Cetacea and Sirenia) are completely aquatic. Nine orders of mammals are native to Australia. These are the monotremes, four orders of marsupials, rodents, bats, carnivores in the form of sea lions and seals, and the dugong. Until recently all marsupials were placed in a single order, Marsupialia, but Wilson and Reeder (1993) reflected the reality of marsupial diversity by recognizing seven orders of marsupials, four of which are Australian. Australia's native mammal families are listed in Table 10.1. Of the approximately 300 mammal species in Australia, about 80 percent are endemic. This is the highest level of mammalian endemism anywhere in the world. When a simplified index of biodiversity is based upon mammals and certain butterfly families, Australia is ranked second in the world in endemism. Only neighboring Indonesia ranks higher, due to its large number of endemic butterflies. The United States is rated ninth. See Sisk *et al.* (1994) for this interesting analysis.

TABLE 10.1 AUSTRALIA'S NATIVE MAMMALS

ORDER	FAMILY	COMMON NAME	AUSTRALIAN GENERA	AUSTRALIAN SPECIES
Monotremata	Tachyglossidae	Echidna	1	1
	Ornithorhynchidae	Platypus	1	1
Dasyuromorphia	Thylacinidae	Tasmanian tiger	1	1
	Dasyuridae	Native 'cats', marsupial 'mice'	13	51
	Myrmecobiidae	Numbat	1	1
Peramelemorphia	Peramelidae	Bandicoots, bilbies	4	10
	Peroryctidae	Spiny bandicoot	1	1
Notoryctemorphia	Notoryctidae	Marsupial mole	1	1
Diprotodontia	Vombatidae	Wombats	2	3
	Phascolarctidae	Koala	1	1
	Burramyidae	Pygmy possums	2	5
	Petauridae	Gliders, possums	3	6
	Pseudocheiridae	Ringtail possums, greater glider	6	8
	Tarsipedidae	Honey possum	1	1
	Acrobatidae	Feathertail glider	1	1
	Phalangeridae	Cuscuses, brushtail possums	4	5
	Potoroidae	Potoroos, bettongs, rat-kangaroos	5	10
	Macropodidae	Kangaroos, wallabies	10	44
Chiroptera	Pteropodidae	Flying foxes	5	12
	Megadermatidae	Ghost bat	1	1
	Vespertilionidae	Common bats	11	30
	Rhinolophidae	Horseshoe bats	1	2
	Hipposideridae	Leafnose bats	2	6
	Molossidae	Mastiff bats	3	6
	Emballonuridae	Sheath-tailed bats	2	7
Rodentia	Muridae	Native mice and rats	14	63
Carnivora	Otariidae	Sea lions, eared seals, fur seals	2	5
	Phocidae	Earless seals, elephant seal	5	5
Sirenia	Dugongidae	Dugong	1	1

• • • • • • • • • •

MONOTREMES

The order Monotremata contains just three species and these are found exclusively in Australia/New Guinea. Only two of the three species are extant in Australia. The word monotremata means 'single hole' and refers to the presence of a cloaca. A cloaca is the opening that receives the discharges from the digestive, excretory, and reproductive systems. This is associated with egg laying and is found in reptiles and birds. Monotremes lack a vagina. Males have internal testes, which is probably related to their relatively low body temperature.

Monotremes are exceedingly primitive mammals. They are warm-blooded with a body temperature of about 31–32°C (89°F) which is 6–7°C lower than that of most mammals. They are egg-layers. Their eggs contain yolk concentrated at the vegetal pole (telolecithal) and the early cleavages are confined to the disc at the animal pole (meroblastic). Bird eggs are also meroblastic. The eggs of monotremes are 15–25 times larger than the eggs of marsupials and placentals respectively. The embryo develops for about 28 days while the egg is in the uterus and continues to develop for about 10 more days of external incubation before hatching.

Contrary to what is written in many popular accounts, nipples are present in monotremes. However, the young may suckle milk exuded onto tufts of fur near the hair-covered nipples. There is a rigid connection between the shoulder joint and the sternum which marsupials and placentals lack. The pectoral girdle exhibits reptilian characteristics (interclavicle and large coracoids) similar to Triassic mammalian fossils called docodonts. The presence of some skull bones (prefrontals and postfrontals) and the absence of others (lacrimals and jugals) also reflect reptilian relationships. The skull sutures are lost early in life. This, along with the elongate toothless rostrum, gives the skull a bird-like appearance. Large epipubic bones, also found in reptiles, project anteriorly from the pubic bones. So different are the monotremes that, until recently, they have not been considered in the direct line of therian evolution. It has been speculated that they have undergone a long and independent evolution and may be more closely related to the extinct multituberculates of the Jurassic than to the other living mammals. This suggests that such mammalian characters as milk production, hair, and thermoregulation may, in fact, have been present in the therapsids. The monotremes provide us with a glimpse of what very early mammals may have been like. In spite of all their primitive features, the monotremes are also highly specialized for their unique lifestyle.

The fossil record of monotremes was practically non-existent until recently, when a platypus-like animal, designated *Obdurodon insignis,* was found from the Australian Miocene (about 22 MYA).

Echidna, *Tachyglossus aculeatus.* Also called spiny anteater, it feeds on ants and on termites by using its strong claws to tear apart hard termite mounds.

Ken Griffiths

Figure 10.1
Echidna,
Tachyglossus aculeatus.

Figure 10.2
The echidna's pouch is a shallow pocket formed by contraction of abdominal muscles.

Australia's first Mesozoic mammal fossil was described in late 1985. *Steropodon galmani,* from Lightning Ridge in New South Wales, is represented by an opalized lower jaw fragment bearing three molar teeth thought to be from a platypus-like monotreme. The describers of the 107 million year old fossil from the early Cretaceous feel that the monotremes are phylogenetically close to the marsupial and placental mammals.

Other reassessments of monotreme-therian relationships also speculate that the two groups shared a common ancestor as recently as 120 MYA. This hypothesis is based upon the fact that monotremes have the same three middle ear bones as placental mammals, but which are absent in the early mammals. In addition, monotremes share several advanced skull and skeleton characters with modern therians that are absent in most early fossil mammals. More fossils are needed to clarify the position of the monotremes in mammalian evolution.

The first monotreme fossil outside the Australian continent was reported in 1992 from Patagonia in southern Argentina. This South American monotreme is represented by an isolated upper right second molar and may have been larger than *Steropodon.* The Paleocene sediments in which the tooth was found have been dated at about 62.5 million years old.

Tachyglossidae—The Echidna (1 species)

The echidna, *Tachyglossus aculeatus* (Figure 10.1), also called the spiny anteater, is widespread throughout Australia. It also occurs in New Guinea. Its dorsal and lateral surfaces are covered with stout spines (modified hairs), and its ventral surface is covered with coarse hair. It is a powerful digger with long, strong claws which it uses to rip apart rock-hard termite mounds. Echidnas feed on termites in dry habitats and ants in moist areas which they slurp up with the long sticky tongue that can extend 180 mm (7 in) from their tubular snout. The echidna's tongue works on the same hydraulic principle as a penis. Blood flows into the tongue which extends and stiffens. The tongue, lubricated by a sticky secretion, can dart in and out 100 times per minute. *Tachyglossus* means 'fast tongue'. When disturbed, the echidna can dig straight down into the soil and completely disappear in less than a minute. If it wedges itself between two rocks, it is practically impossible to dislodge.

Echidnas weigh up to 6 kg (13.2 lb) and are variable in color from light brown to black. In hot climates they have fewer hairs and in cold climates they have dense hair growth that may conceal the spines. The females lay a single 16 mm (0.6 in) leathery egg that is incubated for about 7–10 days in a temporary V-shaped abdominal pouch formed by muscular contraction (Figure 10.2). Unlike the platypus, both ovaries and oviducts are functional. The young hatches without spines and nurses on the mother's rich milk by lapping it from the ventral surface of her skin. When the spines develop after about 55 days, the young leave the pouch but may still nurse from the mother for several more months. The basal metabolic rate of echidnas is about one-quarter that of similar-sized placentals. Dingos and automobiles are the only predators of echidnas.

Zaglossus bruijni is the other species in this family. This long-nosed echidna lives in New Guinea and is found only as a fossil in Australia.

Ornithorhynchidae—The Platypus (1 species)

The platypus, *Ornithorhynchus anatinus* (Figure 10.3), is surely one of the most bizarre animals on earth. When it was first reported in 1799, it was thought to be a hoax. It took a further 80 years after its discovery to demonstrate that it was an egg-layer. The platypus is relatively common in streams, rivers, and lakes of eastern Australia. It is found in streams east and west of the Great Dividing Range, with temperatures from subtropical to near-freezing. Its sensitive duck-like leathery bill is used to probe the stream or lake bottom for aquatic insects, worms, crustaceans, and tadpoles. Electroreceptors and mechanoreceptors in its bill enable the platypus to locate prey in turbid or dark water. It has claws on its feet with which it digs its burrow into stream banks. The winding burrow may be 5–10 m (16.4–32.8 ft) long, 0.5m (20 in) beneath the surface, and the entrance is usually 1–2 m (3.3–6.6 ft) above water level.

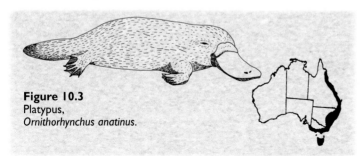

Figure 10.3
Platypus,
Ornithorhynchus anatinus.

Below left: The bill of a platypus is soft and leathery, unlike that of a duck which is hard.
Below: Dorsal view of skull of a platypus (top)—which has a rostrum with widely flaring premaxillae—compared to that of an echidna (bottom)—with a slender rostrum. Adult platypuses do not have teeth, but juveniles have teeth for a short time, which are replaced with horny pads. The location of these pads can be seen on the underside of the platypus skull as the oval patch with small holes.

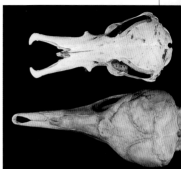

The animal's feet are webbed, which is a useful aquatic adaptation. The webbing folds against the palm when the platypus is digging or walking. The tail is depressed like that of a beaver and used for propulsion. The platypus shows bradycardia (slowing of heart beat) upon diving. The rate may drop from 140 beats/min to 40 beats/min after submersion for 40 seconds. Recovery to 200 beats/min can occur within seconds. The animal is covered with a dense dark brown pelage. The juveniles have molar teeth similar to the Multituberculata, but these are lost and as adults they have only horny pads to crush food. Platypus reach about 2 kg (4.4 lb) and 60 cm (2 ft) at most. Their basal metabolic rate (BMR) is about double that of echidnas, which puts it near the low end of marsupial BMR. This may reflect the greater activity of an aquatic lifestyle. They can maintain a body temperature of 32°C (89°F) for several hours even when surrounded by ice water. Platypus lose very little body heat to the water due to the countercurrent heat exchange system of the blood supply to and from the tail. In the wild, platypus can live to about 12 years old.

Platypus breed in spring after a courtship of several days. The male grabs the female's tail in his mouth, and they swim in circles. Eventually the female usually lays two 13 mm (0.5 in) leathery eggs in the grass-lined chamber at the end of a burrow, about 2–4 weeks after mating. Only the left ovary is functional. Males maintain a separate burrow and have internal testes and therefore lack a scrotum. This is fortunate when you consider how they slide down stream banks. The incubation period is about 11 days, and the female plugs the burrow opening during this time. The eggs are maintained at a temperature of 31.5°C (87°F) even when the outside temperature is much cooler. The 13 mm (0.5 in) young cut their way out of the eggshell with an egg tooth at the tip of their bill. The mother's mammary glands are very large and may extend almost one-third of her length. The young lap iron-rich milk from the mother's two fur-covered nipples and do not leave the burrow until about 17 weeks old and 30 cm (1 ft) long.

Male platypus have a hollow venom spur on the inside of each ankle that is connected to the crural gland in the leg. They will attempt to jab anyone handling them—at least during the breeding season—and there are several cases of injury to humans. The wound is similar to a venomous snake bite: intensely painful with resulting swelling and tenderness. The neurotoxic secretion can kill an animal the size of a dog. The function of the venom spur is not immediately obvious. Various explanations such as establishment and defence of territories, combat between males, prey capture, and holding females during mating have been postulated. The venom gland shows seasonal changes related to the reproduction cycle. Crural gland development, spermatogenesis, and male aggression reach a peak in August. (Male echidnas also have spurs, but they do not seem to be venomous.) Female platypus are hatched with spurs but lose them shortly thereafter.

Platypus were almost hunted to extinction by the fur industry but are now completely protected and common. Water pollution and habitat destruction are their greatest threat today.

MARSUPIALS

Marsupials are characterized by the presence of a marsupium, or pouch, on the female (a few marsupials lack a pouch). Their skull has a small, narrow braincase. The marsupial palate is not solid as in placental mammals but is perforated by openings. The angular process of the lower jaw of marsupials is turned inward. The auditory bullae are formed from the alisphenoid instead of the tympanic bone as in placentals.

Marsupial dentition is unique. The number of upper and lower incisors are not equal (except in wombats) and the number can be as high as 5/4 (upper/lower) on each side (left and right) compared with a maximum of 3/3 in placentals. There may be 3/3 premolars and 4/4 molars compared with the placental arrangement of 4/4 and 3/3 respectively. Total number of teeth in marsupials frequently exceeds the 44 of primitive eutherians (placental mammals).

Marsupials usually have large clavicles but reduced coracoids. There are epipubic bones attached to the pelvis that may function to support the pouch. Some marsupials have very slender second and third toes closely bound in a sheath of skin. This syndactylous arrangement forms a comb used for grooming. Marsupials maintain a lower body temperature than most placentals (35.5° vs 38°C) and have a lower basal metabolic rate, but their temperature-regulating abilities are just as good as those of placental mammals.

The reproductive system of marsupials is their most distinctive feature. The female reproductive tract is double. That is, there are two vaginas and two uteri (Figure 10.4). There is a left and a right lateral vagina each connected to the left or right uterus. Placental mammals have a single vagina and uterus, although two uterine horns may be present in some groups (Figure 10.4). In marsupials the two lateral vaginas join just below the two uteri. At birth a pathway called the pseudovaginal canal forms, and the fetus can pass through to the

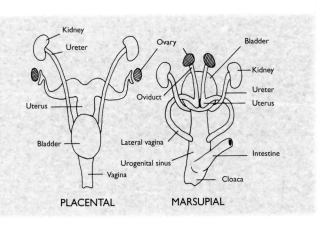

PLACENTAL   **MARSUPIAL**

Figure 10.4
Urogenital system of a generalized female placental mammal and marsupial mammal.

urogenital sinus and then to the outside via the cloaca. In kangaroos this canal is retained, but in other marsupials it closes and must reform with each birth. The testes of male marsupials are anterior to the penis.

Marsupials, except the bandicoots (Perameliae), have a chorio-vitelline placenta instead of the chorio-allantoic placenta found in eutherians and bandicoots. The more primitive marsupial arrangement does not allow the developing blastocyst to implant deeply in the uterine wall. The embryo receives nourishment by excretions of the uterine lining and by limited diffusion from the maternal blood. The very short gestation period of marsupials is probably related to the limitations of the system. It has also been suggested that, since marsupials do not have the placental barrier to fetal antigens, the young must be born soon after its egg membrane ruptures. To retain the embryo longer would invite immunological attack.

The young are born after an extremely short gestation period. They are mere larva-like scraps of pink, blind, naked tissue. However, the forelimbs are very well developed and the newborn climbs from the female's urogenital sinus to the pouch, where it attaches to a teat. The teat swells in its mouth and the mother pumps milk into the young. The young develops in the pouch for a length of time comparable to the gestation period of a placental of the same adult size. The young leaves the pouch at about the same weight as a newborn placental mammal of the same adult size.

Marsupials and placentals approach reproductive strategy from two different directions. The marsupial method of reproduction can result in a greater neonatal death rate, but the female's energy investment in the short gestation period is small, and she may still be able to breed again in the same season. The female marsupial has a greater degree of control over the reproductive processes than does a placental female. The placental mammal's strategy is based upon increased protection for the long-term fetus. However, if the young dies, the placental female has expended a relatively large reproductive and energy effort with no evolutionary result. Both systems offer survival value to their respective groups and demonstrate that natural selection does not always solve the same problem in the same fashion.

Marsupials and placentals probably arose independently at about the same time during the Cretaceous, approximately 130 MYA. Lower Cretaceous marsupials are only known from North America. The oldest marsupial fossil known is a 100 million year old lower jaw named *Kokopellia* from the badlands of Utah. Marsupials probably arose in North America and spread to South America in the late Cretaceous. The oldest South American marsupials date back to about 75 MYA. Following the disappearance of the ruling reptiles, they underwent an explosive radiation in the Cenozoic. Marsupials reached Europe via the Bering land bridge in the Eocene, but died out at the end of the Miocene in both Europe and North America.

South America probably was the source of Australia's marsupial fauna, around 55 MYA in the Eocene. During the Cenozoic a

temperate Antarctica formed a bridge between South America and
Australia over which the marsupials could travel. Both the South
American and Australian marsupial fauna evolved in isolation on island
continents. In the early Tertiary the Isthmus of Panama was under
water, breaking the connection between North and South America.
Likewise Australia began to drift from Antarctica in the Eocene.
North and South America were reunited in the Pliocene. This allowed
the invasion from North to South America of specialized placentals,
including many carnivores, that had already been subject to significant
extinctions and selection. This competition, from animals evolved in a
large diverse continental area of North America, proved too great for
many of the marsupials that had evolved in the smaller, less diverse
island continent of South America. This is not quite the same as
saying that the marsupial method of reproduction is inferior to the
placental method, as many people commonly assume. In fact, marsupials
seem to be able to adjust their reproduction to the vagaries of the
environment more easily than placentals can. A hypothesis has also been
advanced that placental mammals are superior to marsupials in brain
power. Many of the South American marsupials became extinct, but
some survived. The common opossum of North America most likely
reinvaded North America from South America during the Pleistocene.

Australia remained an island, which excluded placental
competition. The marsupials radiated into every possible niche and
remain the dominant mammals in Australia today. Serological studies
show that all extant Australian marsupials are more closely related to
one another than to New World groups.

The above scenario was recently supported by the discovery of
three 40-million-year-old jawbone fragments of a rat-like marsupial,
Polydolops, on Seymour Island in the Weddell Sea, about 48 km
(30 miles) off the northeastern tip of the Antarctic Peninsula. This is
the first discovery of a land mammal in the Antarctic continent. The
jawbones closely resemble those of *Polydolops* from South America.
This is considered strong support for the hypothesis that marsupials
migrated from South America to Australia via Antarctica. Fossils
show that this region had a tropical climate 75–50 MYA.

Similar fossil marsupials have not yet been found in Australia. The
oldest Australian marsupial fossils date back about 23 million years.
More exploration may change that. Likewise primitive placental
mammal fossils also have not been found in Australia. Why should
the marsupials but not the contemporaneous primitive placental
mammals have invaded Australia from South America? This question
cannot, as yet, be answered.

The decline of the Australian marsupial fauna is tragic, but no less
interesting than its origin. Habitat destruction, agricultural use of
large tracts of land, increasing urbanization, pollution, slaughter of
native predators to protect livestock, overharvesting of fur-bearers,
and especially the introduction of exotic predators such as feral cats
and foxes and exotic herbivores such as rabbits have all contributed to

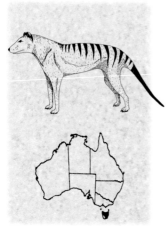

Figure 10.5
Tasmanian tiger,
Thylacinus cynocephalus.

Below: The skull of the
Tasmanian tiger is dog-like, but
the tiger's skull has four more
teeth than a dog.
Below right: Tasmanian tiger,
Thylacinus cynocephalus,
adapted from a John Gould
painting that graces bottles
of Cascade Premium Lager.
The tiger is presumed to be
extinct, but there is a slight
possibility that a few
individuals survive today
in rugged northern parts
of Tasmania.

the reduction of marsupial biodiversity. Since European settlement in 1788, at least ten species of marsupials have become extinct. The list includes several species of bandicoots, wallabies, potoroos, and the Tasmanian tiger. This loss represents about a quarter of the world's mammal extinctions during the last 200 years.

A recent study asked Aboriginals from the central deserts area to comment on various species of marsupials when shown museum specimens. From these interviews it was learned that several species previously thought to have become extinct in the early part of the 20th century actually survived in remote parts of the central deserts until the 1940s and 1950s. Disappearance proceeded from south to north. This pattern coincided with the movement of Aboriginals from their ancestral lands to European settlements. As Aboriginals abandoned their nomadic lifestyle, which included the regular use of fire for hunting and regeneration of food plants, a different fire regime characterized by infrequent, extensive, and very hot wildfires started by lightning replaced the more controlled burns that the Aboriginals had been setting for tens of thousands of years. This disturbance destroyed shelter and food sources and may have been the final blow to some species that were barely hanging on in the last undisturbed areas of Australia.

Order Dasyuromorphia

The carnivorous marsupials, except for the numbat, have an elongate snout with four pairs of needle-like incisors in the upper jaw and three pairs in the lower jaw. They have well-developed canine teeth and serrated premolars and molars. Their front and rear legs are of similar size. The first digit of the hindfoot is small or lacking, and no digits are fused. The tail is not prehensile.

Thylacinidae—The Tasmanian tiger (1 species)

The Tasmanian tiger, wolf, or thylacine, *Thylacinus cynocephalus* (Figure 10.5), is, unfortunately, most likely extinct. This dog-like

Adult thyla...
...thylacine is unrelated to the dog.
...both of which are
...in form,

carnivore once occurred over much of Australia and New Guinea, but died out everywhere except Tasmania about 3000 years ago. This mainland extirpation was probably related to the arrival of the dingo. Thylacines survived in Tasmania, where the dingo never reached, until 1930. As a large predator they had an economic effect on farmers who, along with the government, offered a bounty for their scalps. Between 1888 and 1909 at least 2268 thylacines were killed for the bounty. The actual number was probably larger as some landowners paid higher bounties than the government, and these kills were unrecorded. There was a dramatic decline in the numbers killed in 1905. The last wild specimen was taken in 1930, and the last known thylacine died in the Hobart Zoo on 7 September 1936. There is a slight possibility that a few animals may have escaped destruction in northern Tasmania. 'Sightings' continue to be reported, but no firm evidence of their survival has emerged. A seemingly reliable 1982 sighting was described by a Tasmanian National Parks and Wildlife officer in *Australian Natural History* in 1984, but an extensive search failed to find the animal. *New Scientist,* in the 24 April 1986 issue, carried photographs of an alleged living thylacine from southwest Western Australia.

The thylacine is the largest of the recent marsupial carnivores. It had a digitigrade, loping walk. The thylacine superficially resembled a dog, but it had an oddly tapered rear end with short legs and 15–20 transverse dark stripes across the back and hindquarters. Thyacines had 46 teeth whereas wolves, dogs, and coyotes all have 42 teeth. The average braincase of thylacines was about 53 cc compared to 134 cc for wolves. Body length was about 1.2 m (3.9 ft) and tail length about 0.5 m (1.6 ft). An average weight would be about 25 kg (55 lb). It seemed to prefer open woodlands with rocky outcrops where it established lairs. Thylacines preyed upon wallabies and smaller marsupials, echidnas, birds, reptiles, and of course sheep and chickens, which led to its destruction. Its pouch opened rearward and had four teats. Its tail was rigid like a kangaroo's and could not be wagged.

Thylacines illustrate the principle of convergent evolution beautifully. Their overall body shape and their skull and dentition are very similar to those of a dog or a wolf, which are placental mammals. In convergent evolution similar selective pressures (in this case the fact of being the top hunting carnivore) operate on unrelated animals (marsupials and placentals) and result in similar morphologies (thylacines and wolves).

Dasyuridae—The marsupial 'cats' and 'mice' (51 species)

This interesting marsupial family is closest to the ancestral marsupials that colonized Australia during the late Cretaceous. The members of this family have 42–50 teeth adapted for an insectivorous and carnivorous diet. The variously sized species occupy niches of 'mice', 'rats', and 'cats' which they physically resemble. The pouch of a female dasyurid is easily overlooked outside of the breeding season. It is a depression with teats in it rather than a sac.

Figure 10.6
Quoll or eastern native cat, *Dasyurus viverrinus*.

Figure 10.7
Tasmanian devil, *Sarcophilus harrisii*.

Ken Griffiths

Quoll or eastern native cat, *Dasyurus viverrinus*.

Top: Tasmanian devil, *Sarcophilus harrisii,* a short-limbed, stocky animal found only in Tasmania, although skeletal remains have also been found on Australian mainland. Above: Blade-like canines, crushing molars, and powerful jaw muscles make the Tasmanian devil a formidable scavenger able to crunch even large bones.

Kowari, *Dasyuroides byrnei,* emerging from a burrow in Gilbert Dessert, central Australia.

There are four species of native cats in the genus *Dasyurus.* These attractive animals are usually brown with white spots. They normally feed on insects, mice, and young rabbits, but they can do a great deal of damage if they enter a poultry yard. Their gestation period is about 8–14 days, and the young are weaned after about 4.5 months. Litter size is around six. Gestation, weaning, and litter size vary with the four species. The quoll or eastern native cat, *Dasyurus viverrinus* (Figure 10.6), is capable of producing 10–18 young, but because the pouch of the female contains only six teats, excess young that do not attach die. This phenomenon is called superfetation and also occurs in the American opossum, *Didelphis.* At least one species of *Dasyurus* is found in most regions of Australia.

The Tasmanian devil, *Sarcophilus harrisii* (Figure 10.7), is found only in Tasmania although recent skeletal remains are present on the mainland. Males of this short-limbed, stocky beast average 10 kg (22 lb) and 90 cm (3 ft), counting the tail, while females weigh about 6 kg (13.2 lb). Devils live about six years and are formidable predators as well as highly efficient scavengers. In this respect they may be the ecological equivalent of hyenas. The third and fourth molars of devils slice through flesh while the second molars crush bone. Devils are mostly black with white blotches near the chest and sides. They utter ferocious sounds when fighting, and they will kill and eat lambs or even weakened adult sheep. Normally they eat carrion as well as small animals that they capture by ambush or persistence. They construct a den in hollow logs or rocky crevices. Devils mate in May–June. Females may produce up to 60 eggs, 20 of which develop internally to birth. Usually, however, only four tiny young, 1.25 cm (0.5 in), survive. They remain in the pouch for 15 weeks and then cling to the mother. Weaning occurs at five months.

The kowari, *Dasyuroides byrnei* (Figure 10.8), is a small, brush-tailed, pointy-nosed dasyurid that lives in burrows in the arid center.

D Parer & E Parer-Cook, AUSCAPE International

Some marsupial mice belong to the genus *Antechinus,* and they lack the bushy tail of the kowari but have a sharply tapered snout and a mouse-like body. *Antechinus* occurs throughout Australia and into New Guinea. *A. flavipes* (Figure 10.9) is the best known of the eight species. The number of nipples on female *Antechinus,* and therefore the number of young that can be nursed, varies with habitat. Females in constant climates of the north and coastal areas have fewer nipples. Those in the variable inland have more. Presumably this is related to seasonal fluctuations of insects in the less stable areas. Marsupial mice eat large numbers of insects, especially grasshoppers.

Antichinus stuartii has a strange life cycle. Females give birth to 6–10 young during a highly synchronized two-week period in September. The young leave the nest in January (summer). By June (winter) the young males are reproductively mature. Females come into estrus in August. This triggers a flurry of male-male aggression and male-female copulation which leaves the males exhausted. They lose weight, become sickly, and die by the end of August, apparently of stress. The population at this point consists only of pregnant females, some of which survive to breed again. Males exist only as embryos until birth in September.

Sminthopsis (19 species) is another genus of marsupial mice. They have narrow feet adapted for hopping, whereas *Antechinus* have broad feet more suited for running. *Antechinomys* is a mouse-sized, long-legged relative of *Sminthopsis* that gallops by a leapfrog type of progression. *Planigale ingrami* (Figure 10.10), at 65 mm (2.5 in) not counting the tail, is the smallest known marsupial and perhaps the smallest mammal in the world.

Myrmecobiidae—The numbat (1 species)

The numbat or banded anteater, *Myrmecobius fasciatus* (Figure 10.11), is a beautiful squirrel-sized marsupial currently found only in *Eucalyptus wandoo* or *E. marginata* forests in southwest Western Australia. It is that state's faunal emblem. It was formerly more widespread, but is declining as its habitat is cleared. According to Aboriginals, this species disappeared from the central deserts around 1960. Numbats feed during the day on termites that attack the rotting logs on the forest floor. They utilize hollow logs for shelter. They have a keen sense of smell and a long snout with a very long protrusible tongue. Numbats are colored for camouflage with a black stripe through their eyes and several white and dark bands across the posterior half of their body. The body color is reddish brown. A scent gland on the throat enables them to mark their territory. Their tail is long and is frequently carried erect with the coarse bristles flared out. Numbats have 52 teeth—more than any other terrestrial mammal. Adult females lack a pouch. They carry the young attached to the teats. Juveniles develop a pouch that is lost during maturation. The young are born between January and April, and there is a six month nursing period.

Figure 10.8
Kowari,
Dasyuroides byrnei.

Figure 10.9
Yellow-footed antechinus,
Antechinus flavipes.

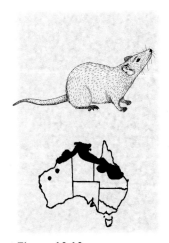

Figure 10.10
Planigale,
Planigale ingrami.

Figure 10.11
Numbat or banded anteater,
Myrmecobius fasciatus.

Figure 10.12
Pig-footed bandicoot,
Chaeropus ecaudatus.

Numbat, *Myrmecobius
fasciatus,* a marsupial with a
reddish brown body, white
and dark stripes on its
posterior, and a long tale that
is often carried erect.

Order Peramelemorphia

Bandicoots and bilbies have four or five pairs of blunt incisors in the upper jaw and three pairs in the lower jaw in an elongate snout. Flattened claws on the second, third, and fourth toes of the forefoot are used for digging. Other digits are small or absent. The hindlimbs have a highly developed fourth toe and are much larger than the forelimbs. The second and third hindlimb toes are syndactylus, and the tail is not prehensile.

Peramelidae—The bandicoots and bilbies (10 species)

Bandicoots range from rat to cat size, and are found in Australia and New Guinea. Bandicoots have an elongate, digitigrade hindlimb specialized for running. The virtually unknown and presumed extinct pig-footed bandicoot, *Chaeropus ecaudatus,* was once widespread in the northern central deserts and survived until the 1950s (Figure 10.12). It has only one functional digit (the fourth) on the hindlimb for running. The forelimb has the second and third digits well developed and clawed, giving a pig-like appearance. Toe reduction is a common adaptation of cursorial animals, but each group has solved the problem in a slightly different anatomical fashion. For example, horses rely on the third digit, antelope retain two digits, and peramelids emphasize the fourth digit. Likewise each of these unrelated running animals support their digits with a different arrangement of bones.

   Isoodon obseulus (Figure 10.13), the southern brown or short-nosed bandicoot, is common in southeastern Australia. There are two other species of *Isoodon,* both of which have rather spiny guard hairs over a soft underfur. Their noses are long and pointy and are used to probe in soil for earthworms and insect larvae. Long-nosed bandicoots, consisting of four species of *Perameles,* have longer snouts than *Isoodon.* Bandicoots are the only marsupials to have a chorio-allantoic placenta, which is the same type found in eutherian mammals. The peramelid placenta is, however, not quite as efficient in transferring materials between maternal and fetal circulation, due to the lack of villi which limits surface area. Other marsupials have the more primitive chorio-vitelline placenta.

Bilbies, *Macrotis lagotis* (Figure 10.14), and the very rare and possibly extinct *M. leucura,* are incredible diggers. They dwell in spiral holes 1–2 m (3.3–6.6 ft) deep and 3 m long, which they dig rapidly with both fore and hind limbs. The burrow affords protection from the heat of the central desert area. Bilbies have long rabbit-like ears. They feed on seeds, bulbs, fruit, fungi and insects along with which they ingest a great deal of soil. Sand composes up to 90 percent of their feces.

Pereoryctidae—Rufous spiny bandicoot (1 species)

This family is common in the rainforests of New Guinea and only one species, *Echymipera rufescens,* reaches Australia's Cape York Peninsula. This species takes its common name from the spiny guard hairs that cover its body. Its tail is naked. It is the only bandicoot in Australia to lack the fifth upper incisor. The snout is very long and somewhat mobile and probably used for searching out insects which it eats. Males are bigger than females and may reach 2 kg (4.4 lb).

Order Notoryctemorphia

The marsupial mole is the only species in this highly specialized order.

Notoryctidae—The marsupial mole (1 species)

The marsupial mole, *Notoryctes typhlops* (Figure 10.15), is another remarkable example of convergent evolution. This 15 cm (6 in) marsupial is very similar to placental moles (Talpidae). These similarities are due to adaptations to a fossorial (burrowing) existence. Their eyes lack lenses and are covered by skin, making the animal blind for all practical purposes. The forelimbs are enlarged and used as spades for digging. The last five cervical vertebrae are fused, which serves to stiffen the spine and allows the animal to force its way through the sandy soil of arid regions. A horny nose shield protects the animals' leading edge. External ears are absent and their white to golden-orange fur is velvety, which reduces resistance in the burrow. Two teats are present in the rear-opening marsupium, which is partially divided, giving each chamber a single nipple. Marsupial moles do not leave a characteristic trail on the surface as do placental moles. Instead, they push the soil behind their body as they advance forward. They eat invertebrate larvae and worms. It is unclear whether *Notoryctes* is related to the dasyuroids or diprotodonts. The dentition and hindfoot structure suggest it is a dasyuroid, but its chromosomes are more like those of the possums and kangaroos.

Order Diprotodontia

This large order of wombats, koala, possums, and kangaroos is united by the presence of only one pair of functional incisors in the lower jaw. The upper jaw may have one to three pairs. The second and third hindfoot toes are syndactylus.

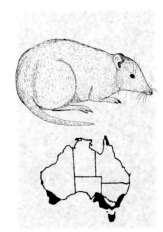

Figure 10.13
Short-nosed bandicoot,
Isoodon obseulus.

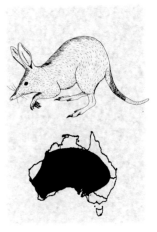

Figure 10.14
Bilby or rabbit-eared
bandicoot, *Macrotis lagotis.*

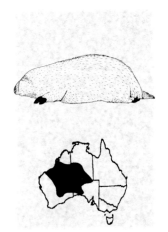

Figure 10.15
Marsupial mole,
Notoryctes typhlops.

Vombatidae—The wombats (3 species)

There are three species of wombats alive today, although at least six extinct species, including a 100 kg (220 lb) giant, are known from the fossil record. The nearly extinct northern hairy-nosed wombat, *Lasiorhinus krefftii*, is found in one small area of central Queensland and numbers only about 40–70 individuals. Its relative, the southern hairy-nosed wombat, *L. latifrons*, is scattered in isolated areas of the Nullarbor Plain and in a few semi-arid localities in South Australia.

Ken Griffiths

Two juvenile wombats, *Vombatus ursinus*. At about 18 months, young wombats must find their own burrows, either by finding a vacant burrow, sharing one, or digging a new burrow.

The common wombat, *Vombatus ursinus* (Figure 10.16), occurs in woodlands and mountainous sclerophyll forests throughout Tasmania, the coastal district and Great Dividing Range of New South Wales in scattered patches from the Queensland border to Adelaide.

Wombats and koalas are related at the subordinal level and may have evolved from a common ancestor about 25 MYA. They share such features as a button-like nose, rudimentary tail, rearward opening pouch, small stomach with very long intestines, and similarly shaped sperm. Wombats, koalas, possums, and kangaroos are diprotodonts with two incisors in the lower jaw, but the wombat is unique among diprotodonts in having only two incisors in the upper jaw. Their 24 teeth are rootless and grow throughout life. Enamel covers only the anterior surface of the incisors, and the chisel-like cutting surfaces are kept sharp by gnawing wood.

Wombats, koalas, and kangaroos also have syndactylous toes on the hindfoot in which the second and third digits are fused. The comb-like structure so formed is used for grooming. Wombats are rodent-like marsupials that superficially resemble woodchucks (*Marmota*) of North America. They have also been compared to large badgers or to small bears from which the specific name *ursinus* is derived. The following information is based upon studies of the common wombat.

Large adults may weigh up to 35 kg (77 lb), be 1 m long and stand 25 cm (10 in) high at the shoulders. They may live to be 26 years old in captivity, but half that time would be a ripe old age in the wild. These stocky, low-slung beasts are built like a tank and resemble a woodchuck on steroids. Their barrel-shaped trunk, powerful shoulders, and bull-like neck are built for burrowing. In spite of all their bulk, they can flatten themselves to squeeze through a small 10 cm (4 in) space under a fence. Their coat color is usually various shades of brown, but beige and black specimens are known. The skin covering the back and rump is especially thick. Wombats, if chased into their burrow by a dog or fox, will plug the burrow entrance with their backsides. A dog can be suffocated or have its skull crushed when its head becomes trapped between the wombat and the burrow wall.

Figure 10.16
Common wombat,
Vombatus ursinus.

The adult wombat's skull is broad and heavy and is often used to force objects such as logs or rocks out of its path. Their short legs are well clawed. There are no obvious sexual dimorphic characters. Females, of course, have a pouch that opens to the rear. This keeps dirt out while digging and moving through the tunnel. A circular muscle (sphincter) guards the opening of the pouch. The male's penis is carried within the body, as in all marsupials, except when in use. The testes hang close to the abdomen in a scrotum.

Wombats are one of the largest burrowing animals. A typical burrow is used repeatedly by many generations of wombats, and is enlarged and extended by each inhabitant. Major burrows may be up to 30 m (98 ft) long and 1.8 m (6 ft) deep, although most burrows are much smaller than this. There may be a network of tunnels that intersect, causing weak spots on the surface, which sometimes collapse under the weight of a heavy animal or a passing vehicle. For this reason wombats are hunted as destructive pests by some landowners. Burrows are usually begun near a rock or tree on a forested hillside. After a more or less straight entrance tunnel, the burrow may twist and turn abruptly. One or two resting chambers, enlarged portions of the tunnel measuring about 1 m by 0.5 m, are usually present. These chambers may be lined with fern fronds, bark, eucalyptus leaves, and grass. Digging for only about one hour per night, a wombat can excavate a 6 m (20 ft) tunnel in sandy soil within a week. Initially dirt is scratched with the strongly clawed front feet, then, as the loosened material accumulates, the rear legs are brought into play to move the soil backwards. A mound of soil accumulates at the mouth of the burrow. Burrows offer excellent protection during most bushfires, although an extremely hot fire can suck the oxygen out of a burrow and suffocate the occupant.

Wombats tend to be solitary animals, but a burrow does not belong exclusively to any individual. If a wombat enters an already occupied burrow, it may be allowed to stay or it may be chased out. Roughly two-thirds of the wombat's life is spent within the burrow. They usually emerge around dusk unless it is very hot. They stay in the cool burrow, near the entrance, until the temperature drops to about 20°C (68°F) around midnight. On cool, overcast days they may emerge in early afternoon. Wombats feed on coarse grasses and other monocotyledonous plants as well as small amounts of dry leaves, bark, and moss. They digest this high-fiber diet with the help of symbiotic intestinal bacteria. They inoculate themselves with these helpful microbes by eating their dry, squarish fecal pellets. Vegetation provides most of their water needs except during the dry season. Wombats may roam several kilometers each night in search of food. Their home range varies from about 5 to 23 ha (12.5 to 57.5 acres), depending upon the density of animals and the richness of the habitat.

If individuals come within a few meters of one another while feeding, a warning may be issued. This involves a low growl, a hiss, and a high-pitched screech. If one animal does not move away other threats, such as calling and the head swinging from side to side, take place. A fight may ensue, with teeth and claws. Wombats frequently show the scars of past battles, with chunks of fur removed from the forehead and muzzle and notches torn from ears.

Before dawn the wombat returns to its burrow to rest and sleep. While sleeping its respiratory rate is halved from the active 30 breaths/minute. Its body temperature falls from about 38°C (100°F) to 35°C (95°F). Even when it is very hot on the surface, 40°C (100°F), the burrow remains cool at about 26°C (77°F). In winter, wombat burrows remain above freezing and the animals spend less time outside, therefore needing and using less food when it is scarce.

As in all marsupials, the gestation period of wombats is very short—about 30 days. The single, hairless newborn is only 15 mm long (0.5 in) at birth and weighs only 0.5 gm. It crawls into the pouch and attaches to a teat within a couple of minutes after birth. The teat swells within the mouth, and the young cannot be removed without injury. By four months the eyes are open, and the pouch young weighs about 400 gm (14 oz). Two months later it weighs around 1 kg (2.2 lb) and is covered with very fine fur. It may poke its head out of the pouch occasionally. At seven months the 2 kg (4.4 lb) young may leave the mother's pouch but remain in the burrow. By eight months the 3 kg (6.6 lb) baby is fully furred and begins to nibble grass. The young leaves the pouch permanently about 9–10 months after birth, but may continue to suckle by placing its head in the pouch for up to 15 months after birth. Wombats are usually independent by 18 months.

As the wombat matures, its pointy snout becomes more rounded and flat, and the body takes on the more robust adult shape. An 18-month-old animal weighs about 20 kg (44 lb). These young animals must now find a vacant burrow, dig a new one, or share one with a willing occupant. Sexual maturity usually occurs at about 2–3 years of age. Males sniff the female's scent to determine if she is in estrus. There is no specific breeding season, but seasonal peaks occur in summer. Mating lasts about 30 minutes while the animals are lying on their side, with the male clasping the female around the thorax from behind. Males play no role in rearing the offspring and do not bond with the female. A pregnant female does not come into heat again until after the young is weaned. Thus wombats usually produce only one offspring every two years.

Wombats have been persecuted by farmers for their destructive digging and a bounty was paid for wombat scalps in the 1920s, when tens of thousands were slaughtered. Now they are protected, and the common wombat species is thriving. They have survived shooting, poisoning, and trapping. Today the most serious threat to wombats is habitat destruction by forest clearing. The car is the most serious predator of wombats. Rural Australians often rear baby wombats orphaned when their mother is killed on a highway. Young wombats make interesting, playful pets and follow their adoptive parents everywhere, but they are none too bright. Young wombats should be nursed on a low-lactose milk diet.

Figure 10.17
Koala,
Phascolarctus cinereus.

Phascolarctidae—The koala (I species)

The koala, *Phascolarctus cinereus* (Figure 10.17), is without a doubt Australia's best goodwill ambassador. Everyone loves koalas and associates them with Australia. This universal appeal may be related to their flat face and forward-directed eyes, conditions shared with humans, rather than the snout-like face and laterally positioned eyes of other animals. Koalas have a patchy distribution in eastern Australia, from eastern South Australia throughout Victoria to eastern New South Wales and northeastern Queensland. About 40 000 years ago *P. cinereus* occurred in the forests of southwestern Australia, but they are now extinct in that region. They have never been found in Tasmania. Koalas tend to be smaller and lighter in coat color in warm regions and larger and darker in the cooler south. Victorian koalas may be twice the size of their Queensland counterparts. Males are up to 50 percent larger than females. They may reach a maximum size of 13.5 kg (30 lb) and 82 cm (2.7 ft). Males tend to have a bulging 'Roman nose', while the female's nose is straight. Koalas are highly specialized marsupials whose closest relatives are wombats, with whom koalas share such features as a rudimentary tail, backward-opening pouch, one pair of teats, and several other unique characters.

Koalas are most common in coastal woodlands, where they dine exclusively on about three dozen of the 600 species of *Eucalyptus* trees in Australia. Their favorite food trees are manna gum, *E. viminalis;* river red gum, *E. camaldulensis;* swamp gum, *E. ovata;* long-leafed box, *E. goniocalyx;* grey gum, *E. punctata;* forest red gum, *E. tereticornis;* Tasmanian blue gum, *E. globulus;* and brush box, *Tristania conferta.* They will eat about a kilogram (2.2 lb) of these oily leaves each night. Most mammals avoid eucalyptus leaves because of the toxic oils, but koalas have circumvented the eucalyptus's defense and can

Koala, *Phascolarctus cinereus*, is undoubtedly one of Australia's best known marsupials.

detoxify the volative oils. They smell of eucalyptus but have few external parasites, which may explain why they groom infrequently.

Koalas are arboreal folivores and have all sorts of adaptations to prove it. Like all herbivores they have a very long alimentary tract, but they also have an extraordinarily long cecum. The cecum is a blind pouch at the junction of the small and large intestine. In the koala it may reach over 2 m (6.5 ft) and hold two liters of thoroughly masticated leaves. This allows the koala to ingest more leaves and speeds up the rate at which nutrients are acquired. In addition, some microbial fermentation of cellulose takes place in the cecum. Young koalas are inoculated with symbiotic cellulose-digesting bacteria by eating emulsified, semi-digested eucalyptus leaves (called pap) voided by the parents. Their nutritionally rather sparse, low-protein diet does not allow the accumulation of body fat.

Other arboreal adaptations include forelimbs with the first two digits opposable to the remaining three, and the hindfoot has one opposable digit and two syndactylous toes used for grooming. Koalas can raise their arms above their shoulders, which is an adaptation shared only with primates. Friction pads are present on the palms and soles, and the digits are strongly clawed. The limbs are long and hook-like, which facilitate hanging in trees even when asleep. Koalas spend most of their time aloft and walk on all fours only to get from tree to tree. Males have a sternal scent gland on the chest and use this to mark trees. Males are highly vocal at night. Their bellow is a combination of 'snore' and 'belch' that lasts up to 30 seconds and can be heard for 800 m (0.5 mile). Females may use this sound to locate the males.

Koalas rarely drink water directly. They get most of their moisture from dew and from the eucalyptus leaves. The cecum can hold large quantities of moist leaves. Koalas are also rather efficient at conserving water by excreting dry feces. Most of their water loss is from exhalation.

The fur of koalas is extremely dense, especially the dorsal fur. This provides excellent insulation for an animal living in windswept trees. The insulating properties of the fur are so good that koalas are among the few tree-dwelling marsupials that do not utilize a shelter. Their luxurious wooly gray-brown fur was almost their undoing, as millions were slaughtered in Queensland during the 1920s for the fur trade. They are now protected. Before Europeans arrived in Australia, Aboriginal hunting, dingo and owl (*Ninox stenua*) predation, and bushfires were the major sources of mortality. Today fires, automobiles, disease, and habitat destruction all take their toll. Bushfires are deadly because koalas have no way of escaping. Of approximately 4000 koalas reported killed annually, about 2500 die as a result of being hit by a car near urban areas.

The rearward-opening pouch of the koala has two teats, but they usually give birth to a single 19 mm (0.7 in) young weighing in at 0.5 gm (0.02 oz) after a 35 day gestation period, usually in summer months. The newborn climbs into the pouch and remains there for about 6–7 months. By then it is fully furred and about 18 cm (7 in) long. The young then come and go from the pouch at will for the next two months. The 1 kg cub is carried on the back of the female for an additional few months. The cub remains with its mother for about one year. By then it weighs over 2 kg. Koalas mature in about two years and usually breed each year. Young females may take up residence near their mother, while males disperse to other areas. Maximum life span in captivity is about eight years, but in the wild it can be 10–15 years.

Koalas have a low metabolic rate, a very small brain that lacks cerebral folds, and a very limited repertoire of behaviors. They lead solitary lives, except for mothers with young. They are mostly nocturnal and spend the larger part of the day sleeping in the crotch of a tree. In fact, observations of wild populations show that they spend about 19 hours each day sleeping or resting, less than five hours feeding, and a few minutes traveling. At night a group can be noisy with grunting and wailing sounds, and males will fight each other if an intruder climbs an occupied tree. Although they are undeniably appealing animals, they are best left alone as they may bite, and their formidable claws can do much damage to human skin. They also seem highly susceptible to cryptococcosis, a fungal disease that can affect people.

The San Diego, Los Angeles, and San Francisco zoos are three of the few places outside Australia where koalas are on permanent display. This is because the proper eucalyptus trees grow in southern California. The power of advertising was vividly demonstrated to me when I overheard a father explaining to his children at the San Diego Zoo that the animals they were watching were called Qantases!

Koalas feed exclusively on 36 of the 600 species of *Eucalyptus*.

A very high proportion of wild koalas, estimated at about 45 percent, suffer from a venereal bacterial infection of *Chlamydia psittaci,* which causes reproductive failure, pneumonia, blindness, and a variety of other ailments. *C. psittaci* causes a cystic condition in koalas similar to pelvic inflammatory disease in humans, which is caused by *C. trachomatis.* Because of the koalas' dependence upon symbiotic bacteria in their intestine for digestion and detoxification of eucalyptus leaves, antibiotics cannot be used to treat the disease. Such treatment would kill the beneficial bacteria, and the koalas would become malnourished and/or poisoned. A vaccine against the chlamydial organism is under development.

Burramyidae—Pygmy possums (5 species)

These tiny possums have strongly prehensile tails that can support their minute weight of 10–50 gm (0.35–1.75 oz). The mountain pygmy possum, *Burramys parvus,* is the only Australian mammal limited to alpine regions where there is snow cover for six months. Likewise, it is the only Australian mammal to hibernate for long periods of time. It lives at the highest altitudes available in Australia,

Ken Griffiths

Eastern pygmy possum, *Burramys nanus.*

1400–2300 m (4600–7500 ft), in southeastern Australia around Mt Kosciuszko, Mt Bogong and Mt Hotham. Hibernation may last as long as seven months during which time body temperature is held at about 2°C (35°F) for up to 20 days before it is interrupted by normal temperatures for less than a day. Hibernation may then resume for another 20 day period. The species' mountainous habitat has many boulders that help buffer temperature extremes and provide hibernation and nestling sites. Insects and other invertebrates form most of their diet, and they deposit fat before hibernation. The other four species in the family are in the genus *Cercartetus.*

Petauridae—The gliders and possums (6 species)

The gliders show striking convergent evolution with the northern hemisphere flying squirrels (*Glaucomys*). Gliders have a fold of skin along the sides of the body from wrists to ankle which, when extended, serves as a parachute. Their furry tail acts as a rudder as they leap from higher to lower branches, with their recurved claws providing a good grip on landing. There is a dorsal stripe that extends to the forehead.

The yellow-bellied glider, *Petaurus australis*, lives in tall wet *Eucalyptus* forests, feeds on plant exudates such as sap and nectar, and reaches 300 mm (1 ft) in body length. It can glide 110 m (360 ft), whereas the smaller sugar glider, *Petaurus breviceps* (Figure 10.18), and squirrel glider, *P. norfolcensis*, manage about 50 m (160 ft). Gliders have scent glands with which they mark their territory and control social interaction.

Leadbeater's possum, *Gymnobelideus leadbeateri* (Figure 10.19), was thought to be extinct until it was rediscovered in mountain ash forests in the central highlands of Victoria in 1961. It resembles a sugar glider but lacks the gliding membrane. It feed on plant exudates and insects. Little is known about this animal.

The striped possum, *Dactylopsila trivirgata* (Figure 10.20), of northeast Queensland rainforests, is a squirrel-like tree dweller with a remarkable similarity to the odd aye aye (*Daubentonia*) of Madagascar. Both have an elongated, probe-like fourth finger and powerful incisors with which they excavate wood-boring insects. This black and white striped marsupial can emit a strong, skunk-like odor.

Pseudocheiridae—Ringtail possums and greater glider (8 species)

The common ringtail possum (Figure 10.21) gets its scientific name, *Pseudocheirus peregrinus*, because the first and second digits of the forefoot oppose the other three—the generic name means 'false hand'. The grip so produced, plus the long prehensile white-tipped tail, make for an agile climber. It builds a nest of woven twigs and

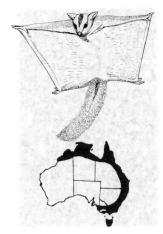

Figure 10.18
Sugar glider,
Petaurus breviceps.

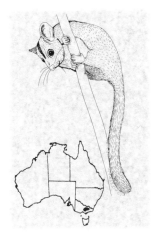

Figure 10.19
Leadbeater's possum,
Gymnobelideus leadbeateri.

Figure 10.20
Striped possum,
Dactylopsila trivirgata.

Sugar glider,
Petaurus breviceps.

Figure 10.21
Ringtail possum,
Pseudocheirus peregrinus.

Figure 10.22
Honey possum,
Tarsipes rostratus.

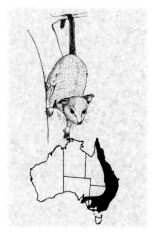

Figure 10.23
Feathertail glider,
Acrobates pygmaeus.

leaves 30 cm (1 ft) in diameter in dense scrub in eastern Australia. It is a nocturnal herbivore. The diet is augmented by coprophagy; that is, it feeds on its own feces. Soft fecal pellets from the cecum are voided once each day and consumed. After the second passage through the gut, undigestible matter is excreted as hard fecal pellets. Coprophagy is probably necessary since this species is one of the very few marsupials that feeds on the nutritionally sparse eucalyptus leaves. Common ringtail possums are familiar to most people living in residential suburbs in the east of Australia. These 1 kg (2.2 lb) animals are frequent visitors to gardens and house roofs during their nocturnal foraging.

The greater glider, *Petauroides volans,* of eastern Australia, is cat-sized and can glide roughly 100 m (328 ft). It is the largest gliding possum. Its gliding membrane (patagium) extends from the elbow, not the wrist as in the Petauridae. Like the common ringtail possum, it feeds on eucalyptus leaves that it breaks down by bacterial fermentation in an enlarged cecum. The greater glider is relatively abundant in undisturbed forests.

Tarsipedidae—The honey possum (1 species)

The honey possum, *Tarsipes rostratus,* (Figure 10.22), is a mouse-sized nectar and pollen feeder from the southwest corner of Western Australia. It has a pointy snout, long prehensile tail, and three black longitudinal strips on its dorsal surface. Its mouth is tubular and the tongue is brush-tipped like the honeyeaters and rainbow lorikeet. These are adaptations for a nectar diet. Its feeding actions probably make it a significant plant pollinator. It has fewer teeth than any other

Honey possum, *Tarsipes rostratus,* is a tiny nectar and pollen eater with modified claws and expanded fingertips for gripping branches.

Australian Marine Photographic Index

marsupial and does not appear to be closely related to other marsupials. Its claws are modified as nails, and its fingertips are expanded as an adaptation for gripping branches. This is reminiscent of the tarsier, a primitive primate—hence the generic name. The young weigh less than 5 mg at birth, the smallest birth weight of any mammal.

Acrobatidae—Feathertail glider (1 species)

The feathertail glider, *Acrobates pygmaeus* (Figure 10.23), lives in eastern Australia. It is mouse-sized and has a feathery tail of stiff bristles that functions in steering and braking during a glide. It has a gliding membrane between the elbows and knees. This smallest of gliders can leap 20 m (66 ft) or more. It is insectivorous and nocturnal.

Phalangeridae—The brushtail possums and cuscuses (5 species)

The brushtail possum, *Trichosurus vulpecula* (Figure 10.24), is familiar to most Australians because this cat-sized marsupial is as at home in the suburbs as it is in the bush. It is widely distributed in sclerophyll forests and open woodlands throughout Australia. It nests in hollow tree branches and has hands and feet adapted for its arboreal existence. It has a prehensile tail and is nocturnal. Brushtail possums feed on leaves and other plant parts as well as insects and birds. They can damage vegetation and have become pests in New Zealand, where they were introduced. The pouch of *Trichosurus* opens forward as in the kangaroos instead of backward as in the wombats and koala. Brushtails have been slaughtered by the millions for their dense gray fur but are still relatively common.

The scaly-tailed possum, *Wyulda squamicaudata* (Figure 10.25), is confined to rugged rainforest areas in northeastern Western Australia in the Kimberley district. Its tail is furred at the base but covered with scales for about 80 percent of the length. Two species of cuscuses, *Phalanger intercastellanus* and *Spilocuscus maculatus*, occur on Cape York Peninsula. They are common in New Guinea. They have large round heads with small ears, a prehensile tail, and are excellent climbers.

Potoroidae—Potoroos, bettongs, rat-kangaroos (10 species)

The tiny musky rat-kangaroo, *Hypsiprymnodon moschatus* (Figure 10.26), is the smallest member of the superfamily Macropodoidea, reaching less than 300 mm (1 ft) in body length. It is one of the oddest. It is the only macropod that retains an opposable big toe on the hindfoot and the only macropod that normally produces two young at a time. It occurs in rainforests of northeast Queensland and runs on all four legs rather than hopping as other kangaroos do. It feeds on insects as well as plant matter and is diurnal, which is another unusual feature for a macropod. *Hypsiprymnodon* may be a primitive link between the macropods and phalangers. It is placed in a separate subfamily, the Hypsiprymnodontinae.

Ringtail possum, *Pseudocheirus peregrinus*. The large eyes are adapted for nocturnal foraging.

Figure 10.24
Brushtail possum, *Trichosurus vulpecula*.

Figure 10.25
Scaly-tailed possum,
Wyulda squamicaudata.

Figure 10.26
Musky rat-kangaroo,
Hypsiprymnodon moschatus.

Figure 10.27
Brush-tailed bettong,
Bettongia penicillata.

Other small macropods include the bettongs (Figure 10.27)—four species of *Bettongia*—that use their prehensile tail to carry grass and sticks for nesting material. Potoroos (Figure 10.28), three rat-like species of *Potorus*, and two other genera form the family Potoroidae.

Macropodidae—The kangaroos (44 species)

The kangaroo family contains about 44 species in ten genera. They vary in size from the eastern gray kangaroo, *Macropus giganteus*, and the red kangaroo, *M. rufus*, which may stand over 2 m (6.6 ft) tall and weigh over 66 kg (145 lb), to the small wallaby, *Petrogale burbidgei*, which is 30 cm (1 ft) tall and weighs just 1.2 kg (2.6 lb).

Kangaroos are the largest living marsupials and are symbolic of Australia. Along with the emu, they are represented on Australia's coat of arms. The larger kangaroos are the ecological equivalents of ungulates of other regions. 'Roos', as they are colloquially called, do not look like antelope, but they function like them in an ecological context. Similarly the wallabies, which are simply small kangaroos, under 20 kg (44 lb), occupy a rabbit-like niche.

Kangaroos and wallabies are highly specialized for saltatorial (jumping) movement. Their forelimbs are small and the hindlegs are elongate. The long, stout tail forms a tripod with the hindlimbs and serves as a balancing organ. The hallux is absent, and the small second and third toes are syndactylous and used for grooming. The fourth and fifth toes are enlarged. When running, the foot is functionally two-toed in most species. Toe reduction is a common adaptation for speed. The larger members of the genus *Macropus* can easily exceed 40 km/h (25 mph) for a couple of kilometers and clear 9 m (30 ft) or more in between leaps reaching a height of 3 m (9.8 ft) off the ground. Speeds of 70 km/h (42 mph) may be possible in short bursts. This enables the larger species to cover a great deal of ground swiftly while searching for isolated patches of forage and water in an arid environment. Red kangaroos may have a home range of about 8.5 km² (3.3 mi²). The erratic leaps of the smaller species enable them to confuse predators and thereby escape. Hopping at moderate speed (20–25 km/h; 12–15 mph) as a means of locomotion is more energetically economical than running for similar-sized mammals, probably because energy is stored from one hop to another in the elastic recoil of large ligaments in the legs, tail, and back. At speeds below 6 km/h (3.6 mph), however, kangaroos use their tail as a fifth leg along with the small forelimbs to support their weight as they move the large hindlimbs forward. This rather strange locomotion is energetically expensive, but less so than hopping at slow speed.

Kangaroos and wallabies have 32 or 34 teeth, with a single pair of blade-like lower incisors that meet a tough pad on the roof of the mouth just behind three pairs of upper incisors. The long axis of the incisors is oriented anterior to posterior, which results in a cropping action like that of an ungulate. A wide gap (diastema) is present between the incisors and the cheek teeth. The largest kangaroos select their food plants carefully, and normally do not eat the same species eaten by sheep and cattle. Gray kangaroos, *Macropus giganteus,* eat grasses, and red kangaroos, *M. rufus,* prefer dicots which they digest by bacterial fermentation in their expanded foregut. Kangaroos normally do not eat mulga, an important sheep forage. In drought years, however, kangaroos and domestic stock do compete for any available vegetation. This has lead to the slaughter of many kangaroos by station owners to protect their forage during bad years. The case can also be made that sheep stations have actually benefited the kangaroo population by converting much of the native vegetation into grasslands with more waterholes. This has allowed kangaroo populations to expand.

Kangaroos are of economic importance for their fur and meat. Where 'roos have overpopulated and are competing with domestic stock, they are killed by licensed hunters. There are about 1600 registered kangaroo shooters in Australia. The hides are valuable, and the meat is used both as pet food and for human consumption. Kangaroos convert grass more efficiently than do sheep of the same weight. A sheep is roughly 27 percent usable muscle, while a red kangaroo is about 52 percent meat with less than 2 percent fat and practically no cholesterol. It would be more efficient to raise kangaroos (rather than cattle) with the sheep, since they eat different plants. In fact, zoologist Gordon Grigg, of the University of Queensland, advocates raising kangaroos instead of either sheep or cattle, whose hard hooves damage the ground and vegetation. The soft, large feet of macropods spread their weight over a greater area and do less damage to the earth and tender plants. Getting the majority of Australians to change their diet to include their national symbol (throw another 'roo on the barbie) will not be easy, however.

Today, the taking of kangaroos by hunters with their mobile refrigerated vans is tightly controlled and based upon an understanding of the population biology of the animals. Quotas for culling kangaroos are determined each year by state faunal

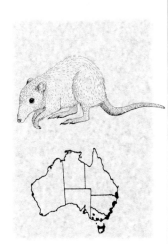

Figure 10.28
Potoroo,
Potorus tridactylus.

Agile wallaby joey, *Macropus agilis,* from Queensland. The white cheek stripe is a diagnostic characteristic of this beautifully colored wallaby.

Figure 10.29
Red kangaroo,
Macropus rufus, with joey.

Figure 10.30
A newborn kangaroo in the
pouch attached to a teat.

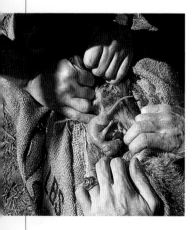

Above: Naked and blind
pouch-young of tammar
wallaby, *Macropus eugenii*.
Right: Red kangaroo,
Macropus rufus.

authorities on the basis of aerial surveys. A harvest level of about 15 percent of the estimated population is considered safe by wildlife biologists. The reason behind the harvest limits is to preserve self-perpetuating populations of each species throughout its range, not to provide a yearly yield for commercial shooters. Recent surveys of kangaroo populations revealed at least 9 million eastern gray, 18 million western gray and 8.4 million red kangaroos. However, kangaroo populations fluctuate greatly from year to year in response to drought and rainfall. The 1984 culling quota for grays was 672 000 and for reds was 900 000. In the early 1990s, about 5 million kangaroos were killed annually.

Kangaroo shooting is an emotional issue in Australia. The United States has recently removed the ban on kangaroo products, and Australian conservationists feel that the lifting of the United States ban is responsible for increased kangaroo slaughter, and that poachers kill far more 'roos than the quotas allow. Australian wildlife biologists generally agree that there are more kangaroos today than before European settlement. This is due to the increased availability of pasture and water. As a reflection of this abundance, the United States Fish and Wildlife Service removed the red, western and eastern gray kangaroos from the list of threatened species in April 1995.

As Australia's dominant form of wildlife, kangaroos, especially the larger economically important species, have been extensively studied. The marvelous adaptations of the reproductive system of the red kangaroo, *Macropus rufus* (Figure 10.29), are well known. This very large species is widely distributed over most of Australia wherever there are large expanses of grassy plains. The gestation period of red kangaroos is a very short 33 days, after which the 0.8 gm (0.03 oz) newborn climbs into the pouch and attaches to a teat (Figure 10.30). The females have a postpartum mating, and the resulting embryo remains dormant in the uterus at the blastocyst stage of 70–100 cells. The blastocyst is only about 0.25 mm (0.01 in) in diameter. The arrested development is called embryonic diapause and the blastocyst, in effect, becomes a 'spare'. The presence of the first pouch-young (joey) suckling the mother prevents the development of the spare embryo due to a hormonal feedback mechanism. The first joey leaves the pouch for short excursions about 6.3 months (190 days) after entering it. It leaves permanently about 7.8 months (235 days) after birth. Even then

it will continue to suckle for up to a year by sticking its head into the pouch. The spare joey is born (exits the vagina) and enters the pouch within 24 hours after the first joey's departure from the pouch.

The female mates and another spare embryo is stored. The female now produces two kinds of milk: one kind for a joey-at-foot, and another kind for the joey in the pouch. If there is a severe drought, the young-at-foot may die and about 16 percent of the large joeys in the pouch will be ejected; even the undeveloped blastocyst may be aborted. If drought conditions are less severe, a percentage of the young-at-foot will die but the pouch-young may be retained. This system has been finetuned by natural selection to provide a continuous stream of young in all but the most severe conditions. When the environment greens up after a severe drought has ended, the females return to breeding condition.

Kangaroos are well adapted for the hot, dry region of Australia. Red kangaroos can survive with a drink of 2–3 liters of water once every week. Sheep require at least four times as much water, and humans under the same conditions would need at least 10 liters of water each day. Kangaroos lose very little moisture through their dry feces, and they allow their body temperature to drop 2–4 degrees lower than normal overnight so that it can rise to normal with the increasing heat of the day. Their metabolic rate is about 30 percent lower than that of placental mammals, and their normal body temperature is a degree or two lower, at 36°C (97°F), than the placentals'. When exercising vigorously, kangaroos will sweat and pant as a cooling mechanism. When at rest, they do not sweat but pant and lick their forelimbs. The resulting evaporation dissipates heat. The forelimbs are highly vascularized, with a superficial network of blood vessels that assist heat transfer. The cessation of sweating at rest conserves water. The animals' dense fur and behavioral posturing protect against the transfer of ambient heat. Blood vessels in the tail also act as a heat radiator.

There are five species of very large kangaroos. The eastern gray (or forester) kangaroo, *Macropus giganteus* (Figure 10.31), is a silvery gray species that lives in forests or woodlands of eastern Australia. A specimen from northeastern Tasmania reportedly measured 2.64 m (8.7 ft) and weighed 82 kg (180 lb). The western gray, *M. fuliginosus*, is lighter gray to brown and occurs from southwestern New South Wales into Western Australia in wooded areas. The wallaroo or euro, *M. robustus* (Figure 10.32), is a powerfully built 'roo that occupies a wide variety of habitats, including rocky ridges, sclerophyll forests, and deserts all over Australia. The antilopine kangaroo, *M. antilopinus,* is confined to plains of north Australia. The red kangaroo, *M. rufus,* was formerly named *Megaleia rufa*. The females of *Macropus rufus* are called blue fliers because of the blue-gray fur. They may reach 39 kg (85 lb). Males are reddish and may, on average, be slightly larger than *M. giganteus.* Of a series of 426 males collected from western New South Wales, the heaviest individual weighed 77 kg (169 lb) and

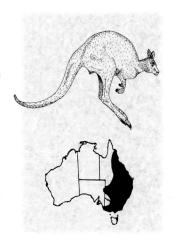

Figure 10.31
Eastern gray kangaroo, *Macropus giganteus*.

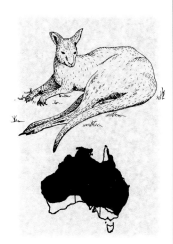

Figure 10.32
Euro, *Macropus robustus*.

Ken Griffiths

Eastern grey kangaroo, *Macropus giganteus*, can be up to 2 m tall and weigh over 66 kg.

Figure 10.33
Tammar wallaby,
Macropus eugenii.

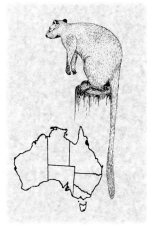

Figure 10.34
Tree kangaroo,
Dendrolagus lumholtzi.

Quokka, *Setonix brachyurus,*
from Rottnest Island off Perth
in Western Australia.
Early European explorers
mistakenly thought that this
animal was a rat.

measured 2.49 m (8.2 ft) along the curve of the body. An extraordinary male red kangaroo may reach 91 kg (200 lb) and live to be 30 years old. Dingos and humans are the main predators of the large kangaroo species.

There are about nine species of large wallabies belonging to the genus *Macropus* (see, for example, Figure 10.33), in addition to *Wallabia bicolor,* and three species of smaller wallabies, called pademelons, in the genus *Thylogale.* Rock wallabies, about 14 species in the genus *Petrogale,* live among piles of rocks, and make incredibly precise leaps from boulder to boulder. Four species of hare-wallabies are in the genus *Lagorchestes.*

The quokka, *Setonix brachyurus,* is a cat-sized wallaby with small rounded ears, from southwestern Western Australia. The species is common on Rottnest Island (off Perth), which the early Dutch explorers mistakenly named after the quokka (rat's nest). They are accustomed to being handfed by tourists who visit Rottnest Island. Quokkas also exist in small colonies in the jarrah and karri forests between Manjimup and Pemberton, but there they are very secretive and hard to find. They prefer to graze on wattle (*Acacia*) at night. Predation by European foxes has seriously reduced their numbers, and habitat alteration due to farming has accelerated their decline.

Tree kangaroos, two species of *Dendrolagus* (Figure 10.34), live in rainforests of northeastern Queensland. They have long, muscular forearms which assist in climbing, and a long, cylindrical, non-prehensile tail is used for balancing. Their feet are shorter than their ground-living relatives and can be twisted from side to side. Unlike other kangaroos, tree kangaroos can move their hind limbs independently of each other. In spite of being agile climbers, these animals hop from the ground into trees and from tree to tree. They back down the tree trunk when descending. They spend the majority of their time sleeping in trees. They eat leaves, too tough for possums, but which they can ferment in their large stomachs. In general, arboreal leaf eaters have a lower metabolic rate than grass eaters, and tree kangaroos are no exception. Their basal metabolic rate is about 30 percent lower than the red kangaroo's. This may represent an adaptation to eating toxin-containing leaves. The lower the metabolic rate the fewer the calories needed and thus the less toxic material ingested. Tree kangaroos tend to lead solitary lives (except for the female and her offspring), interacting only during mating. The young leave the mother's pouch permanently when about 350 days old. This is longer than the 235 days of pouch life for the larger red kangaroos. This slow development may be a reflection of the low metabolic rate of tree kangaroos. About ten species of tree kangaroos are found in the rainforests of New Guinea. The largest Australian species, *D. bennettianus,* weights about 8.6 kg (19 lb) and is about 65 cm (26 in) tall.

Fossil Marsupials and the Megafauna

Three other marsupial families are known from the fossil record. The Wynyardidae, represented by *Wynyardia bassiana* (Figure 10.35) of the late Oligocene, was an arboreal, possum-like creature with some didelphoid characteristics. It is the oldest and most primitive of the Australian marsupial fossils.

Thylacoleo carnifex (Figure 10.36), dubbed the marsupial lion, is relegated to its own family, Thylacoleonidae. It dates to the Miocene and was the largest carnivorous mammal in Australia. It grew to be the size of a leopard, with powerful limbs, clawed feet, and very large, shearing premolars. *Thylacoleo* may have been derived from the herbivorous phalangers. It became extinct in the late Pleistocene, about 18 800 years ago.

The family Diprotodontidae had at least three genera, of which *Diprotodon* (Figure 10.37) is best known. This gigantic wombat-like animal grew to be the size of a rhinoceros during the Miocene. This makes it the largest known marsupial. *Diprotodon* was a grazing animal and, despite its superficial resemblance to wombats, had a tooth cusp pattern very different from theirs. It survived until about 20 000 years ago and was most likely seen by the earlier Aboriginals, who may have hunted it.

There was also a giant kangaroo, *Procoptodon goliath* (Figure 10.38), with a very short muzzle and upturned incisors. *Procoptodon* stood 3 m (9.8 ft) tall and had enormous hindlegs. It could browse on tree leaves, which it crushed with huge molar teeth. It also disappeared in the Pleistocene. This animal and another named *Sthenurus* are put in a subfamily of the Macropodidae, the Sthenurinae. There were other large kangaroos in the subfamily Macropodinae, such as *Protemnodon*, which stood about 1.8 m (5.9 ft) high and died out in the Pleistocene. It is interesting to note that the kangaroos evolved in the Australian Miocene at about the same time as the horses and ruminants of other continents. This was also the time of development of the grasses, which provided a new food source.

The extinction of about 60 species of megafauna (as the gigantic forms have been collectively called) appears to have been substantially completed by about 20 000 years ago (late Pleistocene). Two causes have been proposed: increasing aridity, and the effects of hunting and fire-caused habitat alteration. Climatic change can probably be ruled out because extinction took place throughout

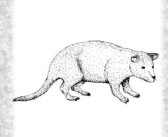

Figure 10.35
Wynyardia bassiana
(a reconstruction of a fossil species).

Figure 10.36
Marsupial lion, *Thylacoleo carnifex* (a reconstruction of a fossil species).

Figure 10.37
Diprotodon optatum
(a reconstruction
of a fossil species).

Figure 10.38
Giant kangaroo, *Procoptodon
goliath* (a reconstruction of
a fossil species).

Australia from the rainforests of the north through the central deserts to the temperate forests of Tasmania. Drought would not affect the entire continent to such an extent that extinctions would be inevitable.

The activities of Aboriginals, using a combination of hunting and fire, could have drastically altered the habitat. There is good evidence that humans and megafauna were contemporary. Sites such as Seton on Kangaroo Island, South Australia (16 000 years ago), Devil's Lair in Western Australia (31 000 years ago), and Menindee in New South Wales (26 300–18 800 years ago) contain both human and megafaunal remains. However, there is a virtual absence of places where megafauna have been killed and butchered. Evidence of breaking, cutting, and burning of megafaunal bones, dating back to 37 000 years ago, is found at Mammoth Cave in Western Australia. Megafaunal species appear to be absent from human sites at Lake Mungo in New South Wales, indicating that extinction occurred prior to 30 000 years ago in that region. The evidence is circumstantial and there is no smoking gun, but many anthropologists favor the hypothesis that overhunting and alterations of the habitat by fire resulted in the extinction of the megafauna.

A new family (Yalkaparidontidae) and order (Yalkaparidontia) of fossil Australian marsupials was described in 1988 from Riversleigh Station, about 200 km (120 miles) north-northwest of Mt Isa, Queensland. Skulls, jaws, and teeth of two species of small marsupials were uncovered from sediment dating to the Miocene, 10–15 MYA. The morphology of this bizarre new mammalian order, named after a northwest Queensland Aboriginal word for boomerang, resembles the marsupial mole (Notoryctidae), but this may reflect convergence rather than ancestry. The habitat at the time was lowland rainforest. Riversleigh Station has yielded more than 200 new species of Tertiary vertebrate fossils, showing that the Australian marsupial fauna was more diverse than previously believed.

PLACENTAL MAMMALS

Bats and Rodents

Bats and rodents arrived in Australia via the Indonesian islands.
A wealth of bat fossils recently found at Riversleigh Station in
northwest Queensland indicates that Australia had a diverse bat fauna
by 25 MYA. Today, there are about 63 species in seven families native
to Australia. The smaller insectivorous bats belong to seven families
(Table 10.1, page 218), and are a joy to study in Australia because
rabies is unknown on that continent. The larger flying foxes or
fruitbats, seven species of *Pteropus* (Figure 10.39) and four other
genera, form the family Pteropodidae and are considered pests by
fruit growers on the east coast. Flying foxes normally feed on
blossoms of eucalyptus and figs. They also eat cultivated fruit. This
habit, combined with their ability to congregate in enormous
numbers, has lead to organized shoots designed to reduce their
numbers around orchards.

Figure 10.39
Flying fox or fruitbat,
Pteropus scapulatus.

Several waves of murid rodents colonized Australia, the first
occurring about 25 MYA. Australia's native rodents, with the
exception of *Rattus,* have a common ancestor and are unique to the
Australasian region. Australia has about 64 species of very interesting
native rodents of the family Muridae. The water rat, *Hydromys
chrysogaster* (Figure 10.40), is in much demand because of its beautiful
pelage (fur), which may be glossy black to golden brown with orange
on the belly. *Hydromys* and the platypus are the only freshwater
mammals in Australia. *Hydromys* feed on fish and crayfish. There are at
least seven species of native bush rats, *Rattus,* plus the three
ubiquitous introduced members of that genus.

The hopping-mice, ten species of *Notomys,* occur in arid regions.
The genus *Pseudomys* has about 23 species of rather typical-looking
mice. *Uromys* and *Melomys* (Figure 10.41) are mosaic-tailed rats, so
called because of the irregular arrangement of scales on their tails.
They belong to a New Guinea group and have established themselves
in northern Australia.

Figure 10.40
Water rat,
Hydromys chrysogaster.

The Dingo

The dingo most likely evolved from the Indian wolf, *Canis lupus
pallipes,* and spread throughout southern Asia. Seafaring Asian traders
transported it to Indonesia, Borneo, the Philippines, and New Guinea,
from which it was brought to Australia about 4000 years ago.

It is difficult to decide if dingos should be considered native or
introduced animals. They are native in the sense that dingos were in
Australia thousands of years before Europeans, but they are
introduced in that they did not evolve in Australia, and they arrived
relatively recently via human intervention. The earliest fossil dingo

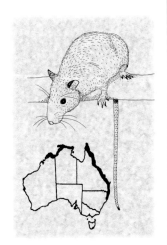

Figure 10.41
Mosaic-tailed rat,
Melomys burtoni.

Figure 10.42
Dingo,
Canis lupus dingo.

remains date to about 3450 years ago, from Madura Cave on the Nullarbor Plain. They never reached Tasmania, as it became separated from the mainland about 12 000 years ago. This allowed thylacines and Tasmanian devils to persist in Tasmania, while competition with dingos exterminated them on the mainland.

Dingos, gray wolves, and domestic dogs are all cross-fertile and considered to be members of the gray wolf species, *Canis lupis.* Domestic dogs belong to the subspecies *C. lupis familiaris.* In older literature domestic dogs and dingos were considered as a species distinct from wolves, *C. familiaris.* Today dingos are classified as *Canis lupis dingo* (Figure 10.42). They differ from domestic dogs in several minor ways. For example, dingos do not bark; they howl. However, they can learn to bark when kept with domestic dogs. They only have one estrus or heat period per year, whereas domestic bitches have two. Dingos mate between April and June (autumn-early winter) and gestation is the same as in domestic dogs—63 days. Dingos probably do not live longer than ten years in the wild. Their courtship behavior and gait are distinctive, and they have a rather aloof attitude. Dingos and dogs interbreed, and the anatomical features of dingos are within the range of variation of dogs. Hybrids often have a dark dorsal stripe or speckling in the white patches. It is feared that within 50–100 years pure dingos will be hybridized out of existence and absorbed into the domestic dog population.

On average, dingos have larger auditory bullae, longer muzzles, bigger carnassial and canine teeth and a larger cranium than do domestic dogs of the same size. The ears of dingos are triangular and permanently erect, and their short, white-tipped tail is bushy. Their feet are white. Coat color can range from black to white but about 88 percent of dingos are yellowish, a genetically dominant color called 'ginger'. Dingos, alone or in pairs, silently stalk their prey. Small marsupials such as wallabies, wombats, and possums, as well as echidnas and rabbits, are their usual prey, but they also kill newborn lambs and calves. This has resulted in the erection of 5614 km (3368 miles) of fences across parts of the Outback, as well as a program of shooting, trapping, and poisoning. However, the dingo is still common over much of Australia, although it has not reached Tasmania. The dingo fences have had the unintended effect of allowing kangaroos to become very abundant within the fenced areas; indeed, in the absence of dingo predation, kangaroo populations have reached plague proportions in some areas. Dingo puppies make interesting, intelligent, and affectionate pets, but when they reach adulthood they are difficult to control and usually do not respond to their owner's commands.

Marine Mammals

Australian marine mammal families (sea lions, seals, and dugong) are listed in Table 10.1 and are not discussed further here. In addition, 36 species of whales have been reported from Australian waters, but

Jean-Paul Ferrero, AUSCAPE International

none can be regarded as 'Australian'. To its credit, Australia is no longer a whaling nation—the last whaling station, at Albany in Western Australia, closed on 31 July 1978. The dolphins of Shark Bay in Western Australia are discussed in Chapter 1.

Dingos, *Canis lupus dingo*, chasing a lace monitor in southeastern New South Wales.

INTRODUCED MAMMALS

The European Rabbit

The European rabbit, *Oryctolagus cuniculus*, was repeatedly introduced into Australia after the arrival of the First Fleet in 1788. However, it was the introduction by one Thomas Austin of 12 wild English rabbits near Geelong, Victoria, in 1859 that changed the face of Australia. From this humble beginning was spawned widespread ecological devastation. By 1880 the rabbits had reached New South Wales and South Australia. Six years later they had spread to Queensland. They were advancing at the rate of 125 km (75 miles) per year. The Western Australia border was crossed by 1894 and the west coast was reached by 1907.

This rapid spread and the ensuing damage to the grasslands prompted governments and property owners to erect huge lengths of 'rabbit-proof fences', some of which were thousands of kilometers long. These failed to halt the rabbits which eventually covered the southern half of Australia. Their spread seemed limited only by the high temperatures north of the Tropic of Capricorn, and now even that barrier has been breached.

The fecundity of rabbits is legendary. They breed, well, like rabbits. Their gestation period is only 30 days and a female can have a litter of 3–12 offspring, depending upon her age and luxuriance of the habitat. Seven litters per year per female are common, and the young become sexually mature at 3–4 months. A single female may produce 38 young in a favorable breeding season.

About 80 percent of the rabbits die within three months, but those that survive to adulthood usually live 2–3 years. In good periods their numbers can increase 12 times in one year. More than 85 percent of the population must die in order to prevent population growth in a favorable year in good rabbit habitat. In less rich areas or in dry years, lower mortality rates can reduce the population.

The principal predators of rabbits are two introduced mammals, the fox and feral cats, as well as dingos, wedge-tailed eagles, other raptors, and some snakes and varanid lizards.

Rabbits are grazers and prefer the choicest green grasses and clovers. Their grazing results in deterioration of pastures and thereby interferes with the sheep and cattle industry. In marginal areas this extra competition has drastically altered the land, causing changes in vegetation and promoting soil drift and erosion. These things have naturally resulted in a great reduction in livestock-carrying capacity. On the plus side, several million dollars worth of rabbit skins and carcasses have been exported each year.

A number of reasons have been identified for the spectacular success of the European rabbit in Australia:

- The rabbit, which originated in Spain, was an ecological generalist and physiologically pre-adapted to the climate of southern Australia.
- Very few of the rabbit's original parasites arrived with it from Europe.
- Rabbits are colonizers of disturbed regions and much of Australia falls under that description due to the conversion of large areas of perennial vegetation into annual grasses for livestock pasture.
- Native Australian predators were depleted by human activity.
- Native Australian grazing mammals produced extensive burrows which were readily exploited by rabbits, especially in arid areas.

After decades of testing, the Commonwealth Scientific Industrial Research Organisation (CSIRO) showed that a pox virus from the South American rabbit, *Sylvilagus brasiliensis,* was deadly to *Oryctolagus cuniculus* but harmless to all other Australian animals and livestock and to humans. The myxomatosis virus is mosquito-borne and in 1950–51 the introduced disease spread spectacularly. It became epidemic and covered much of the rabbits' range by 1952–53. The initial mortality rate was nearly 99.9 percent. The infection is now endemic in wild rabbits and erupts yearly, depending on local conditions.

As any good evolutionary biologist would predict, myxomatosis is only a temporary solution to the rabbit problem. The rabbit populations have been decimated, but they are building again, especially in arid regions. Their numbers are estimated at over 300 million, or 17 rabbits for every Australian. As predicted by evolutionary theory, the rabbits are being selected in the direction of increased genetic resistance to myxomatosis, while the virus is being selected in the direction of less virulence. A new balance between host and parasite is being struck. The 99 percent mortality rate is now closer to 30 percent. A rabbit flea was introduced in 1969 to help spread the disease more efficiently. Meanwhile researchers are trying to stay one jump ahead by developing new strains of the virus.

A new high tech method of control is currently being explored by CSIRO scientists. It involves using a genetically engineered myxoma virus. This recombinant virus is species-specific and is engineered to provoke an auto-immune response in breeding rabbits to specific proteins necessary for reproduction.

Other forms of control, such as poisoning with 1080 (sodium fluoroacetate) or cyanide dust, ripping up the warrens, shooting, or trapping, have proved unsuccessful except on a local basis.

By comparison, the impact of the introduced hare, *Lepus capensis,* has been insignificant. The hare is not susceptible to myxomatosis.

In early 1995 the CSIRO decided to test a new weapon in the biological warfare against the feral European rabbit. An RNA virus called rabbit calicivirus, which causes rabbit hemorrhagic disease, also known as rabbit calicivirus disease (RCD), was released within a quarantined rabbit population on Wardang Island off the South Australian coast. RCD kills via hemorrhaging within 36–48 hours. Preliminary testing had indicated that it affects only European rabbits, not native mammals, birds, reptiles, or livestock. Overseas health screening programs had shown that humans are also unaffected.

By early October 1995, the virus had killed 80 percent of the quarantined rabbits on Wardang Island. Suddenly, and without warning, RCD escaped quarantine and appeared on the mainland in mid-October 1995. By mid-November it had reached Broken Hill in New South Wales, and by December it was in Queensland. It was traveling as fast as 8 km (4.8 miles) daily. In March 1996 RCD reached Victoria. Millions of rabbits died over an area of thousands of square kilometers. Bushflies, active in rabbit warrens, are thought to have fed on infected rabbit carcasses, then flown across the 5 km (3 miles) of sea separating Wardang Island from the mainland, thereby allowing the infection to run amok in rabbit populations throughout Australia. This would be cause for alarm if the virus could cross into other species. At the present time, however, there is no evidence that this is happening or that it is even possible.

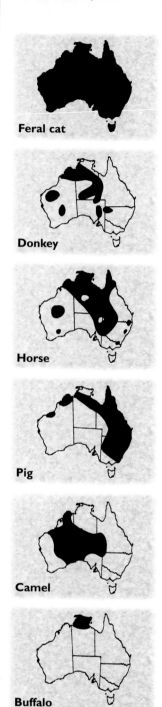

Feral cat

Donkey

Horse

Pig

Camel

Buffalo

Goat

Feral rabbits cause an estimated $115 million a year loss to the wool and meat industries, promote the erosion of millions of acres of topsoil, and are responsible for a great deal of damage to native plants and animals. Over 95 percent of the rabbits are killed where RCD has existed for four or five weeks. This mortality rate is sufficient to overcome the rabbit's enormous fecundity, giving fresh hope to farmers and graziers faced with the problem.

This may be the best time to knock back other introduced predators and encourage native carnivore populations. Efforts are being stepped up to kill foxes and feral cats and dogs for fear that these introduced predators will now kill more native mammal and bird species in the absence of their usual rabbit prey. It is hoped that there will be a resurgence of native plants that have not been seen for decades because their shoots have been grazed mercilessly by the rabbit hordes. In mid-October 1996, following government approval, the CSIRO released the rabbit calcivirus at 280 sites across Australia.

Other Introduced Mammals

Excluding sheep and cattle and failed introductions such as the mongoose (*Herpestes*), alpaca (*Lama pacos*), llama (*L. glama*) and vicuna (*Vicugna vicugna*), there are about 25 exotic mammal species that have become established in Australia (Table 10.2). Of the non-livestock species, the rabbit has had the greatest impact. The introduced house mouse, *Mus musculus,* and rats, *Rattus,* are pests, as they are around the world.

The European red fox, *Vulpes vulpes,* was released in the mid-19th century for sport. Today they number about 5 million and kill medium-sized, native marsupials and birds as well as poultry and lambs. Foxes are in plague proportions in many areas due to a decline in demand for fox skins because of negative publicity aroused by animal rights advocates. The price of pelts has fallen from $50 each to around $4 each. This has made fox shooting uneconomical which, in turn, endangers their prey such as wallabies, bettongs, and numbats. The latter, which formerly covered much of Australia, have been especially decimated and are now confined to a small corner of southwest Western Australia. This range reduction is largely due to fox predation. Foxes do eat rabbits, but considering their destruction of native wildlife they should never have been introduced to Australia. Feral dogs roam in packs and prey upon birds, livestock, and small marsupials, but they are less destructive of the native fauna than are the feral cats, *Felis catus,* which prey on at least 64 species of native mammals. Cats are found throughout Australia.

Herds of wild donkeys, *Equus asinus,* roam sections of central, north, and west Australia. These feral pack animals may number 2–5 million. Their numbers are culled by shooting from helicopters. Wild horses, *E. caballus,* called brumbies, are scattered throughout inland Australia. In some areas brumbies have reached large numbers

TABLE 10.2 **EXOTIC MAMMALS ESTABLISHED IN AUSTRALIA**

ORDER	FAMILY	SCIENTIFIC NAME	COMMON NAME
Lagomorphia	Leporida	*Lepus capenis*	Hare
		Oryctolagus cuniculus	Rabbit
Rodentia	Muridae	*Mus musculus*	House mouse
		Rattus exulans	Pacific rat
		Rattus norvegicus	Norway rat
		Rattus rattus	Black rat
Carnivora	Canidae	*Canis lupus familiaris*	Domestic dog
		Canis lupus dingo	Dingo
		Vulpes vulpes	Red fox
	Felidae	*Felis catus*	Cat
Perissodactyla	Equidae	*Equus asinus*	Donkey
		Equus caballus	Horse
Artiodactyla	Suidae	*Sus scrofa*	Pig
	Camelidae	*Camelus dromedarius*	Camel
	Cervidae	*Axis axis*	Axis deer
		Axis porinus	Hog deer
		Cervus elephus	Red deer
		Cervus unicolor	Sambar deer
		Cervus timorensis	Rusa deer
		Dama dama	Fallow deer
	Bovidae	*Bos javanicus*	Banteng
		Bos indicus	Zebu
		Bos taurus	Cattle
		Bubalus bubalis	Water buffalo
		Capra hircus	Goat
		Ovis aries	Sheep

Source: Based on Strachan (1995).

and are regarded as pests. They destroy pastures and fences and foul water supplies. They are shot or trapped to keep the populations in check. Some wild horses are selected for domestication, others are utilized for pet food or exported for human consumption. There are large numbers of feral pigs, *Sus scrofa,* in eastern Australia and pig-shooting has become a sport as well as a means of control. The razorbacks are escaped domestic stock from the early days of settlement. They are very destructive to native vegetation and soil as well as to crops and small native animals. They destroy water-holding facilities and serve as a reservoir of various livestock diseases such as anthrax and brucellosis.

Considering the large areas of desert in Australia, the introduction of camels, *Camelus dromedarius,* makes more sense than some of the other introductions do. The Victorian government imported 24 dromedaries (one-humped Arabian camels) for use by the ill-fated Burke and Wills expedition in 1860. Other importations followed, and the camel helped explore and develop the arid interior.

About 100 000 of their descendants are living free today in west central Australia. Camels, with their soft food pads, shifting grazing strategy, and relative freedom from the need to drink regularly, do not degrade the desert habitat as other ungulates (sheep, cattle, horses) do. They do compete with livestock for limited grazing in sparsely vegetated areas. The fat in the camel's hump has a very high melting temperature, and used to be cut into small pieces and laced with cyanide or strychnine for use as dingo bait. Its high melting temperature ensured that it remained effective in hot areas for many months. Australian attitudes about camels are changing. They used to be shot on sight. Now camels are viewed as an asset. Some tourism operators offer desert camel rides and safaris to Outback tourists.

Several species of deer have been introduced, especially near Melbourne, where there is limited deer hunting. The various species can damage crops in local situations.

In addition to the domestic cattle and the zebu (or Brahmans) that have gone wild, Australia has wild-living water buffalo, *Bubalus bubalis,* in Arnhem Land—about 100 000 of them. This native Indian species is better adapted to the hot, humid climate than are domestic cattle. It is hunted for hide and meat. Nonetheless, because of the environmental damage it causes, due to erosion and plant destruction, and the cattle disease it carries, an eradication program was carried out which virtually eliminated wild buffalo from the northern wetlands, especially in and near Kakadu National Park. The buffalo is being promoted for redomestication as its meat is less fatty than that of cattle, and it is an excellent milk producer.

The goat, *Capra hircus,* is a serious pest because of its catholic diet. It damages shrub vegetation, which in turn promotes erosion. It also competes with rock wallabies for space in rocky outcrops.

Political and Social Essentials

• • • • • • • • • •

GOVERNMENT

Australia's political institutions and practices reflect both British and North American models of the Western democratic tradition. Basic human rights are guaranteed in law. There are three levels of government: federal, state, and local. Federal and state governments are parliamentary. The federal parliament has two chambers: the House of Representatives (Lower House) and the Senate (Upper House). Parliament meets from February to June and again from August to December. The House of Representatives is required by the Constitution to be approximately twice the size of the Senate. The House has 148 members elected proportionally from the various states. Elections must be held at least every three years. The Senate has 76 members, 12 from each of the six states and two each from the Australian Capital Territory and the Northern Territory. Senators serve six-year terms with half the Senate being returned every three years.

The executive branch of government follows the British system. The party or coalition that achieves a majority in the House of Representatives forms the Government, and the leader of that party becomes Prime Minister. Members of the Cabinet are also members of parliament and are drawn from both houses.

The High Court of Australia is the supreme federal court at the apex of the Australian judicial system. It has appellate jurisdication over all other federal and state courts.

Australia achieved its independent nation status through constitutional processes rather than by unilateral declaration or rebellion. Until passage of the *Australia Act 1986,* Australia, like Canada, gave full allegiance to the monarch of Great Britain. Queen Elizabeth the Second is still formally Queen of Australia, but the *Australia Act* to a large extent severed the constitutional ties with the United Kingdom. The Queen is represented in Australia by the Governor-General and the six state Governors. The national anthem was changed in 1984 from *God Save the Queen* to *Advance Australia Fair*—not *Waltzing Matilda* as many Americans assume. Many citizens want Australia to become a republic by the year 2000, thereby ending the traditional links with the monarchy.

The Governor-General is effectively Head of State and Chief Executive. Among his numerous duties, he summons and dissolves Parliament, approves bills, and appoints ministers and judges. However, these duties are usually performed only on the advice of the Prime Minister and his Government.

A rare occurrence took place in November 1975 when the Governor-General, John Kerr, dismissed the elected Prime Minister, Gough Whitlam (of the left-leaning Labor Party) and appointed the Opposition Leader, Malcolm Fraser (of the right-wing Liberal Party) as caretaker Prime Minister. This startling turn of events shocked the nation and was precipitated by Whitlam's inability to get a Supply Bill through the Senate, which was controlled by a Liberal Party majority. Without the bill being passed, the Government would have run out of money at the end of November 1975. Unfavourable economic conditions and an official loan scandal also contributed to the crisis.

In the following month a new Fraser-led Government was elected which stayed in power until March 1983 when Labor, under the leadership of Bob Hawke, was given a mandate. Labor held power until 1996, with Paul Keating, the former Treasurer in the Hawke Government, becoming Australia's 25th and youngest Prime Minister in late 1991. In March 1996 the pendulum swung again: the Liberal-National Party coalition, under the leadership of John Howard, won office from the Labor Party, with a large majority in the House of Representatiaaves but without control of the Senate.

The state governments are similar to the federal structure: most have a bicameral legislature and a Premier. There are state courts. The states control education, transportation, law enforcement, health service, and agriculture. Local governments also control town planning, local road maintenance, water and sewerage services, maintenance of parks, libraries, and the like.

All elections are by secret ballot and voting is compulsory under penalty of a nominal fine. All adults over 18 years old must vote in state and federal elections and turnouts of nearly 98 percent are not rare. In the 1993 federal elections 96 percent of the eligible voters voted, compared with 55 percent of United States voters who voluntarily exercised their right in the 1992 elections that brought President Clinton to office. Also in 1993, 69 percent of eligible Canadians voted. Australian women were granted voting rights in 1902, nine years after New Zealand and 18 years before women were allowed to vote in the United States.

A preference system is usually used in which the voter ranks the candidates in order of preference. If no candidate receives an absolute majority of first preferences, the candidate with the least votes is excluded and the votes are distributed to the other candidates in accordance with the second preferences. This is repeated until a candidate receives an absolute majority. Voting for the Senate in the federal parliament is by proportional representation. Seats are allocated to competing parties in relation to the proportion of votes each party receives. Smaller parties tend to do relatively well in Senate elections, in part because many Australians have a healthy distrust of their politicians. The Senate is seen in Australia as having a useful 'checks and balances' role.

There are three main political parties: Liberal Party, Australian Labor Party, and National Party. The Liberal Party was formed in 1944. In spite of its name, it is conservative and can be compared with the British Conservative Party. It encourages individual initiative and private enterprise. The ALP (Australian Labor Party) was founded in 1891 and traditionally draws strong support from the trade union movement. It is similar to social democratic parties in Europe. Equality of opportunity for all Australians is its aim, and it advocates use of federal power and finance to reach that goal. The National Party dates from 1918 and usually forms a coalition with the Liberal Party. It was formerly called the Country Party and represented the interests of the rural community by seeking guaranteed prices for primary products and the development of strong manufacturing industry. Under its new name it aims to gain more support in urban areas.

· · · · · · · · · ·

THE PEOPLE, THEIR LIFESTYLE AND STANDARD OF LIVING

There are some 18 million Australians in a land almost the size of the continental United States. That is about 2.3 people per square kilometer, most of whom are clustered along the east and south coasts. By comparison, United States density is 28 per square kilometer. The population of Australia is about the same as the population of Texas. Large areas of Australia are too dry for settlement—the vast arid interior is sparsely settled. Consequently, Australia is one of the most urbanized countries in the world, with some 85 percent of its population living in cities. In 1996, the populations of its major urban centers were:

Sydney 3.8 million	Adelaide 1 million
Melbourne 3.2 million	Canberra 344 900
Brisbane 1.5 million	Hobart 195 300
Perth 1.3 million	Darwin 80 900

Sydney and Melbourne alone account for about 40 percent of the total population of Australia and 70 percent lives in the ten largest cities. The Population Reference Bureau (PRB) in 1994 projected that the Australian population would reach 20.9 million in 2010 and 22.9 million in 2025. Table A1.1 shows selected demographic data for some English-speaking countries.

The Australian population is young, with about half under 25 years of age. About 75 percent of the people are of British and Irish ancestry. Other more recent arrivals include Italians, Greeks,

TABLE A1.1 **POPULATIO**
ENGLISH-SPEAKIN

	Pop. mid-1994 (millions)	Surface area* (sq km)	Pop. density (per sq km)	Crude birth rate (per 1000)	Crude death rate (per 1000)	Annual natural increase (%)
Australia	17.8	7 682 300	2.3	15	7	0.8
Canada	29.1	9 976 139	2.9	14	7	0.7
New Zealand	3.5	268 676	13.0	17	8	0.9
South Africa	41.2	1 221 037	33.7	34	8	2.6
United Kingdom	58.4	244 046	239.3	13	11	0.2
United States	260.8	9 363 123	27.8	16	9	0.7
World	5,607	148 892 000**	37.7	25	9	1.6

† Total Fertility Rate — average number of children born to a woman.
*From the Australian Information Service, Canberra, ACT, Australia.
**From Sverdrup, H.U. and R.H. Fleming (1942), *The Oceans*, Prentice-Hall, New York.
Source: Based on 1994 World Population Data Sheet of the Population Reference Bureau, Inc.

TABLE A1.2 **A SELECTION OF ECONOMIC**
FOR 2

	GDP per head ($) 1991	Inflation annual average 1983–92	Total tax as percent of GDP 1991	Cars per 1000 people 1990†	Secondary school enrollment rate (%)‡ 1990	Life expectancy at birth (years) 1991
United States	22 130	3.8	27	589	92	76
Switzerland	21 780	3.2	32	447	85	78
Germany	19 770	2.2	39	490	97	76
Japan	19 390	1.8	30	285	96	79
Canada	19 320	4.4	36	473	99*	77
Hong Kong	18 520	7.8	11	29	90	78
France	18 430	4.4	41	418	99	77
Sweden	17 490	6.7	50	419	91	78
Italy	17 040	7.4	31	459	79	77
Australia	16 680	6.4	27	435	83	77
Britain	16 340	5.5	36	403	84	75
New Zealand	13 970	7.9	39	455	89	76
Israel	13 460	69.3	40*	145	83	76
Spain	12 670	7.6	35	308	90*	77
Bahamas	11 235	5.1	5*	250*	80	72
South Korea	8 320	5.1	16	27	87	70
Mexico	7 170	59.2	18	65	53	70
Russia	6 920	54.9	40*	50*	80	69
Hungary	6 080	15.2	50*	169	79	70
Brazil	5 240	472.0	75	104	39	66
China	1 680	8.2	45*	2	48	69
India	1 150	9.3	14	2	44	60

† or latest available.
‡ of 12–17 year olds.
* Estimate.
** 1 = most libertarian.

Source: Reported in *The Economist*, 25 December 1993. The rankings in the right-hand columns were subjectively determined by the editorial staff of *The Economist*.

ATISTICS FOR SOME
OUNTRIES

Pop. doubling time (years)	Pop. projected to 2025 (millions)	Infant mortality rate (per 1000)	TFR†	Percent pop. under age 15/over age 65	Life expect. at birth (years)	Urban pop. (%)	Per Capita GNP 1992 (US$)
85	22.9	6.6	1.9	22/12	77	85	17 070
98	36.4	6.8	1.8	21/12	77	77	20 320
76	4.5	7.3	2.1	23/11	75	85	12 060
26	73.2	49.0	4.4	39/4	65	57	2 670
281	62.1	6.6	1.8	19/16	76	92	17 760
98	338.3	8.3	2.1	22/13	76	75	23 120
43	8 378	63	3.2	33/6	65	43	4 340

OCIAL, CULTURAL, AND POLITICAL INDICATORS
OUNTRIES

Infant mortality per 1000 live births 1991	Pop. density: people per 1000 ha 1991	Murders per 100 000 men 1990†	Divorces: Percent of marriages 1990†	TVs per 1000 people 1990	Freedom House Index** 1992: Civil liberties	Rank
9	275	13.3	48	815	1	8
7	1 701	1.4	33	407	1	1
7	2 286	1.0	30	570	2	2
5	3 294	0.7	22	620	2	6
7	29	2.5	43	641	1	14
7	58 121	1.7	12	274	3	10
7	1 036	1.3	31	406	2	11
6	209	1.7	44	474	1	4
8	1 963	3.6	8	424	2	5
8	23	2.7	34	486	1	7
7	2 382	1.0	41	435	2	9
9	128	3.4	37	442	1	12
9	2 396	3.3	18	266	2	13
8	781	1.2	8	396	1	3
25	260	15.3	13	225	2	17
16	4 435	1.3	11	210	3	18
36	452	30.7	8	139	3	16
20	87	16.3	42	283	4	19
16	1 141	3.7	31	410	2	15
58	179	29.4	3	213	3	21
38	1 255	1.0	1*	31	7	20
90	2 902	5.0	1*	32	4	22

Germans, and other Europeans, which make up about 20 percent of the population, and many from southeast Asia have arrived in the last ten years. Australia accepts immigrants who have skills or qualifications that represent a gain to Australia. Many immigrants since 1945 have been refugees, such as Europeans after World War II, Hungarians after the Russian invasion of 1956, and Asians after the Vietnam War. The so-called White Australia Policy was abandoned in 1967. After a two year residence, foreigners may be granted citizenship. Aboriginals and Torres Strait Islanders make up about 1.7 percent of the total Australian population, with 303 300 according to the 1994 census.

The British weekly *The Economist* (25 December 1993) published an analysis of economic, social, cultural, and political indicators of 22 countries (see Table A1.2). Australia was ranked seventh and the United States eighth as desirable places to live. Switzerland, Germany, Spain, Sweden, Italy, and Japan were ranked higher under the subjective criteria of the analysis. Australia ranks with the United States, Switzerland, Canada, Sweden, New Zealand, and Spain as countries most respectful of political freedom and civil liberties. The literacy rate in Australia is 98.5 percent. The infant mortality rate is 6.6/1000 (the United States has 8.3/1000) and life expectancy averages 77 years (76 in the United States)—see Table A1.1.

Lifestyle

Aside from Australians pronouncing some words differently, driving on the left side of the road, and using the metric system, there is little in the way of cultural shock for Americans in Australia. Living conditions, lifestyles, many customs and attitudes, and expectations are similar. Most Australians (70 percent) own their own homes, cars, and television sets. There are 435 cars per 1000 Australians (compared with 589 cars per 1000 Americans). Seatbelt use is compulsory. In Australia in 1990 about 34 percent of marriages ended in divorce; in the United States the figure was 48 percent (Table A1.2).

Australians are frank and friendly, and they dislike pretense and pomposity. The pace of life is somewhat slower than in America, but life in Australian cities is similar to that in American cities. However, the violent crime rate is much lower in Australia. For example, the male homicide rate in the United States in 1990 was 13.3/100 000 head of population; in Australia, in the same year, it was 2.7/100 000 (Table A1.2). In Sydney, with more than 3.7 million people, there were just over 100 homicides in 1992. In Washington DC, with 600 000 people, there were 453 homicides in 1992, a decrease from 489 in 1991. Private handgun ownership is illegal in Australia and, following a tragic mass shooting in Tasmania in 1996, all automatic and semi-automatic rifles and shotguns are now largely outlawed throughout the nation. (A massive gun-surrender scheme, with compensation paid to owners, is currently in place.)

The Legal System

Since American and Australian legal systems were both derived from the British legal system, they have much in common. Some of the similarities are:

- An accused person is presumed innocent until proven guilty beyond a reasonable doubt.
- A person found not guilty may not be tried again for the same offense.
- There is a right to a trial by judge and jury.
- There is a right to apply for release on bail until trial.
- There is a right of *habeas corpus* which protects the accused against arbitrary arrest and imprisonment.

Australia has no capital punishment. Each state maintains its own police force in addition to the Australian Federal Police.

Work

The average Australian works about 40 hours per week, has four weeks of paid vacation each year, and gets another 10 days in public holidays. Australia has been one of the most unionized countries in the world. About 35 percent of the Australian labor force (down from 50 percent in 1982) are members of 157 trade unions, as opposed to about 17 percent of the American labor force that is unionised. Industrial disputes have declined markedly in recent years. In 1988 Australia lost 319 days per 1000 empoyees through disputes, but in 1989–90 this figure dropped to 142 days, and in 1993 the loss was only 39 days, the lowest since 1981. Disputes are often settled by arbitration at state or federal level, but recent changes in the labor relations system have increased the emphasis on mediation. About 75 percent of the Australian work force are employed in the service sector.

Comparing Prices

It is difficult to compare Australian and American prices, because the value of the Australian dollar relative to the US dollar fluctuates daily and can change greatly over time. For example, in June 1982 one Australian dollar was equal to $1.02 in US currency. In August 1997 one A$ was worth US $0.74. The exchange rate is important to a traveller, and while it may not mean a great deal to those living and working in Australia, it does affect the cost of imported goods.

It is my impression that Australian salaries in Australian dollars are similar to American salaries in US dollars for similar jobs, but that almost everything costs more in Australia. The Reebok walking shoes that cost me US$35 in Ohio sold for A$120 in Sydney; Levis 501 jeans that cost US$30 in Ohio sold for A$80 in Sydney; Olympus Stylus 35 mm camera sold for US$99 in United States but cost

A$350 in Sydney; in Ohio I paid US$4.50 for a 5 kg (11 lb) turkey that cost my friends in Victoria A$35. While gasoline was selling for US$1.19 per gallon in Mansfield, Ohio, in August 1997, it cost A$0.72 per liter in Sydney—this is nearly double the United States price, but still cheap by world standards.

Equipping a house or apartment in Australia with major appliances and household electronics (refrigerator/freezer, dishwasher, washer, electric range, microwave, television, VCR, and stereo system) would cost about A$10 000, whereas comparable brands and sizes in the United States would cost US$3500. To set up a house with these items would cost an Australian university lecturer about 21 percent of their annual income. The same items would cost an American assistant professor about 8 percent of the yearly salary. Without converting Australian dollars to US dollars, housing costs are about the same.

Automobiles are very expensive in Australia due to tariff protection. A Honda Accord sedan that sells for US$15 000 would cost about A$30 000.

Australian food prices are also higher than American prices. A grocery basket loaded with 30 staple food items purchased from a supermarket in Sydney would cost A$83.00; the same items purchased at a supermarket in Mansfield, Ohio, sold for US$56.00. Nevertheless, Australians eat well—plenty of beef, lamb, and a wide variety of fruits and vegetables. Meals at a pub are relatively inexpensive, and there are the usual fastfood outlets such as McDonalds (there are 21 McDonalds for every million people in Australia), Pizza Hut, and KFC. Fancy restaurants *are* expensive, but tipping is not compulsory in Australia. Milkbars and delicatessens are more traditional.

Wines and cheeses are the great Australian bargains. You can still get a pretty good Australian wine for under A$10 a bottle. The higher priced wines are superb. Beer, of course, is the great Australian drink and Aussies rate among the highest in the world in beer consumption per capita.

Clothing is expensive, but wool products are plentiful. Department stores and small boutiques are common in major towns and cities, as are the big retail chains.

Opening Hours

In some areas business hours are more restrictive in Australia than they are in the United States. Department stores and other retailers in the cities may open once or twice a week during the evening, but most shops close by 5.30 pm. Weekend trading is becoming more common. Neighborhood milkbars and delicatessens, and some supermarkets, are open longer. Most pubs close at 10 pm.

The Post

Mail service is comparable to that in the United States but slightly more expensive, and there are no Saturday deliveries. Australia Post is a government-owned enterprise that handles 10 million postal items daily.

Education and Science

Education is compulsory to 15 years of age, and my subjective impression is that Australian students are more literate, more knowledgeable and speak better than American children of the same age, from grade school through university level. The school year runs from the end of January to mid-December. About 83 percent of the 12–17 year olds were in secondary school in Australia in 1990. The figure for the United States was 92 percent (Table A1.2, pages 260–61).

There are some 40 universities and there are 285 post-secondary colleges that have a vocational emphasis, such as business management, librarianship, welding, and teacher education. There are ten medical schools and five dental schools at Australian universities. Australia's first private university, Bond University, opened in May 1989. Australian tertiary education was reorganized in 1990 by the mergers of small institutions, as a result of the federal government's economy drive. The federal government pays for most of the cost of university education. Only since 1989 have students been asked to meet part of the cost.

The Commonwealth Scientific and Industrial Research Organization (CSIRO) is a very large government entity with a staff of about 5000 employees at more than 70 laboratories and field stations around the the country. Most fields of scientific research are included in CSIRO activities.

Religion

Religions practised in Australia include Anglican (25 percent), Roman Catholic (25 percent), other Protestant (25 percent). Less than half of 1 percent of those who practice a religion are Jewish. Only about 25 percent of Australians consider religious beliefs to be 'very important', compared with 58 percent of Americans; and 12 percent of Australians say they are non-religious.

Taxation

A social security program provides pensions for the aged, unemployment and sickness benefits, and a health insurance plan. Taxes in Australia, as in the United States, are levied at the federal level (about 80 percent of the taxes), state level (16 percent), and local government level (4 percent). Income tax is deducted from salaries by employers. Taxpayers must file their returns within the four months following the close of the fiscal year on 30 June. The tax load is similar to that in the United States (Table A1.2, pages 260–61). The percent rate of company taxation is 36 percent.

Health

Australia is a very healthy place with a beautiful climate in most areas. Australians love the beach, but they pay the price for sun worship. Australia has the highest death rate related to skin cancer in the world, with about 900 deaths each year. Queensland has twice the skin cancer death rate of Tasmania, as might be predicted from the respective latitudes.

There are virtually no serious tropical or endemic diseases. Rabies is absent from Australia, and importation of pet animals is strictly controlled. Australia is also free of foot and mouth livestock disease and of Newcastle disease which affects birds. In most parts of Australia, water supplies are fluoridated to prevent tooth decay.

Health care and facilities are comparable to those in the United States but are less expensive. In February 1984 a universal health benefits program, Medicare, was introduced by the federal government. It is funded from general revenue and by a 1.5 percent levy on taxable income. Pensioners and low income earners are exempt from the levy. Coverage includes free public hospital care for those without private health insurance, and 85 percent of government-approved charges made by medical practitioners and optometrists.

The Media

There are commercial television stations, the government-owned ABC (Australian Broadcasting Corporation), and the SBS (Special Broadcasting Service, a multicultural service). There are over 140 commercial and public radio stations, and several influential daily newspapers such as *The Australian, The Age* (Melbourne) and *The Sydney Morning Herald.* Interestingly, there is no constitutional guarantee of a free press. Australia also has a large and flourishing book publication industry.

The Arts

The arts are alive and well in Australia. The Australia Council is a statutory authority whose budget provides more than $50 million each year to fund all aspects of the arts. The Australian film industry has recently had one international hit after another with such outstanding films as *My Brilliant Career, Mad Max, The Man from Snowy River, Strictly Ballroom, The Piano, Babe* and many others. Australian rock music is a highly visible and audible export with such groups as INXS and Midnight Oil. Australia has eight professional orchestras of substantial size including six symphony orchestras (one in each state capital).

The Literature Board encourages creative writing by making grants to writers. Two of Australia's best known writers are A.B. 'Banjo' Paterson, known for his bush ballads, and Henry Lawson, whose short stories depict Australians' relationship with the environment. Other prominent writers include Patrick White, Colleen McCullough and Tim Winton, to name only a few. The National Library of Australia holds more than 3.6 million books and over 100 000 periodical titles.

Distinctive Australian painting developed near the end of the 19th century in the works of such artists as Frederick McCubbin. Several living Australian painters enjoy worldwide recognition. These include John Olsen and Bret Whiteley who follow such established names as Sidney Nolan. The Australian National Gallery in Canberra is famous for its collection of Australian art, especially Aboriginal art. Impressive state collections are found in Melbourne's National Gallery of Victoria and Sydney's Art Gallery of New South Wales. There are over 1000 museums in Australia.

Sports

Australians are keen sports fans and support their favorite cricket, rugby, or Australian rules football teams as devotedly as Americans do baseball, football, and basketball. Green and gold are the national sporting colors. Horse racing is widely followed, and the entire country comes to a halt on the afternoon of the first Tuesday in November for the Melbourne Cup. In Victoria itself, the day of this horse race is a public holiday! Weekends may be spent on the beach, or bushwalking, cutting lawns, or at a barbecue.

Transport

Travel within Australia can be by air, rail, or road. Air travel is frequent and regular but it is more expensive than in the United States. The train service, however, is better than in the United States. The Australian domestic airline industry was deregulated on 31 October 1990, a move designed to lower prices by increasing competition. Australia's international airline, Qantas, is the second oldest airline in the world. It grew out of the Queensland and Northern Territory Air Service started in 1920 in outback Queensland. Qantas may be the only word in the English language that begins with a 'Q' and is not followed by a 'U'. In 1993 British Airways acquired 25 percent of Qantas, and Qantas absorbed one of Australia's major domestic carriers, Australian Airlines. Its main domestic competitor is Ansett Airlines.

Many Australian highways have only one lane in each direction, but more four and six lane highways are being built. There are 850 000 km (510 000 miles) of roads, and about 80 percent of all goods are transported by truck. In 1988 the Australian road death total was 2.1 per 100 million vehicle kilometers, compared to 1.5 in the United States, 1.9 in Britain, and 54 in South Korea. Distances and travel times are given in Tables A3.2 and A3.3 (page 281).

· · · · · · · · · ·
THE ECONOMY

By any standard, Australia is a wealthy nation, and is the twelfth largest economy in the world. It has one of the highest ratios of banks to population in the world. Australia has a variety of natural resources and a free enterprise system with considerable federal controls. Economic development was spurred by World War II and by the demands of the post-war immigration program. Sixty-five percent of Australia's land is used for agriculture, most of it for low intensity sheep and cattle grazing. Only about 6 percent of the area is cultivated for crops or used for intensive grazing, lack of water being the major limiting factor. Sheep and wool are Australia's principal rural industry, and Australia leads the world in wool production, producing 33 per-cent of the world's wool. Merino sheep make up 80 percent of the total. Sheep farming is concentrated west of the eastern highlands and in the southeast and southwest regions of the continent. There are about eight sheep for every Australian.

Cattle farming is also very important, especially in the semi-arid and savanna regions of northern and northwest Australia. Australia is the world's largest exporter of beef and veal, much of which goes to the United States for beef patties in fast-food restaurants. This range-fed beef is leaner than the grain-fed United States beef, and is less desirable to the American palate for steaks. American quotas on imported beef are a concern to Australian graziers and to the government. Dairy production is also high.

Wheat is a major grain crop in the southeast and southwest, in conjunction with sheep. Only the United States and Canada export more wheat than does Australia, whose markets include China, Egypt, Pakistan, Japan, Iran, and Russia. Sugar and cotton are important crops in Queensland and northern New South Wales. Sugar prices are supported by a tariff on foreign sugar, which otherwise would undersell the domestic product. Australia may soon become the world's largest raw sugar producer and exporter.

Australia's range of climate allows a variety of fruit production: bananas and pineapples in subtropical areas, apples in the southeastern states, citrus in the irrigated inland area around the Murray River, stone fruits near the state capitals, and wine grapes in the Barossa Valley (SA), Hunter Valley (NSW), and many other areas. More than 400 million liters of wine are produced annually from vineyards in South Australia and New South Wales. The United Kingdom is the leading export destination (48.5 million liters), the United States imported 11.5 million liters of Australian wine in 1994, and Canada imported another 7 million liters. Fresh vegetables are grown throughout the year. Although Australia has the lowest percentage of cultivable area among all continents, it is one of the world's largest food producers. This is due to advances in agricultural technology, farm mechanization, and market expansion.

Prawns (shrimp) and rock lobsters are principal marine exports.

Natural forests occur mostly in the coastal and highland areas of eastern and northeastern Australia. The forests are dominated by *Eucalyptus* and are deficient in softwood. Hence the softwood, *Pinus radiata,* has been introduced from California and supplies about 30 percent of the forest products.

Australia is one of the major producers of minerals and metals. It is the world's second largest exporter of iron ore and third largest producer of iron ore, mostly from Western Australia. Australia is the world's largest coal exporter, with coal being Australia's biggest export earner (12 percent), followed by wool, wheat, beef, and iron ore. Australia is also the largest bauxite and alumina producer in the world. Lead and zinc are major exports, and copper, nickel, uranium, and phosphate rock are also important. Australia is the world's fourth largest gold producer, following South Africa, Russia, and the United States.

Oil refining is of great importance. Australia meets most of its own gasoline requirements, but must import crude oil for the manufacture of lubricating and fuel oils. Overall, Australia produces about 80 percent of its domestic petroleum demand.

Motor vehicle production, shipbuilding, paper production, chemicals, textiles, and food canning are a few of the important manufacturing industries.

Nine of Australia's top 12 trading partners are Asian countries (Table A1.3, page 272). Japan is Australia's major customer, receiving 25 percent of Australia's exports. Australia is one of only a handful of countries that consistently run a trade surplus with Japan. Japan is the biggest buyer of Australian wool, iron ore, coal, and other raw materials. New Zealand is the largest market for manufactured goods. The United States is Australia's largest source of imports and second largest export market. It is also Australia's largest market for beef. Formerly Britain was the largest source of Australia's imports (40–50 percent), but it has now been replaced by the United States and Japan. About 22 percent of the imports now come from United States and 19 percent from Japan. Australia exported to the United States in 1994 about US$3.4 billion worth of goods and imported from there goods worth about US$11 billion. Australia's three-to-one trade deficit with the United States is significantly higher than the United States' trade deficit with Japan. American trade barriers and export subsidies are responsible for some of the Australian deficit. Australia ranks 20th among the trading nations of the world, even though it ranks only 40th in population. The two major constraints on Australia's economic growth are the small population size and great distances—distance from trading partners as well as distances within Australia. These economic 'problems' are, however, ecological benefits in my opinion.

As one might expect from the low average rainfall and limited areas of high-terrain relief, Australia is not one of the world's major producers of hydro-electric power. Tasmania, though, has reliable rainfall and a number of high-gradient rivers have been dammed. The island state accounts for about half of Australia's total hydro-electric output. The price paid for this, of course, has been the degradation of free-flowing rivers and the concomitant loss of biodiversity. The second largest producer of hydro-electricity in Australia is the Snowy

Mountain Scheme. This project was begun in 1949 and completed in 1974 at a cost of A$810 million. This enormous engineering feat in southeastern Australia impounds the south-flowing waters of the Snowy River and diverts them inland through two tunnel systems into the Murray and Murrumbidgee rivers. The scheme includes 16 dams that operate seven power stations with a generating capacity of 3.7 million kilowatts. After power generation, the water is used for irrigation of the dry interior region. Victoria and New South Wales derive a substantial portion of their electric power from the Snowy Mountains. Electricity costs in Australia are comparable to those in the United States—among the cheapest in the world. Most of Australia's electric power is produced by coal-fired generators, but solar and wind power sources are growing in importance.

· · · · · · · · · ·
CONSERVATION

Australia was one of the first of the 150 countries to sign and ratify the conventions on conservation of biodiversity and climate change at the Earth Summit in Rio de Janeiro in June 1992. A conservation ethic is becoming increasingly strong in Australia, and more 'green policies' are being put into effect. Australia has a national strategy to eliminate ozone-depleting chemicals, and recycling and water usage reduction are encouraged. The federal government considers the environmental implications of its foreign aid. State governments are responsible for pollution control, which must meet national guidelines.

By 1991 about 6.4 percent of Australia's total land area (49.9 million ha) was protected as national parks and other types of conservation reserves. An additional 43.6 million ha is protected in marine reserves that form 4.9 percent of Australia's total marine area. There are 3587 nature conservation areas in Australia. The largest terrestrial area is Kakadu National Park, at 20 000 km², in the Northern Territory. The Royal National Park south of Sydney was established in 1879. This 7284 ha (18 000 acre) reserve was Australia's first national park and the world's second after Yellowstone in the United States.

There are ten World Heritage areas in Australia. These are regions of unique natural and cultural features of worldwide significance. They are the Great Barrier Reef, Kakadu National Park, the Willandra Lakes region in New South Wales, the Tasmanian wilderness, Lord Howe Island Group, the Australian east coast temperate and subtropical rainforest parks, Uluru (Ayers Rock) National Park in the Northern Territory, a wet tropics area of Queensland, Shark Bay in Western Australia, and Fraser Island in Queensland.

ASPECTS OF INTERNATIONAL RELATIONS

Australia has had a longstanding close relationship with the United States, deriving from common culture and language. This relationship was initiated when Australian Prime Minister John Curtin asked for and received United States help to defend Australia against a possible Japanese invasion during World War II. The Battle of the Coral Sea took place between 5 and 8 May 1942 off the northern coast of Queensland. This first major battle between Allied and Japanese fleets stemmed Japan's advance in the South Pacific and thwarted its plans to invade Port Moresby (in Papua New Guinea) and Australia. This ongoing alliance promotes stability in the Pacific. The joint United States-Australian base at Pine Gap, near Alice Springs (NT), is a site for the collection of intelligence. Nurrungar, near Woomera (SA), is a key link in the United States defense support program that monitors launching and detonation of nuclear weapons. The Pine Gap and Nurrungar bases played an important role in detecting Iraqi Scud missile launches and eavesdropping on Iraqi communications during the 1991 Gulf War. A base in the North West Cape region (WA) plays a part in the United States nuclear submarine strategy. Command of this base was handed over to the Royal Australian Navy in 1992. The submarine communications station will become a wholly Australian facility by the year 2000. These bases were a source of concern to some Australians, who pointed out that the facilities made Australia a target in the event of nuclear war. However, with the collapse of communism and the end of Cold War tensions, this is no longer much of a problem. Australia is one of the few countries that fought alongside the United States in all five wars in which the United States was involved during the 20th century.

Australia has an especially close relationship with nearby New Zealand that covers a wide variety of political, economic, and social issues. Papua New Guinea, as a former Australian trust territory, also has close links with Australia. Papua New Guinea is a major recipient of Australian foreign aid.

As Table A1.3 shows, Asia has become the principal Australian market. There is a growing awareness in Australia of the political, economic and strategic importance of Asian countries. This has led Australia to establish closer relations with Asian nations as well as its South Pacific neighbors.

TABLE A1.3 **AUSTRALIAN TRADE STATISTICS FOR 1992–93**

Major markets	Percent exports to	Percent imports from
Japan	25.0	18.7
United States	8.1	21.8
Korea, Republic of	6.5	2.8
Singapore	6.2	2.5
New Zealand	5.5	4.7
Taiwan	4.4	3.7
Hong Kong	4.3	1.3
United Kingdom	3.9	5.7
China	3.7	4.3
Indonesia	2.8	2.2
Malaysia	2.2	1.6
Thailand	2.0	1.3
Germany	1.6	5.7
Italy	1.4	2.3
France	1.4	2.5
Papua New Guinea	1.4	2.1
Other countries	19.6	16.8

Source: Statistical Services, October 1993, Department of Foreign Affairs and Trade, Canberra.

Australian
Idiomatic Language

Just as related organisms may speciate while geographically isolated, so too may languages diverge from the parental stock. Australians and Americans both claim to speak English, but neither speak the same mother tongue our British forefathers brought to our respective continents. (Do you ever wonder why it isn't 'father tongue' and 'foremothers'?)

Australian English has evolved to include words of Aboriginal origin, words for things that are endemically Australian, and all sorts of slang, as well as words from British English. To make matters more complicated, Australians have incorporated many Americanisms into their vocabulary, due to contact with American GIs during World War II and the Vietnam War, and to the barrage of American movies and television programs as well as other cultural insults.

Some Australians have the attractively peculiar habit of adding an 'ie' or 'y' to the end of many nouns, thus making the word a diminutive. It sounds like baby talk when grown men talk about shopping for 'Chrisie presies' instead of Christmas presents. A college student might say, 'I made a screaming uie into the uni to watch the footie on the tele with a tinny on me belly'. What he is trying to communicate is that he made a U-turn into the university to watch a football game on television while drinking from a can of beer.

Australians do not seem to have major geographically varying accents as do Americans. However, there are three forms of Australian pronunciation, which vary with social class and education rather than region of the country. *Cultivated Australian* is used by about 10 percent of the population and sounds somewhat like southern British English. *Broad Australian* is easy to recognize and is used by about 35 percent of the population. In Broad Australian 'praise' sounds like 'prize' and 'buy' may be heard as 'boy'. The other 45 percent of the population speak *General Australian*, which lies between the other two forms.

The following are just a few of the more common words or expressions used by many Australians. Some are uniquely Australian, some are Aboriginal, and some are British. If a few of the selections seem a bit salacious, such is the lot of an American field biologist in Australia (especially when his technical assistant is Robert Barton). These are the words and expressions I heard most often from my friends. These are certainly not all of the Aussieisms and many of the words have other meanings as well as those listed.

(The) Alice—Alice Springs.

Are you right?—equivalent to a sales clerk saying, 'May I help you?'

arse—rear end; a stupid person.

arvo—afternoon.

Aussie Rules—Australian Rules football.

avago—have a go, try it.

back o' Bourke—a very faraway place, similar to 'Woop Woop'; as in 'He's gone fishing back o' Bourke'.

banana bender—someone from Queensland.

barbie—barbecue.

barrack—to cheer for or against a team or person; see 'root', which polite people do not do in public.

bastard—in the friendly sense, a good ol' boy; in the unfriendly sense, an obnoxious person.

beaut or beauty—great, terrific, nice; broadly pronounced 'bewdy'.

billabong—a stagnant pool in the bed of an intermittent stream.

billy—metal can for boiling tea over a campfire.

Biro—ballpoint pen.

biscuit—cookie or cracker.

bloke—guy, man, fellow.

bloody—the great Australian expletive; as in 'The bloody bastard stole my car'.

blow—damn, as in 'I'll be blowed'.

bludger—someone who works as little as possible; as in 'dole-bludger'.

blue—a fight.

bluey—a bundle; a cattle dog.

bob—a shilling in the days before decimal currency.

bonnet—hood of a car.

bonzer—very good, excellent.

boomer—anything big; an adult male kangaroo.

boot—trunk of a car.

brolly—umbrella.

Buckley's chance—no chance at all.

bugger—term for a man viewed with affection or humor as in 'silly old bugger'.

bugger all—nothing, as in 'He knows bugger all about sheilas'.

bugger up—to break or ruin.

bum—arse.

bush—the countryside.

bush lawyer—person unqualified to give legal advice, but who does so anyway.

bushranger—outlaw living in the bush; someone who robs through high prices.

caravan—house-trailer pulled by a car.

cask—lined cardboard container in which some cheaper brands of wine are sold.

cheeky—impudent.

cheese and kisses—wife (rhymes with missus; from Cockney slang).

chemist—druggist, pharmacist.
chips—French fries.
chook—chicken.
chunder—to throw up, vomit.
cobber—friend, mate.
cockie—farmer.
come the raw prawn—attempt to gain advantage, to con someone.
compo—workers' compensation.
cordial—liquid fruit drink concentrate.
corroboree—a celebration, an Aboriginal dance.
crook—sick, broken, no good.
cuppa—cup of tea.

dag—an eccentric, oddball or slob; from daglock, a dung-encrusted
    lump of wool near a sheep's bottom.
damper—unleavened bush bread cooked on campfire.
dear—expensive.
dero—a derelict.
didgeridoo—Aboriginal musical instrument.
Digger—an Australian soldier.
dill—idiot.
dingo—Australian wild dog, *Canis lupus dingo*; coward or cheat.
dinkum—genuine.
dob in—to inform against someone; report to the police.
docket—receipt or bill.
dole—unemployment benefit.
drongo—slow-witted person; a bird, *Dicrurus bracteatus*, that
    sometimes migrates to colder regions during winter.
drover—one who drives cattle or sheep to (usually) a distant
    destination.
dunny—lavatory, outhouse.

earbash—to talk incessantly.
Esky—portable ice box.

fair dinkum—genuine or true, as in 'fair dinkum Aussie'.
fair go—fair chance.
flagon—half-gallon of wine in glass container.
flaming—an informal intensifier, as in 'You flaming idiot'.
flat out—as fast as possible.
flog—to sell.
footy—Australian rules football.
fossick—to search for something, especially minerals, gems, gold, etc.
first floor—floor above ground floor (2nd floor in the United
    States).

galah—a small Australian cockatoo, *Cacatua roseicapilla*; a fool or
    simpleton.
garbo—garbage man.
gibber—stone or boulder.
gaol—jail.

g'day—good day.

good on you—well done, a term of congratulations.

grazier—sheep or cattle rancher.

grotty—dirty, unkempt.

ground floor—at ground level (1st floor in the United States).

gum tree—any Australian eucalyptus tree.

hire purchase—buying over time.

humping your bluey—to carry a pack on your back (traditional).

in it—to do something or go along with it; as in: 'Hey, mate, let's go sink some tubes' (drink some beer). 'Sure, I'll be in it!'

Itie—an Italian (pronounced eye-tie).

jackeroo—young male management trainee on a sheep or cattle station.

jelly—gelatin, Jello.

joey—young kangaroo or other marsupial.

jumbuck—sheep.

jumper—sweater.

Kiwi—New Zealander.

knockback—refusal or rejection.

layby—down payment on an itme to be collected when paid in full.

lemonade—carbonated lemon-lime drink like 7 UP.

lolly—candy.

lollywater—any sweet, colored softdrink.

mallee—scrub in arid region composed of low-growing, multiple stem *Eucalyptus.*

mate—form of address for males, as in 'G'day, mate'; friend.

matilda—a swag or pack of belongings.

me—my; as in 'Me mates and me grog'.

middy—half-pint glass of beer; size varies in the different Australian states.

mongrel—unpleasant person; frequently used before 'bastard'.

mossie—mosquito.

mulga—scrub or bush country in arid areas, composed of dense stands of *Acacia.*

nappie—diaper.

Never-never (the)—remote desert country, as in central and northern Australia.

New Australian—an immigrant, usually from southern Europe (the term is now rare).

no-hoper—loser, failure, or useless person.

nong—dill, drongo, or galah; a stupid or incompetent person.

nosh—tucker, food.

ocker—uncultivated Australian; good ol' boy; redneck.

offsider—assistant or partner.

Outback—remote bush country.
Oz—Australia.

paddock—any area of fenced land.
petrol—gasoline.
pinch—steal.
piss—beer or booze.
piss off—go away, get lost.
plasticine—modelling clay.
plonk—very bad white wine (a corruption of *vin blanc*); see 'red
    Ned'.
point the bone—wish bad luck, curse.
pokies—slot machines, poker machines.
Pommy—Englishman; sometimes shortened to Pom, mildly
    offensive, frequently used between 'bloody' and 'bastard';
    probably derived from 'Prisoner of Mother England' stamped on
    clothes of transported convicts.
poof, pooftah or poofter—an effeminate male or a homosexual man.
prang—auto accident.
pull your head in—shut up, be quiet!
punt—to bet or gamble.

quid—money, formerly a pound (20 shillings); as in 'not the full
    quid'; mentally deficient.

Rafferty's rules—no rules at all.
randy—lustful, horny.
ratbag—troublemaker.
red Ned—bulk claret, really bad red wine.
ripper—beaut, exclamation of approval.
roo—kangaroo.
root—slang for sexual intercourse, noun or verb.
rubber—eraser.

scone—biscuit.
screw—prison guard; salary or earnings.
septic—an American (goes with tank, which rhymes with Yank).
sheila—girl or woman.
she's apples—it will be alright; don't worry.
she'll be right—that's OK, reassurance.
shoot through—leave in a hurry.
shout—turn at buying drinks, as in 'It's my shout'.
sickie—day of sick leave from work.
skite—brag, boast.
sleeper—railroad tie.
smoko—short work break for a cigarette or tea.
snag—sausage.
souvenir—to steal or purloin something.
spanner—a wrench.
spunk—attractive male or female.
station—large sheep or cattle ranch.

stickybeak—nosey person or a look around.
stockman—Australian cowboy.
straight away—immediately.
strewth!—exclamation of suspicion or dismay, like 'good grief!'
stubby—small bottle of beer.
swag—pack containing personal belongings.
swagman—vagrant worker, also called swaggie.

ta—thank you (pronounced 'tah').
tall poppies—successful people, who some Australians feel compelled
    to 'knock'; this is known as 'the tall poppy syndrome.
tank—dam formed by excavation, for water storage; very large
    container for water storage.
Tassie—Tasmania.
tea—dinner, evening meal, as in 'What's for tea, mum?'; (if you're
    invited to tea in the bush don't eat dinner first!).
technicolor yawn—chunder; vomit.
tinny—can of beer.
togs—swimming costume.
too right!—yes sir!
torch—flashlight.
tube—can of beer.
tucker—food.
tucker box—container for food.

uni—university.
ute—pickup truck, utility truck.

waltzing matilda—travelling the road, carrying one's swag.
walkabout—periodic nomadic excursion into the bush by an
    Aboriginal, as in 'He's gone walkabout'.
walloper—policeman.
wank, wanker—to masturbate, a masturbater; self-indulgent or
    egotistical person.
whinge, whinger—to complain, chronic complainer.
wog—flu, as in 'I've got a wog and I'm feeling pretty crook. Guess
    I'll take a sickie'; also means a foreigner, typically a non-white or
    non-English-speaking one (derogatory).
Woop Woop—a name for any backward or remote town.
wowser—straightlaced person.
wrapped (rapped)—very pleased.

yabbie—freshwater crayfish.
yakka—work.
Yank—an American.
You wouldn't read about it!—incredible.

Z—the last letter of the alphabet; pronounced 'zed'.

Some Facts and Figures for Travelers

· · · · · · · · · ·

TRAVELING FROM THE UNITED STATES AND EUROPE

The nonstop flight from Los Angeles to Sydney takes 14.5 hours. This is a long time to be cooped up in an airplane. It is a good idea to get up and walk around once an hour. This promotes blood circulation and helps prevent swelling of the feet. Since cabin air is very dry you should drink lots of liquids such as fruit juice or water. Alcoholic beverages should be avoided because alcohol inhibits the secretion of anti-diuretic hormone from the pituitary gland and, therefore, is dehydrating. You will excrete more liquid than you ingest in beer, wine, or liquor. The extra liquid comes from your tissue fluids. Colas and coffee with caffeine are also dehydrating.

Such a long journey is bound to disturb your sleep/wake patterns. This disruption is commonly known as jet lag. My personal experience has been such that the trip from the United States to Australia has produced very little disturbance, because the journey is westward. This lengthens the day. On arrival, even though you may be tired, it is best to carry on with the day according to local time. If you arrive in the morning, have breakfast and undertake a full day's schedule. Exposure to bright lights helps reset your biological clock. A visit to the beach is in order. Eat dinner at the usual dinner time and go to bed when it is dark.

Returning to the United States from Australia is another matter. Traveling eastward compresses the day/night cycle and produces serious jet lag. One way to deal with this is to begin shifting your sleep/wake cycle before you leave Australia. Go to bed and get up an hour later each day for three days before departure. Exposure to several hours of sunlight upon arrival is helpful. It is also a good idea to break up the flight, if convenient—spend a few days in Hawaii.

Travelers from Europe have an even longer flight, and can expect the same biological clock disturbance. Determine whether you want to be less exhausted on arrival in Australia or on arrival back home, then plan your route accordingly.

- - - - - - - - - -
TIME ZONES AND DISTANCES

The standard time zone system is based on the division of the world into 24 zones, each of 15° longitude. The 'zero' time zone is centered at the Greenwich meridian with longitudes 7.5°W and 7.5°E as its western and eastern limits. The time in this zone is Greenwich Mean Time. The 12th time zone is divided in two by the 180th meridian (the International Date Line) and the two parts are designated the plus 12 and minus 12 zones. Table A3.1 should enable you to calculate the time in Australia or in America given local time in one of the two places. Both countries utilize daylight-saving time during part of the year. This is not shown in Table A3.1.

TABLE A3.1 **LOCAL TIME CONVERSION CHART (24 HOUR CLOCK)**

Hawaii	North America (Pacific)	North America (Mountain)	North America (Central)	North America (Eastern)	Greenwich Mean Time	Perth, WA; Philippines	Australian E.T; Papua New Guinea	New Zealand; Fiji
1500	1700	1800	1900	2000	0100	0900	1100	1300
1600	1800	1900	2000	2100	0200	1000	1200	1400
1700	1900	2000	2100	2200	0300	1100	1300	1500
1800	2000	2100	2200	2300	0400	1200	1400	1600
1900	2100	2200	2300	2400	0500	1300	1500	1700
2000	2200	2300	2400	0100	0600	1400	1600	1800
2100	2300	2400	0100	0200	0700	1500	1700	1900
2200	2400	0100	0200	0300	0800	1600	1800	2000
2300	0100	0200	0300	0400	0900	1700	1900	2100
2400	0200	0300	0400	0500	1000	1800	2000	2200
0100	0300	0400	0500	0600	1100	1900	2100	2300
0200	0400	0500	0600	0700	1200	2000	2200	2400
0300	0500	0600	0700	0800	1300	2100	2300	0100
0400	0600	0700	0800	0900	1400	2200	2400	0200
0500	0700	0800	0900	1000	1500	2300	0100	0300
0600	0800	0900	1000	1100	1600	2400	0200	0400
0700	0900	1000	1100	1200	1700	0100	0300	0500
0800	1000	1100	1200	1300	1800	0200	0400	0600
0900	1100	1200	1300	1400	1900	0300	0500	0700
1000	1200	1300	1400	1500	2000	0400	0600	0800
1100	1300	1400	1500	1600	2100	0500	0700	0900
1200	1400	1500	1600	1700	2200	0600	0800	1000
1300	1500	1600	1700	1800	2300	0700	0900	1100
1400	1600	1700	1800	1900	2400	0800	1000	1200

PREVIOUS DAY

FOLLOWING DAY

Most Australian states observe daylight-saving time from November to early March, during which period the clocks are advanced one hour. But it can be complicated. The times for 1996 and into 1997 were as follows:

March–October		November–March	
Vic, NSW, Tas, ACT, Qld	12 noon	Vic, NSW, Tas, ACT	12 noon
SA	11:30 am	SA	11:30 am
		QLD	11:00 am
NT	11:30 am	NT	10:30 am
WA	10:00 am	WA	9:00 am

The setting of daylight-saving times is somewhat contentious in parts of Australia and is subject to change. If appropriate, you should check on current arrangements upon arrival.

Table A3.2 shows distance and travel time from Sydney to major cities, Table A3.3 road distances between cities, and Table A3.4 metric measurement conversions.

TABLE A3.2 **DISTANCES AND TRAVEL TIMES**

Sydney to:	Distance (km)	Travel time (hrs: mins) Air	Rail	Bus
Adelaide	1422	1.50	25.15	23.40
Alice Springs	2960	5.20	64.30	56.25
Brisbane	1027	1.10	15.30	17.45
Canberra	304	0.30	5.00	4.45
Darwin	4095	6.30	—	92.50
Hobart	1145	2.30	18.00	19.30
Melbourne	893	1.10	13.00	14.30
Perth	4135	5.20	65.45	72.15

TABLE A3.3 **ROAD DISTANCES BETWEEN CITIES (KILOMETERS)**

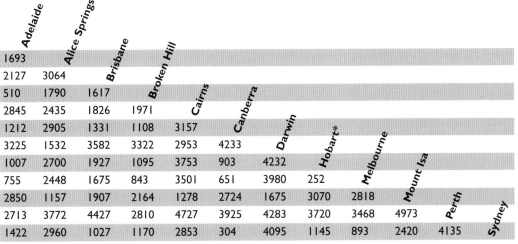

Adelaide	Alice Springs	Brisbane	Broken Hill	Cairns	Canberra	Darwin	Hobart*	Melbourne	Mount Isa	Perth	Sydney
1693											
2127	3064										
510	1790	1617									
2845	2435	1826	1971								
1212	2905	1331	1108	3157							
3225	1532	3582	3322	2953	4233						
1007	2700	1927	1095	3753	903	4232					
755	2448	1675	843	3501	651	3980	252				
2850	1157	1907	2164	1278	2724	1675	3070	2818			
2713	3772	4427	2810	4727	3925	4283	3720	3468	4973		
1422	2960	1027	1170	2853	304	4095	1145	893	2420	4135	

* Excludes Melbourne – Devonport (Bass Strait) ferry crossing

· · · · · · · · · ·

METRIC CONVERSIONS

TABLE A3.4 **METRIC CONVERSION TABLES**

APPROXIMATE CONVERSIONS TO METRIC MEASURES

Symbol	When you know...	Multiply by...	To find...	Symbol
LENGTH				
in	inches	2.54	centimeters	cm
ft	feet	30.5	centimeters	cm
yd	yards	0.9	meters	m
mi	miles	1.6	kilometers	km
AREA				
in^2	square inches	6.5	square centimeters	cm^2
ft^2	square feet	0.09	square meters	m^2
yd^2	square yards	0.8	square meters	m^2
mi^2	square miles	2.6	square kilometers	km^2
	acres	0.4	hectares	ha
MASS (weight)				
oz	ounces	28	grams	g
lb	pounds	0.45	kilograms	kg
	short tons (2000 lb)	0.9	tonnes	t
VOLUME				
tsp	teaspoons	5	milliliters	ml
Tbsp	tablespoons	15	milliliters	ml
fl oz	fluid ounces	30	milliliters	ml
c	cups	0.24	liters	l
pt	pints	0.47	liters	l
qt	quarts	0.95	liters	l
gal	gallons	3.8	liters	l
ft^3	cubic feet	0.03	cubic meters	m^3
yd^3	cubic yards	0.76	cubic meters	m^3

APPROXIMATE CONVERSIONS FROM METRIC MEASURES

Symbol	When you know...	Multiply by...	To find...	Symbol
LENGTH				
mm	millimeters	0.04	inches	in
cm	centimeters	0.4	inches	in
m	meters	3.3	feet	ft
m	meters	1.1	yards	yd
km	kilometers	0.6	miles	mi
AREA				
cm^2	square centimeters	0.16	square inches	in^2
m^2	square meters	1.2	square yards	yd^2
km^2	square kilometers	0.4	square miles	mi^2
ha	hectares (10 000m^2)	2.5	acres	ac
MASS (weight)				
g	grams	0.035	ounces	oz
kg	kilograms	2.2	pounds	lb
t	tonnes (1000 kg)	1.1	short tons	
VOLUME				
ml	milliliters	0.03	fluid ounces	fl oz
l	liters	2.1	pints	pt
l	liters	1.06	quart	qt
l	liters	0.26	gallons	gal
m^3	cubic meters	35	cubic feet	ft^3
m^3	cubic meters	1.3	cubic yards	yd^3

Information Sources

REFERENCES AND FURTHER READING

Chapter 1: Great South Land

Anderson, I. (1994) Dolphins pay a high price for handout. *New Scientist,* vol. 144, pp. 5.

Anon. (1985) *The Australian almanac,* Angus & Robertson, Sydney. 793 p.

Australian Bureau of Statistics (1984) *Australia at a glance,* ABS, Canberra.

Australian Information Service (1981–82) Fact sheets on Australia, Department of Administrative Services, Canberra.

Bayly, B. (1984) *Australia in brief,* Australian Information Service, Canberra. 80 p.

Blume, H. (1992) *Colour atlas of the surface forms of the Earth,* Harvard University Press, Cambridge, MA. 140 p.

Breeden, S. and B. Wright (1989) *Kakadu,* Simon & Schuster, Sydney. 208 p.

Buick, R., J.R. Thornett, N.J McNaughton, J.B. Smith, M.E. Barley, and M. Savage (1995) Record of emergent continental crust—3.5 billion years ago in the Pilbara craton of Australia. *Nature,* vol. 375, pp. 574–77.

Bunce, P. (1988) *The Cocos (Keeling) Islands,* Jacaranda Press, Milton, Qld. 144 p.

Clark, C.M.H. (1968) *A history of Australia,* 5 vols, Melbourne University Press, Melbourne.

———(1969) *A short history of Australia,* Mentor, New York. 280 p.

Conlon, D. (1987) *Pocket history of Australia from 1770 to federation,* Child & Associates, Sydney. 222 p.

Coppell, B. (1994) *Australia in facts and figures,* Penguin Books, Ringwood, Victoria. 502 p.

Dayton, L. (1994) Outback oases. *New Scientist,* vol. 143, pp. 30–34.

Grotzinger, J.P. and D.H. Rothman (1996) An abiotic model for stromatolite morphogenesis. *Nature* vol. 383, pp. 423–25.

Hill, B. (ed) (1983) *Australia handbook 1983–84,* Australian Information Service, Canberra. 160 p.

Hughes, R. (1987) *The fatal shore,* Knopf, New York. 688 p.

Jeans, D.N. (ed) (1977) *Australia: a geography,* Sydney University Press, Sydney. 571 p.

Kerle, A. (1995) *Uluru, Kata Tjuta and Watarrka,* University of New South Wales Press, Sydney. 202 p.

Leeper, G.W. (ed) (1970) *The Australian environment,* CSIRO and Melbourne University Press, Melbourne. 163 p.

McNamara, K. (1990) Survivors from the primordial soup. *New Scientist,* vol. 128, pp. 50–52.

———(1992) *Stromatolites,* Western Australian Museum, Perth. 27 p.

Molony, J. (1987) *History of Australia,* Viking Penguin Books, Ringwood, Victoria. 422 p.

Population Reference Bureau (1994) *World population data sheet,* Washington, DC.

Schall, J.J. and E.R. Pianka (1978) Geographical trends in number of species. *Science,* vol. 201, pp. 679–86.

Shaw, A.G.L. (1967) *A short history of Australia,* Praeger, New York. 332 p.

Skivington, R. (ed) (1980) *Australian handbook 1980–81,* Australian Information Service, Canberra. 160 p.

Smith, R.F. (ed) (1992) *Australia,* Australian Government Publishing Service, Canberra. 126 p.

Smolker, R.A., J. Mann, and B.B. Smuts (1993) Use of signature whistles during separations and reunions by wild bottlenose dolphin mothers and infants. *Behavioral Biology and Sociobiology,* vol. 33, pp. 393–402.

Smuts, B.B., R. Connor, J. Mann, A. Richards, and R. Smolker (1991) Dolphins of Shark Bay. *Yearbook of Science and the Future,* Encyclopedia Britannica.

Stevens, J.D. (1994) Whale sharks at Ningaloo Reef, northern Western Australia. *Chondros,* vol. 5, pp. 1–3.

Steven, M. (1988) *First impressions: The British discovery of Australi,* British Museum (Natural History), London. 96 p.

Various authors (1983) The weather 'Down Under'. *Weatherwise,* vol. 36, no. 3.

Voice, N. (1983) Drought. *Australian Natural History,* vol. 21, pp. 2–9.

Wilks, G. (1994) *Australia in brief,* Australian Government Publishing Service, Canberra. 112 p.

Chapter 2: The Gondwana Legacy

Alexander, T. (1975) A revolution called plate tectonics has given us a whole new Earth. *Smithsonian,* vol. 5, no. 10, pp. 30–40 and vol. 5, no. 11, pp. 38–47.

Anon. (1972) Australia and continental drift. *Australian Natural History,* vol. 17, no. 8, p. 245–84.

Ballance, P.F. (ed) (1980) Plate tectonics and biogeography in the southwest Pacific: The last 100 million years. *Palaeogeography, Palaeoclimatology, and Palaeocology,* vol. 31, no. 2–4, pp. 101–372.

Ben-Avraham, Z. (1989) (ed) *The evolution of the Pacific Ocean margins,* Oxford University Press, New York. 234 p.

Berra, T.M. (1990) *Evolution and the myth of creationism,* Stanford University Press, Stanford, CA. 198 p.

Brunnschweiler, R.O. (1984) *Ancient Australia* (3rd ed.), Angus & Roberston, Sydney. 342 p.

Colbert, E.D. (1973) *Wandering lands and animals,* Dutton, New York. 323 p.

Darlington, P.J., Jr (1957) *Zoogeography: The geographical distribution of animals,* Wiley, New York. 675 p.

The Earl of Cranbrook (1981) The vertebrate faunas, pp. 57–69 in T.C. Whitmore (ed) *Wallace's line and plate tectonic,* Clarendon Press, Oxford. 91 p.

Glen W. (1975) *Continental drift and plate tectonics,* Merrill, Columbus, OH. 188 p.

Harland, W.B., A.V. Cox, P.G. Llewellyn, C.A. Pickton, A.G. Smith, and R. Walters (1982) *A geologic time scale,* Cambridge University Press, Cambridge. 131 p.

Mayr, E. (1944) Wallace's Line in light of recent zoogeographical studies. *Quarterly Review of Biology,* vol. 19, pp. 1–14.

Rich, P.V. and E.M. Thompson (1982) *The fossil vertebrate record of Australasia,* Monash University, Melbourne. 759 p.

Smith, A.G. and J.C. Briden (1977) *Mesozoic and Cenozoic paleocontinental maps,* Cambridge University Press, Cambridge. 63 p.

Tarling, D.H. (1980) Continental drift and biological evolution. *Carolina Biology Reader,* no. 113. 32 p.

Tarling, D. and M. Tarling (1975) *Continental drift,* Anchor Books, Garden City, New York. 142 p.

Truswell, E.M. and G.E. Wilford (1985) The setting: Changing Australian environments throughout ecological time, pp. 59–92 in *Kadimakara,* P.V. Rich, and G.F. van Tets (eds), Pioneer Design Studio, Lilydale, Victoria. 284 p.

Vickers-Rich, P. and T.H. Rich (1993) *Wildlife of Gondwana,* Reed, Sydney. 276 p.

Wallace, A.R. (1876) *The geographical distribution of animals,* Macmillan, London. 2 vols.

Wegener, A. (1966) *The origin of continents and oceans,* Dover Publications, New York. 246 p.

Whitmore, T.C. (1981) *Wallace's Line and plate tectonics,* Clarendon Press, Oxford. 91 p.

Wilson, J.T. (ed) (1973) *Continents adrift: Readings from Scientific American,* Freeman, San Francisco. 172 p.

Chapter 3: The First Australians

Abbie, A.A. (1969) *The original Australians,* American-Elsevier, New York. 288 p.

Allen, J., J. Golson, and R. Jones (eds) (1977) *From Sunda to Sahel: Prehistoric studies in southeast Asia, Melanesia, and Australia,* Academic Press, New York.

Altman, J.C. (1987) *Hunter-gathers today: An Aboriginal economy in north Australia,* Australian Institute of Aboriginal Studies, ABS Canberra. 251 p.

Anderson, I. (1995) The sickness at Australia's heart. *New Scientist,* vol. 146, pp. 15–16.

Anon. (1985) *Australian Aboriginal culture,* Australian Government Publishing Service, Canberra. 48 p.

Australian Bureau of Statistics (1996) *1996 Year Book Australia,* ABS, Canberra.

Australian Information Service (1980) *The Australian Aboriginals,* Australian Government Publishing Service, Canberra. 20 p.

——(1984) *The Australian Aboriginals,* Australian Government Publishing Service, Canberra. 12 p.

Bellwood, P.S. (1978) *Man's conquest of the Pacific,* Oxford Univiversity Press, Oxford. 462 p.

——(1980) The peopling of the Pacific. *Scientific American,* vol. 243, pp. 174–85.

Berndt, R.M. and C.H. Berndt (1988) *The world of the first Australians,* Aboriginal Studies Press, Canberra. 608 p.

Berra, T.M. (1990) *Evolution and the myth of creationism,* Stanford University Press, Stanford, CA. 198 p.

Bowdler, J.M., R. Jones, J. Allen, and A.G. Thorne (1970) Pleistocene human remains from Australia: A living site and a cremation from Lake Mungo, Western NSW. *World archaeology,* vol. 2, pp. 39–60.

Burenhult, G. (ed.) (1994) *Traditional peoples today,* Harper Collins, New York. 239 p.

Chaloupka, G. (n.d.) *Burrunguy Nourlangie Rock,* Northart, Darwin. 40 p.

Davies, D. (1973) *The last of the Tasmanians,* Frederick Muller, London. 284 p.

Dayton, L. and J. Woodford (1996) Australia's date with destiny. *New Scientist,* vol. 152, pp. 28–31.

Elkin, A.P. (1979) *The Australian Aborigines* (5th ed.), Angus & Robertson, Sydney. 397 p.

Flood, J. (1983) *Archaeology of the Dreamtime,* Collins, Sydney. 288 p.

Forge, A. (ed) (1973) *Australia and Melanesia,* vol. 1 of *Peoples of the Earth,* Danbury Press, Danbury, Conn. 144 p.

Fullagar, R.L.K., D.M. Price, and L.M. Head (1996) Early human occupation of northern Australia: Archaeology and thermoluminescence dating of Jinmium rock-shelter, Northern Territory. *Antiquity,* vol. 70, pp. 751–73.

Gould, R.A. (1969) *Yiwara: Foragers of the Australian desert*, Charles E. Scribner's Sons, New York. 239 p.

Habgood, P.J. (1985) The origin of the Australian Aborigines: An alternative approach and view, p. 367–80 in *Hominid evolution: Past, present and future*, P.V. Tobias (ed) Alan R. Liss, Inc. New York.

Hallam, S. (1975) *Fire and hearth*, Australian Institute of Aboriginal Studies, Canberra.

Hart, C.W.M. and A.R. Pilling (1979) *The Tiwi of North Australia*, Holt, Rinehart & Winston, New York. 140 p.

Hess, F. (1968) The aerodynamics of boomerangs. *Scientific American*, vol. 219, pp. 124–36.

Horton, D.R. (1978) The extinction of the Australian megafauna. *Australian Institute of Aboriginal Studies newsletter*, vol. 9, pp. 72–75.

Isaacs, J. (1980) *Australian dreaming: 40,000 years of Aboriginal history*, Lansdowne Press, Sydney. 304 p.

———(1989) *Australian Aboriginal paintings*, Dutton Studio Books, New York. 192 p.

Kirk, R.L. and A.G. Thorne (eds) (1976) *The origin of the Australians*, Human Biology Series no. 6, Australian Institute of Aboriginal Studies, Canberra. 449 p.

Leakey, M. (1995) The farthest horizon. *National Geographic*, vol. 188, pp. 38–51.

Leakey, M., C.S. Feibel, I. McDougall, and A. Walker (1995) New four-million-year-old hominid species from Kanapoi and Allia Bay, Kenya. *Nature*, vol. 376, pp. 565–71.

Lofgren, M.E. (1975) *Patterns of life*, Western Australian Museum Information Series no. 6, Perth. 72 p.

Maddock, Kenneth (1982) *The Australian Aborigines*, Penquin Books, Ringwood, Vic. 198 p.

McCarthy, F.D. et al. (1977) Aborigines, pp. 3–66 in *The Australian Encyclopedia*, vol. 1, The Grolier Society of Australia, Sydney.

Manon, S. (1997) Art in Australia 60,000 years ago. *Discover*, vol. 18, pp. 33–34.

Moorehead, A. (1968) *The fatal impact*, Penguin Books, Melbourne. 283 p.

Morell, V. (1994) An anthropological culture shift. *Science*, vol. 264, pp. 20–22.

———(1995) The earliest art becomes older—and more common. *Science*, vol. 267, pp. 1908–1909.

———(1995) Who owns the past? *Science*, vol. 268, pp. 1424–26.

Morgan, S. (1987) *My place*, Fremantle Arts Centre Press, Fremantle, WA. 358 p.

Mulvaney, D. (1975) *The prehistory of Australia* (2nd ed.), Pelican Books, Baltimore.

Mulvaney, D.J. and J.P. White (eds) (1987) *Australians to 1788*, Fairfax, Syme & Weldon Associates, Sydney. 476 p.

Rowley, C.D. (1970) *The destruction of Aboriginal society*, Australian National University Press, Canberra. 430 p.

———(1971) *The remote Aborigines*, Australian National University Press, Canberra. 379 p.

———(1971) *Outcasts in white Australia*, Australian National University Press, Canberra. 472 p.

Shipman, P. (1993) On the origin of races. *New Scientist*, 16 January, pp. 34–37.

Simpson, C. (1952) *Adam in ochre*, Angus & Robertson, Sydney. 221 p.

Sutton, P. (ed) (1988) *Dreamings: The art of Aboriginal Australia*, Viking, Melbourne. 266 p.

Swisher, C.C., G.H. Curtis, T. Jacob, A.G. Getty, A. Suprijo and Widiasmoro (1994) Age of the earliest known hominids in Java, Indonesia. *Nature*, vol. 263, pp. 1118–21.

Swisher, C.C., W.J. Rink, S.C. Antón, H.P. Schwarcz, G.H. Curtis, A. Suprijo, and Widiasmoro (1996) Latest *Homo erectus* of Java: Potential contemporaneity with *Homo sapiens* in Southeast Asia. *Science*, vol. 274, pp. 1870–74.

Tattersall, I. (1993) *The human odyssey*, Prentice Hall, New York. 191 p.

Thomas, J. (1983) Why boomerangs boomerang. *New Scientist*, vol. 9, no. 1376, pp. 838–43.

Thorne, A.G. and M.H. Walpoff (1992) The multiregional evolution of humans. *Scientific American*, vol. 266, pp. 76–83.

Tindale, N.B. (1974) *Aboriginal tribes of Australia*, University of California Press, Berkeley, CA. 2 vols.

Tindale, N.B. and B. George (1978) The Australian Aborigines, pp. 67–124 in *Incredible Australia*, Budget Books, Melbourne.

Tonkinson, R. (1978) *The Mardudjara Aborigines*, Holt, Rinehart & Winston, New York. 149 p.

Turnbull, C. (1948) *Blackwar: The extermination of the Tasmanian Aborigines*, Lansdowne, Melbourne. 274 p.

White, J.P. (1993) The settlement of ancient Australia, pp. 146–69 in *The first humans*, vol. 1, G. Burenhult (ed), Harper, San Francisco, CA.

White, J.P. and J.F. O'Connell (1979) Australian prehistory: New aspects of antiquity. *Science*, vol. 203, pp. 21–28.

———(1983) *A prehistory of Australia, New Guinea, and Sahal*, Academic Press, New York. 286 p.

White, T.D., G. Suwa, and B. Asfaw (1994) *Australopithecus ramidus*, a new species of early hominid from Aramis, Ethiopia. *Nature*, vol. 371, pp. 306–12.

White, T.D., G. Suwa, and B. Asfaw (1995) Corrigendum, *Nature*, vol. 375, pp. 88.

Wilford, J.N. (1996) In Australia, signs of artists who predate *Homo sapiens*. *New York Times*, 21 September, pp. 1, 5.

Wilson, A.C. and R.L. Cann (1992) The recent African genesis of humans. *Scientific American*, vol. 266, pp. 68–73.

Wolpoff, M. and R. Caspari (1997) *Race and human evolution*, Simon and Schuster, New York. 462 p.

Wolpoff, M.H., W.X. Zhi, and A.G. Thorne (1984) Modern *Homo sapiens* origins: A general theory of hominid evolution involving the fossil evidence from East Asia, pp. 411–83 in *The origins of modern humans*, F.H. Smith and F. Spencer (eds), Alan R. Liss, New York.

Chapter 4: A Floral Mosaic

Adam, P. (1992) *Australian rainforests,* Clarendon Press, Oxford. 308 p.

Anderson, I. (1995) Australia's growing disaster. *New Scientist,* vol. 147, pp. 12–13.

Beadle, N.C.W. (1981) *The vegetation of Australia,* Cambridge University Press, Cambridge. 690 p.

Blombery, A.M. (1979) *Australian native plants,* Angus & Robertson, Sydney. 481 p.

Burbridge, N.T. and M. Gray (1970) *Flora of the Australian Capital Territory,* Australian National University Press, Canberra. 447 p.

Carnahan, J.A. (1977) Vegetation, pp. 175–95 in *Australia: A geography,* D.N. Jeans (ed), Sydney University Press, Sydney. 571 p.

Erickson, R., A.S. George, N.G. Marchant, and M.K. Morcombe (1983) *Flowers and plants of Western Australia,* Reed, Sydney. 231 p.

George, A.S. (1987) *The Banksia book* (2nd ed.), Kangaroo Press, Sydney. 240 p.

Groves, R.N. (ed) (1981) *Australian vegetation,* Cambridge University Press, Cambridge. 449 p.

Herbert, D.A. and J.H. Willis (1977) Plants, pp. 37–44 in *The Australian Encyclopaedia,* vol. 5, The Grolier Society of Australia, Sydney.

Harris, T.Y. (1979) *Wildflowers of Australia,* Angus & Robertson, Sydney. 207 p.

Holden, C. (1995) Ancient trees Down Under. *Science,* vol. 267, p. 334.

Jacobs, M.R. (1955) *Growth habits of eucalypts,* Commonwealth of Australia Forestry and Timber Bureau, Canberra. 262 p.

Jessop, J. (ed) (1982) *Flora of central Australia,* Reed, Sydney. 537 p.

Jones, W.G., K.D. Hill, and J.M. Allen (1995) *Wollemia nobilis,* a new living Australian genus and species in the Araucariaceae. *Telopea,* vol. 6, pp. 173–76.

Millett, M. (1969) *Australian eucalypts,* Periwinkle Books, Melbourne. 112 p.

Pate, J.S. and A.J. McComb (eds) (1981) *The biology of Australian plants,* University of Western Australia Press, Nedlands, WA. 412 p.

Pryor, L.D. (1976) *The biology of eucalyptus,* The Institute of Biology's Studies in Biology, no. 61, Edward Arnold, London.

Pryor, L.D. and L.A.S. Johnson (1971) *A classification of the eucalyptus,* Australian National University, Canberra. 192 p.

Pyne, S.J. (1991) *Burning bush: A fire history of Australia,* Henry Holt and Co., New York. 520 p.

Sharkey, T.D. and E.L. Singsaas (1995) Why plants emit isoprene. *Nature,* vol. 374, p. 769.

Specht, R.L. (1981) Foliage projective cover and standing biomass, pp. 10–21 in *Vegetation classification in Australia,* A.N. Gillison and D.J. Anderson (eds), CSIRO/Australian National University Press, Canberra.

White, M.E. (1990) *The flowering of Gondwana,* Princeton University Press, Princeton, NJ. 256 p.

Woodford, J. (1994) Found: Tree from the dinosaur age, and it's alive. *Sydney Morning Herald,* 14 December, pp. 1, 8.

Chapter 5: Dangerous Waters

Baldridge, H.D. (1974) *Shark attack,* Berkley Publications, New York. 263 p. (See also *Contributions from the Mote Marine Laboratory,* vol. 1, no. 2, pp. 1–98.)

Bennett, I. (1971) *The Great Barrier Reef,* Lansdowne, Melbourne. 183 p.

Berra, T.M. (1997) Some 20th century fish discoveries, *Environmental biology of fishes,* vol. 50, pp. 1–12.

Berra, T.M. and B. Hutchins (1990) A specimen of megamouth shark, *Megachasma pelagios* (Megachasmidae) from Western Australia. *Records of the Western Australian Museum,* vol. 14, no. 4, pp. 651–56.

——(1991) Natural history notes on the megamouth shark, *Megachasma pelagios,* from Western Australia. *Western Australian Naturalist,* vol. 18, pp. 224–33.

Carcasson, R.H. (1977) *A field guide to the coral reef fishes of the Indian and West Pacific Oceans,* Collins, London. 320 p.

Coleman, N. (1974) *Australian marine fishes,* Reed, Sydney. 108 p.

——(1978) *Australian fisherman's guide,* Bay Books, Sydney. 143 p.

——(1981) *Australian sea fishes north of 30°S,* Doubleday, Sydney. 297 p.

Coppleson, V.M. (1962) *Shark attack* (2nd ed.), Angus & Robertson, Sydney. 269 p.

Dakin, W.J. (1968) *The Great Barrier Reef,* Walkabout Pocket Books, Sydney. 176 p.

——(1969) *Australian seashores,* Angus & Robertson, Sydney. 372 p.

Edmonds, C. (1978) *Dangerous marine animals of the Indo-Pacific region,* Wedneil Publications, Newport, Victoria. 235 p.

Ellis, R. (1976) *The book of sharks,* Grossett & Dunlap, New York. 320 p.

Endean, R. (1983) *Australia's Great Barrier Reef,* University of Queensland Press, Brisbane. 348 p.

Fautin, D.G. and G.R. Allen. (1992) *Field guide to anemone fishes and their host sea anemones,* Western Australian Museum, Perth. 157 p.

Gilbert, P.W. (ed) (1963) *Sharks and survival,* Heath, Boston. 578 p.

Gillett, K. (1967) *The Great Barrier Reef and adjacent isles,* Coral Press, Sydney. 209 p.

——(1968) *The Australian Great Barrier Reef,* Coral Press, Sydney. 112 p.

Grant, E.M. (1982) *Guide to fishes* [of Queensland] (5th ed.), Department of Harbours and Marine, Brisbane. 896 p.

Green, J. (1976) *Recorded shark attacks in Australian waters,* published by author. 15 p.

Halstead, B.W. (1978) *Poisonous and venomous marine animals of the world,* Darwin Press, Princeton, NJ. 1043 p.

Healy, A. and J. Yaldwyn (1970) *Australian crustaceans,* Reed, Sydney. 112 p.

Hopley, D. (1982) *The geomorphology of the Great Barrier Reef,* Wiley-Interscience, New York. 454 p.

Hughes, R. (1985) *Australia's underwater wilderness,* Weldon, Sydney.

Hutchins, B. (1992) Megamouth: Gentle giant of the deep. *Australian Natural History*, vol. 23, pp. 910–17.

Klimley, A.P. and D.G. Ainley (eds) (1996) *Great white sharks*, Academic Press, San Diego. 517 p.

Krogh, M. (1994) Spacial, seasonal and biological analysis of sharks caught in the New South Wales protective beach meshing programme. *Aust. J. Mar. Freshwater Res.*, vol. 45, pp. 1087–160.

Last, P.R. and J.D. Stevens (1994) *Sharks and rays*, CSIRO, East Melbourne. 507 p.

Leis, J.M. and D.C. Rennis (1983) *The larvae of Indo-Pacific coral reef fishes*, University of New South Wales Press, Sydney. 269 p.

Marshall, T.C. (1965) *Fishes of the Great Barrier Reef and coastal waters of Queensland*, Livingston Publications, Narberth, Pennsylvania. 566 p.

McGregor, C. (1975) *The Great Barrier Reef*, Time-Life Books, Amsterdam. 184 p.

Munro, I.S.R. (1967) *The fishes of New Guinea*, Department of Agriculture, Stock and Fisheries, Port Moresby, Papua New Guinea. 651 p.

Oceanus (1986) The Great Barrier Reef: Science and management. *Oceanus*, vol. 29, no. 2, pp. 1–124.

Paterson, R.A. (1990) Effects of long-term anti-shark measures on target and non-target species in Queensland, Australia. *Biological Conservation*, vol. 52, pp. 147–59.

Polland, J. (ed) (1980) *G.P. Whitley's handbook of Australian fishes*, Pollard Publications, Sydney. 629 p.

Potts, D.C. (1981) Crown-of-thorns starfish—man-induced pest or natural phenomenon? pp. 54–86 in *The ecology of pests: Some Australian case histories*, R.L. Kitching and R.E. Jones (eds), CSIRO, Melbourne. 254 p.

Pownall, P. (1979) *Fisheries of Australia*, Fishing News Books, Surrey, England. 149 p.

Randall, J.E., G.R. Allen, and R.C. Steene (1990) *Fishes of the Great Barrier Reef and Coral Sea*, University of Hawaii Press, Honolulu. 507 p.

Reader's Digest (1988) *Visitor's guide to the Great Barrier Reef*, Reader's Digest, Sydney. 168 p.

Roughley, T.C. (1966) *Wonders of the Great Barrier Reef*, Angus & Robertson, Sydney. 279 p.

Russell, B.C. (1983) *Annotated checklist of the coral reef fishes in the Capricorn-Bunker Group, Great Barrier Reef of Australia*, GBR Marine Park Authority, Townsville, Qld.

Saville-Kent, W. (1893) *The Great Barrier Reef of Australia*, Allen, London. 387 p.

Stead, D.G. (1963) *Sharks and rays of Australian seas*, Angus & Robertson, Sydney. 211 p.

Thomson, J.M. (1974) *Fish of the ocean and shore*, Collins, Sydney. 208 p.

——(1977) *A field guide to the common sea and estuary fishes of non-tropical Australia*, Collins, Sydney. 144 p.

Tinker, S.W. (1965) *Pacific crustacea*, Charles E. Tuttle, Reuland, Vermont. 134 p.

Veron, J.E.N. (1986) *Corals of Australia and the Indo-Pacific*, Angus & Robertson, Sydney. 644 p.

West, J.G. (1993) The Australian shark attack file with notes on preliminary analysis of data from Australian waters, pp. 93–101 in J. Pepperell, J. West, and P. Woon (eds) *Shark conservation*, Zoological Parks Board of NSW, Sydney.

Whitley, G.P. (1940) *The fishes of Australia, Part I: Sharks, rays, etc.*, Royal Zoological Society of New South Wales, Sydney. 280 p.

——(1962) *Marine fishes of Australia*, Jacaranda Press, Brisbane. 2 vols. 287 p.

——(1963) Shark attacks in Australia, pp. 329–39 in *Sharks and survival*, P.W. Gilbert (ed) Heath, Boston. 578 p.

Chapter 6: Some Interesting Invertebrates

Barrett, C. and A.N. Burns (1951) *Butterflies of Australia and New Guinea*, N.H. Seward, Melbourne. 187 p.

Bayly, I.A.E. and W.D. Williams (1973) *Inland waters and their ecology*, Longman, Melbourne. 314 p.

Burns, A. and E.R. Rotherham (1969) *Australian butterflies in color*, Reed, Sydney. 112 p.

Child, J. (1968) *Australian pond and stream life*, Cheshire-Lansdowne, Melbourne. 94 p.

Common, I.F.B. (1964) *Australian butterflies*, Jacaranda Press, Brisbane. 131 p.

——(1966) *Australian moths*, Jacaranda Press, Brisbane. 129 p.

CSIRO (eds) (1991) *The insects of Australia*, Melbourne University Press, Melbourne, and Cornell University Press, Ithaca, NY. 2 vols.

Groombridge, B. (ed) (1992) *Global biodiversity: Status of the Earth's living resources*, World Conservation Monitoring Centre, Chapman & Hall, London. 585 p.

Healy, A. and J. Yaldwyn (1970) *Australian crustaceans in color*, Reed, Sydney. 112 p.

Heath, J. (1989) *The fly in your eye*, J & E Publishing, Perth. 20 p.

Hughes, R.D. (1981) The Australian bushfly: A climate-dominated nuisance pest of man, pp. 176–91 in *The ecology of pests*, R.L. Kitching and R.E. Jones (eds), CSIRO, Melbourne.

Main, B.Y. (1967) *Spiders of Australia*, Jacaranda Press, Brisbane. 125 p.

May, R.M. (1988) How many species are there on Earth? *Science*, vol. 241, pp. 1441–49.

Merrick, J.R. and C.N. Lambert (1991) *The yabby, marron and red claw production and marketing*, J.R. Merrick Publishing, Sydney. 180 p.

Mukerjee, M. (1995) Giving your all. *Scientific American*, vol. 273, p. 32.

Smith, B.J. and R.C. Kershaw (1980) *Field guide to non-marine molluscs of southeastern Australia*, Australian National University Press, Canberra.

Waterhouse, D.F. (1974) The biological control of dung. *Scientific American*, vol. 230, no. 4, pp. 100–109.

Watson, J.A.L. (1982) A truly terrestial dragonfly larva from Australia (Odonata: Corduliidae). *J. Aust. Ent. Soc,* vol. 21, pp. 309–11.
Weatherley, A.H. (ed) (1967) *Australian inland waters and their fauna,* Australian National University Press, Canberra. 287 p.
Williams, W.D. (1980) *Australian freshwater life,* Macmillan, Melbourne. 321 p.

Chapter 7: The Freshwater Fishes

Allen, G.R. (1978) A review of the archerfishes (family Toxotidae). *Records of the Western Australian Museum,* vol. 6, pp. 355–78.
——(1982) *A field guide to inland fishes of Western Australia,* Western Australian Museum, Perth. 86 p.
——(1989) *Freshwater fishes of Australia,* T.F.H. Publishing, Neptune City, NJ. 240 p.
Allen, G.R. and T.M. Berra (1989) Life history aspects of the West Australian salamanderfish, *Lepidogalaxias salamandroides* Mees. *Records of the Western Australian Museum,* vol. 14, pp. 253–67.
Allen, G.R. and N.J. Cross (1982) *Rainbow fishes of Australia and Papua New Guinea,* T.F.H. Publishing, Neptune City, NJ. 141 p.
Barlow, C.G. and A. Lisle (1987) Biology of the Nile perch *Lates niloticus* (Pisces: Centropomidae) with reference to its proposed role as a sport fish in Australia. *Biological Conservation,* vol. 39, pp. 269–89.
Berra, T.M. (1981) *An atlas of distribution of the freshwater fish families of the world,* University of Nebraska Press, Lincoln. 197 p.
——(1982) Life history of the Australian grayling *Prototroctes maraena* (Salmoniformes: Prototrotidae) in the Tambo River, Victoria. *Copeia,* 1982, pp. 795–805.
Berra, T.M. and G.R. Allen (1989) Burrowing, emergence, behavior, and functional morphology of the Australian salamanderfish, *Lepidogalaxias salamandroides.* Fisheries, vol. 14, no. 5, pp. 2–10.
——(1991) Population structure and development of *Lepidogalaxias salamandroides* (Pisces: Salmoniforms) from Western Australia. *Copeia,* 1991, pp. 844–49.
Berra, T.M., L.E.L.M. Crowley, W. Ivantsoff, and P.A. Fuerst (1996) *Galaxias maculatus:* An explanation of its biogeography. *Marine and Freshwater Research,* vol. 47, pp. 845–49.
Berra, T.M. and B.J. Pusey (1997) Threatened fishes of the world: *Lepidogalaxias salamandroides* Mees, 1961 (Lepidoglaxiidae). *Environmental Biology of Fishes,* vol. 50, pp. 201–202.
Berra, T.M., and A.H. Weatherley (1972) A systematic study of the Australian freshwater serranid fish genus *Maccullochella.* Copeia, 1972, pp. 53–64.
Cadwallader, P.L. and G.N. Backhouse (1983) *A guide to the freshwater fish of Victoria.* Victorian Government Printing Office, Melbourne. 249 p.
Grant, E.M. (1982) *Guide to fishes* [of Queensland] (5th ed.), Department of Harbours and Marine, Brisbane. 896 p.
Kailoa, P.J., M.J. Williams, P.C. Stewart, R.E. Reicbelt, A. McNee, and C. Grieve (1993) *Australian fisheries resources,* Bureau of Resources and the Fisheries Research and Development Corporation, Canberra. 422 p.
Lake, J.S. (1971) *Freshwater fishes and rivers of Australia,* Nelson, Melbourne. 61 p.
——(1978) *Australian freshwater fishes,* Nelson, Melbourne. 160 p.
Last, P.R., E.O.G. Scott, and F.H. Talbot (1983) *Fishes of Tasmania,* Tasmanian Fisheries Development Authority, Hobart. 563 p.
Lloyd, L.N. and J.F. Tomasov (1985) Taxonomic status of the mosquitofish, *Gambusia affinis* (Poeciliidae), in Australia. *Aust. J. Mar. Freshw. Res.,* vol. 36, pp. 447–51.
Martin, K.L.M., T.M. Berra, and G.R. Allen (1993) Cutaneous aerial respiration during forced emergence in the Australian salamanderfish, *Lepidogalaxias salamandroides.* Copeia, 1993, pp. 875–79.
McDowall, R.M. (1996) (ed.) *Freshwater fishes of south-eastern Australia* (2nd ed.), Reed, Sydney. 247 p.
——(1980) Freshwater fishes and plate tectonics in the southwest Pacific, in *Plate tectonics and biogeography in the southwest Pacific: The last 100 million years,* P.F. Ballance (ed). *Paleogeography, Palaeoclimatology, and Palaeocology,* vol. 31, nos 2–4, pp. 101–372.
——(1981) The relationships of Australian freshwater fishes, pp. 1253–73 in *Ecological Biogeography of Australia,* A. Keast (ed), Junk, The Hague.
McDowall, R.M. and R.S.Frankenberg (1981) The galaxiid fishes of Australia. *Records of the Australian Museum,* vol. 33, no. 10, pp. 443–605.
McKay, R.J. (1984) Introduction of exotic fishes in Australia, pp 177–94 in *Distribution, biology, and management of exotic fishes,* W.R. Courtnay Jr and J.R. Stauffer Jr (eds), Johns Hopkins University Press, Baltimore.
Merrick, J.R. and G.E. Schmida (1984) *Australian freshwater fishes: Biology and management,* J.R. Merrick, School of Biological Sciences, Macquarie University, Sydney. 409 p.
Munro, I.S.R. (1967) *The fishes of New Guinea,* Department of Agriculture, Stocks and Fisheries, Port Moresby, Papua New Guinea. 651 p.
Nelson, J.S. (1994) *Fishes of the world* (3rd ed.), John Wiley & Sons, New York. 600 p.
Pollard, J. (ed) (1969) *Australian and New Zealand fishing,* Paul Hamlyn, Sydney. 952 p.
Roughley, T.C. (1966) *Fish and fisheries of Australia,* Angus & Robertson, Sydney. 328 p.
Rowland, S.J. (1993) *Maccullochella ikei,* an endangered species of freshwater cod (Pisces: Percichthyidae) from the Clarence River system, NSW, and *M. peelii mariensis,* a new subspecies from the Mary River system, Qld. *Records of the Australian Museum,* vol. 45, pp. 121–45.

Scott, T.D., C.J.M. Glover, and R.V. Southcott (1974) *The marine and freshwater fishes of South Australia* (2nd ed.), Government Printer, Adelaide. 392 p.

Taylor, W.R. (1964) Fishes of Arnhem Land. *Records of the American-Australian Scientific Expedition to Arnhem Land,* vol. 4, pp. 45–307.

Vari, A.P. (1978) The terapon perches (Percoidei, Teraponidae): A cladistic analysis and taxonomic revision. Bull. *Amer. Mus. Nat. Hist.* vol. 159, no. 5, pp. 175–340.

Waters, J.M. (1996) Aspects of the phylogeny, biogeography and taxonomy of galaxioid fishes. Ph.D. thesis, University of Tasmania, Hobart. 159 p.

Weatherley, A.H. (ed) (1967) *Australian inland waters and their fauna,* Australian National University Press, Canberra. 287 p.

Whitley, G.P. (1964) *Native freshwater fishes of Australia,* Jacaranda Press, Brisbane. 127 p.

COASTAL MARINE FISHES EXCLUDING THE GREAT BARRIER REEF

Allen, G.R. and R. Swainston (1988) *The marine fishes of northwestern Australia,* Western Australian Museum, Perth. 201 p.

Coleman, N. (1980) *Australian sea fishes south of 30°S,* Doubleday, Sydney. 302 p.

——(1986) *Australian sea fishes north of 30°S,* Doubleday, Sydney. 297 p.

Edgar, G.J., P.R. Last, and M.W. Wells (1982) *Coastal fishes of Tasmania and Bass Strait,* Cat & Fiddle Press, Hobart. 176 p.

Gloerefelt-Tarp, T. and P.J. Kailola (1984) *Trawled fishes of southern Indonesia and northwestern Australia,* Australian Development Assistance Bureau, Canberra. 406 p.

Gomon, M.F., J.C.M. Glover, and R.H. Huiter (1994) *The fishes of Australia's south coast,* State Printer, Adelaide. 992 p.

Hutchins, B. and R. Swainston (1986) *Sea fishes of southern Australia,* Swainston Publishing, Perth. 180 p.

Hutchins, B. and M. Thompson (1983) *The marine and estuarine fishes of southwestern Australia,* Western Australian Museum, Perth. 103 p.

Kailola, P.J. et al. (1993) *Australian fisheries resources,* Bureau of Resource Sciences and the Fisheries Research and Development Corporation, Canberra. 422 p.

Kuiter, R.H. (1993) *Coastal fishes of southeastern Australia,* University of Hawaii Press, Honolulu. 437 p.

May, J.L. and J.G.H. Maxwell (1986) *Field guide to trawl fish from temperate waters of Australia,* CSIRO Division of Fisheries Research, Hobart. 492 p.

Sainsbury, K.J., P.J. Kailola, and G.G. Leyland (1985) *Continental shelf fishes of northern and northwestern Australia,* Clouston & Hall and Peter Pownall Fisheries Information Service, Canberra. 375 p.

Chapter 8: Warts, Scales, Frills, and Thorns

Anon. (1993) Leatherback sea turtle. *Currents,* no. 50, pp. 4–7.

Auffenberg, W. (1981) *The behavioral ecology of the Komodo monitor,* University Press of Florida, Gainesville. 406 p.

Banks, C.B. and A.A. Martin (1982) *Proceedings of the Melbourne Herpetological Symposium, 1980,* Zoological Board of Victoria, Melbourne. 199 p.

Barker, J. and G. Grigg (1977) *A field guide to Australian frogs,* Rigby, Adelaide. 229 p.

Biddell, G. and C. Stringer (1988) *Crocodiles of Australia,* Adventure Publications, Casuarina, Northern Territory. 80 p.

Burbidge, A. (1983) A very rare Australian: The western swamp tortoise. *Australian Natural History,* vol. 21, pp. 14–17.

Bustard, R. (1970) *Australian lizards,* Collins, Sydney. 162 p.

Bustard, R. (1972) *Sea turtles,* Taplinger, New York. 220 p.

Caldwell, M.W. and M.S.Y. Lee (1997) A snake with legs from the marine Cretaceous of the Middle East. *Nature,* vol. 386, pp. 705–709.

Cann, J. and J. Legler (1994) The Mary River tortoise: A new genus and species of short-necked chelid from Queensland, Australia (Testudines: Pleurodira). *Chelonian Conserv. Biol.,* vol. 1, pp. 81–96.

Cogger, H.G. (1992) *Reptiles and amphibians of Australia* (5th ed.), Cornell University Press Ithaca, NY. 775 p.

Cogger, H.G., E.E. Cameron, and H.M. Cogger (1983) of *Amphibians and reptiles,* vol. 1 *Zoological catalogue of Australia,* Australian Government Publishing Service, Canberra. 313 p.

Daugherty, C.H., A. Cree, J.M. Hay, and M.B. Thompson (1990) Neglected taxonomy and continuing extinctions of tuatara (*Sphenodon*). *Nature,* vol. 347, pp. 177–79.

Duellman, W.E. (1993) *Amphibian species of the world: Additions and corrections,* Special Publication no. 21, Museum of Natural History, University of Kansas.

Dunson, W.A. (ed) (1975) *The biology of sea snakes,* University Park Press, Baltimore. 550 p.

Edwards, H. (1988) *Crocodile attack in Australia,* Swan Publishing, Sydney. 192 p.

Ernst, C.H. and R.W. Barbour (1989) *Turtles of the world,* Smithsonian Institution Press, Washington, DC. 313 p.

Frost, D.R. (ed) (1985) *Amphibian species of the world,* Allen Press and the Association of Systematic Collections, Lawrence, Kansas. 732 p.

Goin, C.J., O.B. Goin and G.R. Zug (1978) *Introduction to herpetology* (3rd ed.), Freeman, San Francisco, CA. 378 p.

Goode, J. (1967) *Freshwater tortoises of Australia and New Guinea,* Lansdowne Press, Melbourne. 154 p.

Gow, G.F. (1983) *Snakes of Australia*, Angus & Robertson, London. 118 p.

Green, B. and D. King (1993) *Goanna: The biology of varanid lizards*, University of New South Wales Press, Sydney. 102 p.

Grigg, G., R. Shine, and H. Ehman (eds) (1985) *Biology of Australasian frogs and reptiles*, Surrey Beatty & Sons, Sydney. 527 p.

Guggisberg, C.A.W. (1972) *Crocodiles: Their natural history, folklore, and conservation*, David & Charles, Newton Abbot, England. 202 p.

Halliday, T. and K. Adler (1986) *The encyclopedia of reptiles and amphibians*, Facts on File, New York. 143 p.

Heatwole, H. (1987) *Sea snakes*, University of New South Wales Press, Sydney. 85 p.

Horgan, J. (1990) *Bufo abuse. Scientific American*, vol. 263, pp. 26–27.

Houston, T.F. (1978) *Dragon lizards and goannas of South Australia*, South Australian Museum Special Edition Bulletin, Adelaide. 84 p.

King, F.W. and R.L. Burke, (eds). (1989) *Crocodilian, tuatara, and turtle species of the world*, Association of Systematics Collections, Washington, DC. 216 p.

Kinghorn, J.R. (1967) *The snakes of Australia*, Angus & Robertson. 197 p.

Márquez, M.R. (1990) *Sea turtles of the world. FAO Fisheries Synopsis*, no. 125, vol. 11, Rome. 81 p.

Marshall, J.T. (1985) *Guam: A problem in avian conservation. Wilson Bulletin*, vol. 97, pp. 259–62.

Martin, A.A. and M.J. Littlejohn (1982) *Tasmanian amphibians*. Fauna of Tasmania Handbook no. 6, University of Tasmania, Hobart. 52 p.

Meyer, A. (1995) Molecular evidence on the origin of tetrapods and the relationships of the coelacanth. *Trends in Ecology and Evolution*, vol. 10, pp. 111–16.

Minton, Jr., S.A. and M.R. Minton (1969) *Venomous reptiles*, Charles Scribner's Sons, New York. 275 p.

Mirtschin, P. and R. Davis (1992) *Dangerous snakes of Australia* (rev. ed.), Hew Holland, London. 208 p.

Neill, W.T. (1971) *The last of the ruling reptiles: Alligators, crocodiles, and their kin*, Columbia University Press, New York. 486 p.

Pianka, E.R. (1986) *Ecology and natural history of desert lizards*, Princeton University Press, Princeton, NJ. 208 p.

Priede, I.G. (1990) The sea snakes are coming. *New Scientist*, 10 Nov., pp. 29–33.

Sabath, M.D., W.C. Boughton, and S. Easteal (1981) Expansion of the range of the introduced toad *Bufo marinus* in Australia from 1935 to 1974. *Copeia*, no. 3, pp. 679–80.

Schmida, G. (1985) *The cold-blooded Australians*, Doubleday, Sydney.

Schwenk, K. (1994) Why snakes have forked tongues. *Science*, vol. 263, pp. 1573–77.

Shine, R. (1991) *Australian snakes: A natural history*, Cornell University Press, Ithaca, NY. 223 p.

Sordino, P., F. van der Hoeven, and D. Duboule (1995) Hox gene expression in teleost fins and the origin of vertebrate digits. *Nature*, vol. 375, pp. 678–81.

Sutherland, S.K. (1983) *Australian animal toxins: The creatures, their toxins and care of the poisoned*, Oxford University Press, Oxford. 480 p.

Storr, G.M., L.A. Smith, and R.E. Johnstone (1983) *Lizards of Western Australia II: Dragons and monitors*, Western Australian Museum, Perth. 113 p.

——(1986) *Snakes of Western Australia*, Western Australian Museum, Perth. 187 p.

Tyler, M.J. (1979) *The status of endangered Australian wildlife*, Royal Zoological Society of South Australia, Adelaide. 210 p.

——(ed) (1983) *The gastric brooding frog*, Croom Helm, London/Canberra. 163 p.

——(1989) *Australian frogs*, Viking O'Neil, Ringwood, Vic. 220 p.

Webb, G. and C. Manolis (1988) *Australian freshwater crocodiles*, G. Webb Pty Ltd, Winnellie, NT. 33 p.

——(1988) *Australian saltwater crocodiles*, G. Webb Pty Ltd, Winnellie, NT. 33 p.

Webb, G.,C. Manolis, and P.J. Whitehead (1987) *Wildlife management: Crocodiles and alligators*, Surrey Beatty & Sons, Sydney. 552 p.

Wilson, S.K. and D.G. Knowles (1988) *Australia's reptiles*, Collins, Sydney. 447 p.

Worrell, E. (1969) *Dangerous snakes of Australia and New Guinea*, Angus & Robertson, Sydney. 65 p.

——(1970) *Reptiles of Australia*, Angus & Robertson, Sydney. 169 p.

Chapter 9: Aerial Australians

Blackers, M., S. Davies, and P. Reilly (1984) *The atlas of Australian birds*, Melbourne University Press, Melbourne. 738 p.

Borgia, G. (1986) Sexual selection in bowerbirds. *Scientific American*, vol. 254, pp. 92–100.

Cayley, N. (1968) *What bird is that?* (5th ed.), Angus & Robertson, Sydney. 348 p.

Chisholm, A.H. (1960) *The romance of the lyre bird*, Angus & Robertson, Sydney. 156 p.

Christidis, L. and W. Boles (1994) *The taxonomy and species of birds of Australia and its territories*, Monograph no. 2, Royal Australian Ornithologists Union, Melbourne.

Eastman, M. (1969) *The life of the emu*, Angus & Robertson, Sydney. 72 p.

Eastman, W. (1970) *The life of the kookaburra and other kingfishers*, Angus & Robertson, Sydney. 64 p.

Everett, M. (1978) *The birds of paradise*, G.P. Putnam's, New York. 144 p.

Faaborg, J. (1988) *Ornithology: An ecological approach*, Prentice-Hall, Englewood Cliffs, NJ. 470 p.

Forshaw, J.M. (1969) *Australian parrots*, Lansdowne Press, Melbourne. 306 p.

Frith, H.J. (1962) *The mallee fowl,* Angus & Robertson, Sydney.

——(1967) *Waterfowl in Australia,* Angus & Robertson, Sydney. 328 p.

——(ed) (1969) *Birds in the Australian high country,* Reed, Sydney. 481 p.

Hoser, R. (1993) *Smuggled: The underground trade in Australia's wildlife,* Apollo Books, Sydney. 149 p.

Leach, J.A. (1968) *An Australian bird book* (9th ed.), Whitcombe & Tombs, Melbourne. 224 p.

Monroe, B.L. and C.G. Sibley (1993) *A world checklist of birds,* Yale University Press, New Haven, CN. 393 p.

Mulder, R.A. (1994) Faithful philanderers. *Natural History,* vol. 103, pp. 56–63.

Pizzey, G. (1983) *A field guide to the birds of Australia,* Collins Australia, Sydney.

Reader's Digest (1979) *Reader's Digest complete book of Australian birds,* Reader's Digest Pty Ltd, Sydney. 615 p.

Reilly, P. (1988) *The lyrebird,* University of New South Wales Press, Sydney. 93 p.

Rowley, I. (1994) 'What a Galah!' *Australian Natural History,* vol. 24, pp. 22–29.

Serventy, D.L. and H.M. Whittell (1967) *Birds of Western Australia,* Lamb Publications, Perth. 440 p.

Seymour, R.S. (1991) The brush turkey. *Scientific American,* vol. 265, no. 6, pp. 108–14.

Sibley, C.G. and J.E. Ahlquist (1985) The phylogeny and classification of the Australo-Papuan passerine birds. *Emu,* vol. 85, pp. 1–14.

——(1990) *Phylogeny and classification of birds,* Yale University Press, New Haven, CN. 976 p.

Sibley, C.G. and B.L. Monroe (1990) *Distribution and taxonomy of birds of the world,* Yale University Press, New Haven, CN. 1111 p.

Simpson, K., N. Day, and P. Trusler (eds) (1996) *The Princeton field guide to the birds of Australia* (5th ed.), Princeton University Press, Princeton, NJ. 400 p.

Slater, P. (1970) *A field guide to Australian birds,* vol. 1, *Non-passerines,* Rigby, Adelaide. 428 p.

——(1974) *A field guide to Australian birds,* vol. 2, *Passerines,* Rigby, Adelaide. 309 p.

Slater, P., P. Slater and R. Slater (1986) *The Slater field guide to Australian birds,* Rigby, Sydney. 343 p.

Stahel, G. and R. Gales (1987) *Little penguin.* University of New South Wales Press, Sydney, 117 p.

Storr, G.M. and R.E. Johnstone (1985) *Field guide to the birds of Western Australia,* Western Australian Museum, Perth. 214 p.

Tuck, G.S. and H. Heinzel (1980) *A field guide to the seabirds of Australia and the world,* Collins, Sydney. 276 p.

Tyler, M.J. (ed) (1979) *The status of endangered Australian wildlife,* Royal Zoological Society of South Australia, Adelaide. 210 p.

Welty, J.C. (1975) *The life of birds* (2nd ed.), Saunders, Philadelphia. 623 p.

Wolters, H.E. (1982) *Die Vogelarten der Erde / The bird species of the world,* Paul Parey, New York. 748 p.

Chapter 10: Mostly Marsupials

Anderson, S. and J.K. Jones (eds) (1967) *Recent mammals of the world,* Ronald Press, New York. 453 p.

Anon. (1981) *Kangaroo management,* Fact sheet on Australia, Australian Information Service, Canberra.

Archer, M. (ed) (1982) *The carnivorous marsupials,* Royal Zoological Society of NSW. 2 vols. 1300 p.

——(1988) A new order of Tertiary Zalambdodont marsupials. *Science,* vol. 239, pp. 1528–31.

Archer, M., T.F. Flannery, and G.C. Grigg (1985) *The kangaroo,* Kevin Weldon, Sydney. 263 p.

Archer, M., T.F. Flannery, A. Ritchie, and R.E. Molnar (1985) First Mesozoic mammal from Australia—an early Cretaceous monotreme. *Nature,* vol. 318, pp. 363–66.

Augee, M.L.(ed) (1978) Monotreme biology. *Australian Zoologist,* vol. 20, no. 1, pp. 1–257.

Barrett, C. (1944) *The platypus,* Robertson & Mullens, Melbourne. 62 p.

Bennett, M.B. and G.C. Taylor (1995) Scaling of elastic strain energy in kangaroos and the benefits of being big. *Nature,* vol. 378, pp. 56–59.

Beresford, Q. and G. Bailey (1981) *Search for the Tasmanian tiger,* Blubber Head Press, Hobart. 54 p.

Breckwoldt, R. (1988) *The dingo,* Angus & Robertson, Sydney. 283 p.

Brown, S. and F. Carrier (1985) Koala disease breakthrough. *Australian Natural History,* vol. 21, pp. 314–17.

Burrell, H. (1927) *The platypus,* Angus & Robertson, Sydney. 227 p.

Burbidge, A.A., K.A. Johnson, P.J. Fuller, and R.I. Southgate (1988) Aboriginal knowledge of the mammals of the central deserts of Australia. *Australian Wildlife Research,* vol. 15, pp. 9–39.

Campbell, J. (1985) She'll go a-waltzing Australia with you. *Sports Illustrated,* vol. 62, no. 6, pp. 102–31.

Corbett, L. (1995) *The dingo in Australia and Asia,* University of New South Wales Press, Sydney, 200 p.

Corbett, L. and A. Newsome (1975) Dingo society and its maintenance: A preliminary analysis, pp. 369–79 in Fox, M.W. (ed) *The wild canids,* Van Nostrand Reinhold, New York.

Dagabriele, R. (1980) The physiology of the koala. *Scientific American,* vol. 243, no. 1, pp. 110–17.

Dawson, T.J. (1977) Kangaroos. *Scientific American,* vol. 237, no. 2, pp. 78–89.

——(1983) *Monotremes and marsupials: The other mammals,* Edward Arnold, London. 90 p.

——(1995) *Kangaroos: Biology of the largest marsupials,* University of New South Wales Press, Sydney. 162 p.

——(1995) Red kangaroos, the kings of cool. *Natural History,* vol. 104, pp. 38–45.

Diamond, J. (1984) Big bang reproduction. *Discover,* vol. 5, pp. 53–55.

Domico, T. (1993) *Kangaroos: The marvelous mob,* Facts on File, New York. 202 p.

Douglas, A.M. (1986) Tigers in Western Australia. *New Scientist,* vol. 110, pp. 44–47.

Eadie, R. (1935) *The life and habits of the platypus,* Stillwell & Stephen, Melbourne. 78 p.

Eaton, J.G. (1993) Marsupial dispersal. *National Geographic Research and Exploration,* vol. 9, pp. 436–43.

Eisenberg, J.F. (1981) *The mammalian radiations,* University of Chicago Press, Chicago. 610 p.

Fleay, D. (1944) *We breed the platypus,* Robertson & Mullens, Melbourne. 44 p.

——(1980) *Paradoxical platypus,* Jacaranda Press, Brisbane, 150 p.

Frith, H.J. and J.H. Calably (1969) *Kangaroos,* Cheshire, Melbourne. 209 p.

Godthelp, H., M. Archer, and P.R. Vail (1992) Earliest known Australian tertiary mammalian fauna. *Nature,* vol. 356, pp. 514–15.

Grant, T. (1984) *The platypus,* University of New South Wales Press, Sydney. 76 p.

Green, R.H. (1973) *The mammals of Tasmania,* Mary Fisher Bookshop, Launceston, Tas. 74 p.

Griffiths, M. (1968) *Echidnas,* Pergamon Press, Oxford. 282 p.

——(1979) *The biology of the monotremes,* Academic Press, New York. 376 p.

——(1988) The Platypus. *Scientific American,* vol. 258, pp. 84–91.

Grigg, G. (1984) 'Roo harvesting. *Australian Natural History,* vol. 21, pp. 122–29.

Gunderson, H.L. (1976) *Mammalogy,* McGraw-Hill, New York. 483 p.

Hume, I.D. (1982) *Digestive physiology and nutrition of marsupials,* Cambridge University Press, Cambridge. 256 p.

Janke, A., N.J. Gemmell, G. Feldmaier-Fuchs, A. von Haeseler, and S. Pääbo (1996) The mitochondrial genome of a monotreme—the platypus (*Ornithorhynchus anatinus*). *Journal of Molecular Evolution,* vol. 42, pp. 153–59.

Janke, A., X. Xu, and U. Arnason (1997) The complete mitochondrial genome of the wallaroo (*Macropus robustus*) and the phylogenetic relationship among Monotremata, Marsupialia, and Eutheria. *Proceedings of the National Academy of Science USA,* vol. 94, pp. 1276–81.

Jones, F.W. (1923–25) *The mammals of South Australia,* Parts I–III, Government Printer, Adelaide. 458 p.

Jones, M. (1995) Dinning with the devil. *Australian Natural History,* vol. 24, pp. 32–41.

Keast, A., F.C. Erk, and B. Glass (eds) (1972) *Evolution, mammals, and southern continents,* State University of New York Press, Albany. 543 p.

Kirsch, J.A.W. (1977) The comparative serology of marsupials and a classification of marsupials. *Australian Journal of Zoology,* suppl. series no. 52. 152 p.

Laycock, G. (1966) *The alien animals,* Natural History Press, Garden, NY. 240 p.

Lee, A. and R. Martin (1988) *The koala: A natural history,* University of New South Wales Press, Sydney, 102 p.

Lee, A.K. and A. Cockburn (1985) *Evolutionary ecology of marsupials.* Cambridge University Press, New York. 274 p.

Lee, A.K., K.A. Handasyde, and G.D. Sanson (eds) (1990) *Biology of the koala,* Surrey Beatty & Sons, Sydney. 336 p.

Le Souef, A.S., and H. Burrell (1926) *The wild animals of Australasia,* George G. Harrap, London. 388 p.

Lucas, J. (ed) (1971) Marsupials in captivity, pp 1–54 and 351–54 in *International Zoo Yearbook,* vol. 11, Zoological Society of London, London.

Lyne, G. (1967) *Marsupials and monotremes of Australia,* Angus & Robertson, Sydney. 72 p.

Macintosh, N.W.G. (1975) The origin of the dingo: An enigma, pp. 87–106 in Fox, M.W. (ed)*The wild canid,* Van Nostrand Reinhold, New York.

McKnight, T.L. (1969) *The camel in Australia,* Melbourne University Press, Melbourne. 154 p.

McNamara, K. and P. Murray (1985) *Prehistoric mammals of Western Australia,* Western Australian Museum, Perth. 32 p.

Marlow, B.J. (1965) *Marsupials of Australia.* Jacaranda Press, Brisbane. 141 p.

Merrilees, D. and J.K. Porter (1979) *Guide to the identification of teeth and some bones of native land mammals occurring in the extreme southwest of Western Australia,* Western Australian Museum, Perth. 152 p.

Modney, N. (1984) Tasmanian tiger sighting casts marsupial in new light. *Australian Natural History,* vol. 21, pp. 177–80.

Murray, P. (1984) *Australia's prehistoric animals,* Methuen, Sydney. 32 p.

Nowak, R.M. and J.L. Paradiso (1983) *Walker's mammals of the world* (4th ed.), vol. 1. Johns Hopkins University Press, Baltimore. 568 p.

Pratt, A. (1937) *The call of the koala,* Robertson & Mullens, Melbourne. 120 p.

Procter-Gray, E. (1990) Kangaroos up a tree. *Natural History,* vol. 99, pp. 60–67.

Proske, U. (1996) Hopping mad. *Australia Nature,* vol. 25, pp. 56–63.

Quirk, S. and M. Archer (eds) (1983) *Prehistoric animals of Australia,* Australian Museum, Sydney. 80 p.

Ratcliffe, F. (1947) *Flying fox and drifting sand,* Angus & Robertson, Sydney. 332 p.

Rich, P.V. and E.M. Thompson (eds) (1982) *The fossil vertebrate record of Australia,* Monash University, Melbourne. 759 p.

Ride, W.D.L. (1970) *A guide to the native mammals of Australia,* Oxford University Press, Melbourne. 249 p.

Rolls, E.C. (1969) *They all ran wild,* Angus & Robertson, Sydney. 444 p.

Roots, C. (1976) *Animal invaders,* Universe Books, New York. 203 p.

Schmidt-Nielson, K., L. Bolis, and C.R. Taylor (eds) (1980) *Comparative physiology: Primitive mammals,* Cambridge University Press, Cambridge. 338 p.

Sharland, M. (1962) *Tasmanian wild life,* Melbourne University Press, Melbourne. 86 p.

Sisk, T.D., A.E. Launer, K.R. Switky, and P.R. Ehrlich (1994) Identifying extinction threats. *BioScience,* vol. 44, pp. 592–604.

Smith, A. and I. Hume (eds) (1984) *Possums and gliders,* Surrey Beatty & Sons, Sydney. 800 p.

Stodart, E. and I. Parer (1988) *Colonisation of Australia by the rabbit,* CSIRO Division of Wildlife and Ecology, Report no. 6, pp. 1–21.

Stonehouse, B. and D. Gilmore (eds) (1977) *The biology of marsupials,* University Park Press, Baltimore. 494 p.

Strahan, R. (1981) *A dictionary of Australian mammal names,* Angus & Robertson, Sydney. 191 p.

——(1983) *Complete book of Australian mammals,* Australian Museum, Sydney. 530 p.

——(ed) (1995) *Mammals of Australia,* Smithsonian Institution Press, Washington, DC. 756 p.

Szabo, M. (1995) Australia's marsupials — going, going, gone. *New Scientist,* vol. 145, pp. 30–35.

Taylor, J.M. (1984) *Mammals of Australia,* Oxford University Press, Melbourne. 148 p.

Thompson, H.V. and C.M. King (eds) (1994) *The European rabbit: The history and biology of a successful colonizer,* Oxford University Press, Oxford. 245 p.

Triggs, B. (1988) *The wombat,* University of New South Wales Press, Sydney, 141 p.

Troughton, E. (1967) *Furred animals of Australia* (9th ed.) Angus & Robertson, Sydney. 384 p.

——(19770 *Kangaroos and wallabies,* vol. 3, *Australian Encyclopaedia* (3rd ed.), The Grolier Society of Australia, Sydney, pp. 440–44.

——(1977) *Mammals, marsupials, monotremes,* vol. 4, *Australian Encyclopaedia* (3rd ed.), The Grolier Society of Australia, Sydney, pp. 108–11, 131–33, 222–23.

Tyler, M.J. (ed). (1979) *The status of endangered Australian wildlife,* Royal Zoological Society of South Australia, Adelaide. 210 p.

Tyndale-Biscoe, H. (1973) *Life of marsupials,* American Elsevier, New York. 254 p.

Vaughn, T.A. (1972) *Mammalogy.* Saunders, Philadelphia. 463 p.

Williamson, H.D. (1975) *The year of the koala,* Charles Scribner's Sons, New York. 209 p.

——(1977) *The year of the kangaroo,* Charles Scribner's Sons, New York. 187 p.

Wilson, D.E. and D.M. Reeder (1993) *Mammal species of the world: A taxonomic and geographical reference* (2nd ed.), Smithsonian Institution Press, Washington, DC. 1206 p.

Zinsmeister, W.J. (1986) Fossil windfall at Antarctica's edge. *Natural History,* vol. 95, pp. 60–66.

Appendix 1: Political and Social Essentials

Anon. (1993) Where to live: Nirvana by numbers. *The Economist,* vol. 329, no. 7843, pp. 39–42.

Anon. (1994) *Australia in the world,* Departments of the Prime Minister and Cabinet, Treasury, and Foreign Affairs and Trade, Canberra. 84 p.

Buckely, A. and P. McLaughlin (1996) *Australia and USA,* Office of Public Affairs, Embassy of Australia, Washington, D.C. 50 p.

Coppell, B. (1994) *Australia in facts and figures,* Penguin Books, Ringwood, Vic. 502 p.

Fein, R. (1992) Health care reform. *Scientific American,* vol. 267, no. 5, pp. 46–53.

Figgis, P., G. Mosley, and L. Meier (1988) *Australia's wilderness heritage,* Weldon Publishing, Sydney. 2 vols.

Keneally, T., P. Adam-Smith, and R. Davidson (1987) *Australia: Beyond the Dreamtime,* Facts on File, New York. 248 p.

Kurian, G.T. (1990) *National profiles: Australia and New Zealand,* Facts on File, New York. 216 p.

McDonald, J. (1987) *Pocket facts of Australia,* Child & Associates, Sydney. 320 p.

Nathan, J.A. (1983) Dateline Australia: America's foreign Watergate? *Foreign Policy,* vol. 49, pp. 168–85. (See also *Time Magazine,* 13 Dec. 1982, p 49; 10 Jan. 1983, p. 4.)

Population Reference Bureau (1994) World Population Data Sheet, PRB, Washington, DC.

Smith, R.E. (ed) (1992) *Australia,* Australian Government Publishing Service, Canberra. 126 p.

Wilks, G. (1994) *Australia in brief,* Australian Government Publishing Service, Canberra. 112 p.

Appendix 2: Australian Idiomatic Language

Baker, S.J. (1978) *The Australian language,* Currawong Press, Sydney.

Delbridge, A. (ed) (1988) *The Macquarie dictionary* (2nd ed.), Macquarie Library, Sydney. 2009 p.

——(ed) (1984) *Aussie talk,* Macquarie Library, Sydney. 353 p.

Lauder, A., (1965) *Let stalk strine,* Ure Smith, Sydney. 47 p.

Mitchell, A.G. and A. Delbridge. (1978) *The pronunciation of English in Australia,* Angus & Robertson, Sydney.

O'Grady, J. (1965) *Aussie English,* Ure Smith, Sydney. 104 p.

Papps, E.H. (1965) *Aboriginal words of Australia,* Reed, Sydney. 144 p.

Ramson, W.S. (1966) *Australian English,* Australian National University Press, Canberra.

Reed, A.W. (1969) *Place-names of New South Wales,* Reed, Sydney. 156 p.

Turner, G.W. (1972 *The English language in Australia and New Zealand,* Longmans, Melbourne.

Wilkes, G.A. (1978) *A dictionary of Australian colloquialisms,* Sydney University Press, Sydney.

——(1979) Australian English, pp. xix–xxii in *Collins Dictionary of the English Language,* Collins, Sydney.

• •

OTHER BOOKS ON NATURAL HISTORY

Archer, M. and G. Clayton (eds) (1984) *Vertebrate zoogeography and evolution in Australasia,* Hesperian Press, Perth. 1203 p.

Barker, W.R. and P.J.M. Greenslade (eds) (1982) *Evolution of the flora and fauna of arid Australia,* Peacock Publishing, Frewville, SA. 392. p.

Bergamini, D. (1980) *The land and wildlife of Australia (*rev. ed.), Time-Life Books, Alexandria, Va. 198 p.

Breeden, S. (1970) *Tropical Queensland,* Collins, Sydney. 262 p.

——(1972) *Australia's south east,* Collins, Sydney. 256 p.

——(1975) *Australia's north,* Collins, Sydney. 208 p.

Brunschweiler, R.O. (1984) *Ancient Australia (*3rd ed.), Angus & Robertson, Sydney. 349 p.

Darlington, P.J. (1957) *Zoogeography: The geographical distribution of animals,* Wiley, New York. 675 p.

——(1965) *Biogeography of the southern end of the world,* Harvard University Press, Cambridge, MA. 236 p.

Dyne, G.R. and D.W. Walton (eds) (1987) *Fauna of Australia: General articles,* vol. 1A, Australian Government Publishing Service, Canberra. 339 p.

Evans, H.E. and M.R. Evans (1983) *Australia: A natural history,* Smithsonian Institution Press, Washington, DC. 208 p.

Finney, C.M. (1984) *To sail beyond the sunset,* Rigby, Adelaide, SA. 206 p.

Groves, R.H. and W.D.L. Ride (eds) (1982) *Species at risk: Research in Australia,* Springer-Verlag, Berlin. 216 p.

Jeans, D.N. (ed) (1977) *Australia: A geography,* Sydney University Press, Sydney. 571 p.

Keast, A. (1966) *Australia and the Pacific islands: A natural history,* Random House, New York. 298 p.

——(ed) (1981) *Ecological biogeography of Australia,* vol. 41, Monographiae Biologicae, Junk, The Hague. 2182 p.

Keast, A., R.L. Crocker, and C.S. Christian (eds) (1959) *Biogeography and ecology in Australia,* vol. 8, Monographiae Biologicae, Junk, The Hague. 640 p.

Knox, B., P. Ladiges, and B. Evans (eds) (1994) *Biology,* McGraw-Hill Book Co., Sydney. 1067 p.

Leeper, G.W. (ed) (1970) *The Australian environment (*4th ed.), CSIRO and Melbourne University Press, Melbourne. 163 p.

Morrison, R. and M. Morrison. (1988) *Australia: The four billion year journey of a continent,* Facts on File, New York. 334 p.

Quirk, S. and M. Archer (eds) (1983) *Prehistoric animals of Australia,* Australian Museum, Sydney. 80 p.

Raymond, R. and R. Morrison (1979) *Australia: The greatest island,* Ure Smith, Sydney. 352 p.

Recher, H.F., D. Lunney, and I. Dunn (1979) *A natural legacy: Ecology in Australia,* Pergamon Press. 276 p.

Rich, P.V. and E.M. Thompson (eds) (1982) *The fossil vertebrate record of Australasia,* Monash University, Melbourne. 759 p.

Rich, P.V. and G.F. van Tets, (1985) *Kadimakara,* Pioneer Design Studio, Lilydale, Vic. 284 p.

Smith, J. (ed) (1992) *The unique continent,* University of Queensland Press, Brisbane. 282 p.

Tyler, M.J., C.R. Twidale, and J.K. Ling (1979) *Natural history of Kangaroo Island,* Royal Society of South Australia, Adelaide. 184 p.

Vandenbeld, J. (1988) *Nature of Australia,* Facts on File, New York. 292 p.

Vickers-Rich, P. and T.H. Rich (1993) *Wildlife of Gondwana,* Reed, Sydney. 276 p.

Williams, W.D. (ed) (1974) *Biogeography and ecology in Tasmania,* vol. 25, Monographiae Biologicae, Junk, The Hague. 498 p.

• •

NATIONAL GEOGRAPHIC ARTICLES

In February 1988 the *National Geographic* celebrated Australia's Bicentennial with an issue devoted exclusively to Australia. It includes eight articles and a large map.

Abercrombie, T.J. (1982) Perth. 161: 638–667. May.

Adams, M.P.G. (1924) Australia's wild wonderland. 45: 329–356. Mar.

Arden, H. (1991) Journey into Dreamtime. 179: 8–41. Jan.

Ballard, W.J. (1905) Australia's future. 16: 570–571. Dec.

Barrett, C. (1930) The Great Barrier Reef and its isles. 58: 355–384. Sept.

Breeden, K. and S. (1973) Rock paintings of the Aborigines. 143: 174–187. Sept.

——(1973) Eden in the outback. 143: 189–203. Feb.

Breeden, S. (1988) The Australians. 173: 266–289. Feb.

Canby, T.Y. (1984) El Nino's ill wind. 165: 144–183. Feb.

Cayley, N.W. (1945) The fairy wrens of Australia. 88: 488–498. Oct.

Chaffer, N. (1961) Australia's amazing bowerbirds. 120: 866–873. Dec.

Davidson, R. (1978) Alone across the outback. 153: 580–611. May.

Doubilet, D. (1991) Lord Howe Island: Australian haven in a distant sea. 180: 126–146. Oct.

——(1991) Australia's magnificent pearls. 180: 108–123. Dec.

——(1997) Beneath the Tasman Sea 191:82–101. Jan.

Ellis, R. (1987) Australia's southern seas. 171: 286–319. Mar.

Ellis, W.S. (1986) Queensland. 169: 2–39. Jan.

Ellison, N. (1932) Shark fishing — an Australian industry. 62: 369–386. Sept.

Everingham, J. (1988) Children of the First Fleet. 173: 232–245. Feb.

Fisher, A.C. (1971) Australia's pacesetter state, Victoria. 139: 218–253. Feb.

Fleay, D. (1958) Flight of the platypuses. 114: 512–525. Oct.

———(1963) Strange animals of the island continent. 124: 388–411. Sept.

Fox, R. (1991) The sea beyond the outback. 179: 42–73. Jan.

Gore, R. (1997) Expanding worlds. 191: 84–109. May.

Green, P. (1977) Australia's feathered playboy: the satin bowerbird. 152: 865–872. Dec.

Gregory, H.E. (1916) Lonely Australia: the unique continent. 30: 473–568. Dec.

Hamner, W.M. (1994) Australia's box jellyfish: A killer down under. 186: 116–130. Aug.

Hauser, H. (1984) Exploring a sunken realm in Australia. 165: 128–142. Jan.

Holmes, C.H. (1939) Australia's patchwork creature, the platypus. 76: 273–282. Aug.

Johnson, D.H. (1955) The incredible kangaroo. 108: 487–500. Oct.

Judge, J. (1979) The tragic journey of Burke and Wills. 155: 152–191. Feb.

———(1988) Child of Gondwana. 173: 170–177. Feb.

Kahl, M.P. (1987) The royal spoonbill. 171: 280–284. Feb.

Keltie, J.S. (1897) The great unmapped areas on the earth's surface awaiting the explorer and geographer. 8: 251–266. Sept.

Levathes, L.E. (1985) The land where the Murray flows. 168: 252–278. Aug.

Lewis, F. (1931) The koala, or Australian teddy bear. 60: 346–355. Sept.

MacLeish, K. (1968) Queensland, young titan of Australia's tropic north. 134: 593–639. Nov.

———(1972) Diving with sea snakes. 141: 565–578. April.

———(1973) Exploring Australia's coral jungle. 143: 742–779. June.

———(1973) The top end of down under. 143: 145–174. Feb.

———(1975) Western Australia: The big country. 147: 150–187. Feb.

Mark, M.E. (1988) Sydney's changing face. 173: 246–265. Feb.

Moore, K. (1976) Coober Pedy: Opal capital of Australia's outback. 150: 560–571. Oct.

Moore, W.R. (1935) Capital cities of Australia. 68: 667–722. Dec.

———(1936) Beyond Australia's cities. 70: 709–747. Dec.

Mountford, C.P. (1946) Earth's most primitive people: A journey with the Aborigines of central Australia. 89: 89–112. Jan.

———(1949) Exploring Stone Age Arnhem Land. 96: 745–782. Dec.

National Geographic Society (1902) The completion of the cable between Canada and Australia. 13: 410. Nov.

———(1948) The society maps a new Australia. 93: 431–432. Mar.

———(1956) Sports-minded Melbourne host to the Olympics. 110: 688–693. Nov.

———(1963) Australia on maps — old and new. 124: 386–387. Sept.

O'Brian, M. (1988) Australians. 173: 212–231. Feb.

O'Neill, T. (1997) Travelling the Australian dog fence. 191: 18–37. April.

Patterson, C.B. (1983) A walk and ride on the wild side [Tasmania]. 163:676–692. May.

Payne, O. (1995) Koalas: Out on a limb. 188: 36–59. April.

Roughley, T.C. (1940) Where nature runs riot: On Australia's Great Barrier Reef. 77: 823–850. June.

Scollay, C. (1980) Arnhem Land Aboriginals cling to dreamtime. 158: 644–664. Nov.

Smith, L.H. (1955) Lyrebird, Australia's meistersinger. 107: 849–857. June.

Smith, R. (1921) From London to Australia by aeroplane. 39: 229–339. Mar.

Sharman, G. (1959) Kangaroos: That marvelous mob. 155: 192–210. Feb.

Sisson, R.F. (1973) Life cycle of a coral. 143: 780–793. June.

Starbird, E. (1979) Sydney: Big, breezy, and a bloomin' good show. 155: 211–236. Feb.

Terrill, R. (1988) Australia at 200. 173: 181–211. Feb.

Vesilind, P.J. (1984) Monsoons: The life breath of half the world. 166: 712–747. Dec.

Vessels, J. (1992) The Simpson outback. 181: 64–93. April.

Watkins, T.H. (1996) Sir Joseph Banks. 190: 28–53. Nov.

Wolinsky, C. (1995) Wild flowers of Western Australia. 187: 68–89. Jan.

Wright, B. and S. Breeden. (1988) Living in two worlds. 173: 290–294. Feb.

Zwingle, E. (1986) The tea and sugar: Lifeline in Australia's outback. 169: 736–757. June.

General Index

S*cientific* N*ames* I*ndex*